高等学校国际化人才培养土木工程专业系列教材

# BASIC PRINCIPLES OF STEEL STRUCTURES
# 钢结构基本原理

Lu Linfeng
卢林枫　主编

中国建筑工业出版社
China Architecture & Building Press

图书在版编目（CIP）数据

钢结构基本原理 = BASIC PRINCIPLES OF STEEL STRUCTURES：英文 / 卢林枫主编. — 北京：中国建筑工业出版社，2022.1
高等学校国际化人才培养土木工程专业系列教材
ISBN 978-7-112-26769-9

Ⅰ.①钢… Ⅱ.①卢… Ⅲ.①钢结构－高等学校－教材－英文 Ⅳ.①TU391

中国版本图书馆CIP数据核字（2021）第211091号

责任编辑：聂 伟 王 跃
责任校对：李美娜

高等学校国际化人才培养土木工程专业系列教材
# BASIC PRINCIPLES OF STEEL STRUCTURES
## 钢结构基本原理
Lu Linfeng
卢林枫 主编

\*

中国建筑工业出版社出版、发行（北京海淀三里河路9号）
各地新华书店、建筑书店经销
北京红光制版公司制版
北京建筑工业印刷厂印刷

\*

开本：787毫米×1092毫米 1/16 印张：21½ 字数：534千字
2022年5月第一版 2022年5月第一次印刷
定价：**56.00元**（赠教师课件）
ISBN 978-7-112-26769-9
（38597）

**版权所有 翻印必究**
如有印装质量问题，可寄本社图书出版中心退换
（邮政编码100037）

## Book Description

This book is compiled following the design provisions and descriptions of the current *Standard for Design of Steel Structures* GB 50017. It fully complies with the *Guiding Professional Standards for Undergraduate Civil Engineering in Colleges and Universities*, strictly follows the basic principles of preparing professional specifications, and fully covers the core knowledge points proposed by the professional specifications. Through the core knowledge points, the basic design provisions and basic principles of GB 50017 are introduced to the greatest extent. Thus, international students can get familiar with and master the basic theories and basic design methods of the Chinese steel structure design system to the greatest extent, which makes it easier to approach the actual project. This book introduces the basic principles of steel structures. The content arrangement considers the internal logic of steel structure courses: student-oriented, concise, practical, and readable. It highlights the principles and focuses on analysis and problem-solving skills.

This book is suitable for international students and teachers of civil engineering major, teachers and students who carry out bilingual teaching, as well as technicians engaging in the design and construction of steel structure projects overseas.

In order to better support the teaching of corresponding courses, we provide courseware to teachers who use this book as teaching materials. Anyone who needs it can contact the publisher. Jian Gong Shu Yuan: http://edu.cabplink.com. Email: jckj@cabp.com.cn; 2917266507@qq.com. Tel: (010) 58337285.

本教材根据现行《钢结构设计标准》GB 50017 的设计条文与条文说明，按照《高等学校土木工程本科指导性专业规范》编写。严格遵循专业规范编制的基本原则，对专业规范提出的核心知识点做到了全面覆盖。通过核心知识点，介绍我国《钢结构设计标准》GB 50017 的基本设计条文和基本原则，最大程度地让留学生了解和掌握我国钢结构设计体系的基础理论和基本设计方法，易于与实际工程靠近。本教材介绍钢结构的基本原理，内容安排考虑钢结构课程的内在逻辑规律，遵循以学生为主、简明实用、可读性强的编写原则，突出原理，注重分析问题和解决问题的思路。

本教材适用于土木工程专业国际留学生和授课教师、开展双语授课的师生，以及对外从事钢结构工程设计与施工的技术人员。

为了更好地支持相应课程的教学，我们向采用本书作为教材的教师提供课件，有需要者可与出版社联系。建工书院：http://edu.cabplink.com，邮箱：jckj@cabp.com.cn，2917266507@qq.com，电话：(010) 58337285。

# PREFACE

With the adjustment and improvement of the education policy for international students in China, the number of international students majoring in civil engineering is also steadily increasing. Many international students are enrolled in the curriculum in English, which puts forward new requirements for the education quality of international students. Many international students comes from non-English countries. Their domestic civil engineering development is still very different from that of China, which poses a new challenge of constructing an all-English curriculum for international students majoring in civil engineering.

The course "Basic Principles of Steel Structures" in English is a professional knowledge development course for international students majoring in civil engineering. One of the main goals of training international students is to make them familiar with and master the structural design system of China. Therefore, the compilation of the content is in full compliance with the *Guiding Professional Standards for Undergraduate Civil Engineering in Colleges and Universities*. The book introduces the basic design provisions and basic principles of China's steel structure professional code—*Standard for Design of Steel Structures* GB 50017. International students can understand and master the basic theories and basic design methods of China's steel structure design system, which makes it easier to approach the actual project. This book introduces the basic principles of steel structure. The content arrangement considers the internal logic of steel structure courses, follows the student-oriented, concise, practical, and readable writing principles, highlights the principles, and analyzes and solves problems.

Participants in the writing of this book are Lu Linfeng (Editor-in-Chief, Chapters 1-7, Appendix), Wu Hanheng (Chapter 5), Liu Yan (Chapter 6, Appendix), and Guan Yu (Chapter 7, Appendix). In addition, many graduate students of Chang'an University (Sui Lu, Li Jinnan, Yan Hongwei, Jiang Taisheng et al.) drew the figures in the book, calculated and proofread the example questions, exercises, and table data in the Appendix. Thanks to Professor Nie Shaofeng, Dr. Chao Sisi, and Dr. Lan Guanqi for their support in the book compilation.

The choice of content in this book is in full compliance with the core knowledge requirements of the *Guiding Professional Standards for Undergraduate Civil Engineering in Colleges and Universities*. It is closely related to the *Standard for Design of Steel*

*Structures* GB 50017. The use of professional terms is as international as possible, which is the feature of this book. However, due to the author's language level limitation, readers are kindly requested to provide valuable comments on the errors and areas that need to be improved in the book.

June 2021<br>
Xi'an

# 前　言

随着我国对来华留学生教育政策的调整和完善，土木工程专业国际留学生的数量也在稳步上升，很多留学生是按照全英文授课方式招生入学，这对我国留学生教育质量提出了新的要求。国际留学生很多来自非英语国家，且其国内土木工程的发展程度与我国还有很大差别，这对我国土木工程专业留学生全英文课程的建设提出了新的挑战。

全英文课程《钢结构基本原理》是土木工程专业国际留学生的专业发展课程。我国培养国际留学生的主要目标之一是让其熟悉和掌握中国结构设计体系的知识。因此，教材编写完全遵从《高等学校土木工程本科指导性专业规范》，最大程度地介绍我国钢结构专业规范——《钢结构设计标准》GB 50017 的基本设计条文和基本原则，最大程度地让国际留学生了解和掌握我国钢结构设计体系的基础理论和基本设计方法，而且易于与实际工程靠近。本教材介绍钢结构的基本原理，内容安排考虑钢结构课程的内在逻辑规律，遵循以学生为主、简明实用、可读性强的编写原则，突出原理，注重分析问题和解决问题的思路。

参加本书编写工作的有：卢林枫（主编，第 1～7 章，附录），吴函恒（第 5 章），刘岩（第 6 章，附录），管宇（第 7 章，附录）。长安大学多名研究生（隋璐、李劲男、严宏伟、姜太升等）绘制了书中的插图，对例题、习题、附表数据进行了计算与校对。感谢聂少锋教授、晁思思博士、兰官奇博士在教材编写中给予的支持。

本教材在内容取舍上完全符合《高等学校土木工程本科指导性专业规范》对核心知识点的要求，与《钢结构设计标准》GB 50017 联系紧密，在专业术语的使用上尽可能国际化，这是本书的特色。但由于作者的语言水平限制，对于书中的错误和需要完善的地方，敬请读者提出宝贵的意见。

2021 年 6 月
于西安

# CONTENTS

**Chapter 1  INTRODUCTION** ............... 1
  1.1  Characteristics of Steel Structures ............... 1
    1.1.1  Advantages ............... 1
    1.1.2  Disadvantages ............... 2
  1.2  Development History of Steel Structures ............... 2
  1.3  Applications of Steel Structures ............... 5
  1.4  Classification of Structural Steel Members ............... 10
  1.5  The Basic Design Method of Steel Structures ............... 12
    1.5.1  The purpose of steel structure design ............... 12
    1.5.2  Probabilistic limit state design method ............... 13
    1.5.3  Design expression ............... 14
  1.6  Chapter Summary ............... 15
  Questions ............... 16

**Chapter 2  STRUCTURAL STEEL** ............... 17
  2.1  Introduction ............... 17
  2.2  Several Important Performance Indexes of Structural Steel ............... 18
    2.2.1  Strength performance ............... 18
    2.2.2  Plastic performance ............... 20
    2.2.3  Steel physical performance index ............... 20
    2.2.4  Cold bending performance ............... 21
    2.2.5  Impact toughness ............... 21
    2.2.6  Structural steel weldability ............... 22
  2.3  Stress-Strain Relationship of Structural Steel under Different Loadings ............... 22
  2.4  Performance of Steel under Complex Stress ............... 23
  2.5  Main Factors Affecting the Performance of Structural Steel ............... 24
    2.5.1  Chemical compositions of steel ............... 24
    2.5.2  Production process ............... 25
    2.5.3  Steel hardening ............... 26
    2.5.4  Temperature ............... 26
    2.5.5  Stress concentration ............... 27
  2.6  Types and Specifications of Structural Steel ............... 27
    2.6.1  Classification of steel in China ............... 27

  2.6.2 The choice principles of structural steel ............... 29
  2.6.3 Steel selection recommendations ............... 30
 2.7 Structural Steel Products in China ............... 31
  2.7.1 Hot-rolled steel products in China ............... 31
  2.7.2 Cold-formed steel shapes in China ............... 36
 2.8 Chapter Summary ............... 37
 Questions ............... 38

## Chapter 3 POSSIBLE FAILURE OF STEEL STRUCTURES ............... 39
 3.1 Introduction ............... 39
 3.2 The Strength Failure of Steel Structures ............... 40
  3.2.1 The strength failure of tension members ............... 40
  3.2.2 The strength failure of bending members ............... 40
 3.3 Overall Instability Failure of Steel Structures ............... 41
  3.3.1 Aspects that must be considered ............... 41
  3.3.2 Overall instability of steel structures ............... 42
  3.3.3 Overall instability of steel members ............... 42
 3.4 Local Instability Failure of Steel Structures ............... 44
  3.4.1 Local instability of steel structures ............... 44
  3.4.2 Local instability of steel members ............... 45
  3.4.3 Classification of cross-section ............... 45
 3.5 The Fatigue Failure of Steel Structures ............... 48
  3.5.1 The concept of fatigue failure ............... 48
  3.5.2 Factors affecting fatigue failure ............... 49
  3.5.3 Fatigue calculation and anti-brittle fracture design ............... 51
 3.6 The Deformation Failure of Steel Structures ............... 51
  3.6.1 The deformation failure of steel members ............... 51
  3.6.2 The deformation failure of steel structures ............... 52
 3.7 Brittle Fracture of Steel Structures ............... 52
  3.7.1 Examples of brittle fracture of steel structures ............... 52
  3.7.2 The main cause of brittle fracture ............... 53
  3.7.3 Measures to prevent brittle fracture ............... 54
 3.8 Chapter Summary ............... 55
 Questions ............... 56

## Chapter 4 CONNECTIONS OF STEEL STRUCTURES ............... 57
 4.1 Introduction ............... 57
 4.2 Methods and Characteristics of Welded Connection ............... 58
  4.2.1 The method of welding and the form of the weld ............... 58
  4.2.2 Weld defects and inspection ............... 61

  4.2.3 Residual welding stress and residual deformation ······················ 62
4.3 Construction Details and Calculation of Groove Welded Connection ············ 66
  4.3.1 Construction details of groove welded connection ······················ 66
  4.3.2 Calculation of full penetration groove welded connection ················ 68
4.4 Construction Details and Calculation of Fillet Welded Connection ················ 70
  4.4.1 Construction details of fillet welded connection ······················ 70
  4.4.2 Force performance of fillet weld ·································· 73
  4.4.3 The basic formula for calculating the strength of a right-angle fillet weld ········ 74
  4.4.4 Calculation of right-angle fillet welded connection ····················· 75
4.5 Form of Bolted Connection and Construction Detail Requirements ················ 79
  4.5.1 Characteristics and form of bolted connections ······················· 79
  4.5.2 Forms of bolted connections ····································· 80
  4.5.3 Construction detail requirements of bolted connections ·················· 81
4.6 Calculation of Bolted Connection ········································· 83
  4.6.1 Calculation of unfinished/ordinary bolt connections ····················· 83
  4.6.2 Calculation of high-strength bolt friction-type connection ················· 89
  4.6.3 Calculation of high-strength bolt bearing-type connection ················ 92
4.7 Working Examples ···················································· 93
4.8 Chapter Summary ···················································· 104
Questions ································································ 105
Exercises ································································ 106

# Chapter 5 AXIALLY LOADED MEMBERS ······························· 110

5.1 Introduction ························································· 110
5.2 Strength and Stiffness of Axially Loaded Members ··························· 112
  5.2.1 Strength calculation of axially loaded members ······················· 112
  5.2.2 Calculation of stiffness ········································· 114
5.3 Stability of Axially Loaded Compression Member ··························· 115
  5.3.1 Calculation of overall stability ···································· 116
  5.3.2 The calculation of local stability ·································· 127
5.4 Design of Axially Loaded Columns ······································ 132
  5.4.1 Design of solid web column ····································· 132
  5.4.2 Laced column design ·········································· 136
  5.4.3 Diaphragm of columns ········································· 144
5.5 Working Examples ··················································· 145
5.6 Chapter Summary ··················································· 155
Questions ································································ 156
Exercises ································································ 157

## Chapter 6  FLEXURAL MEMBER (BEAM) ... 159
### 6.1  Introduction ... 159
### 6.2  The Strength and Stiffness of the Beam ... 161
#### 6.2.1  Bending strength ... 161
#### 6.2.2  Shear strength ... 162
#### 6.2.3  Local compressive strength ... 165
#### 6.2.4  Reduced stress ... 166
#### 6.2.5  The stiffness of the beam ... 167
### 6.3  Torsion of the Beam ... 170
#### 6.3.1  Free torsion ... 170
#### 6.3.2  Constrained torsion of open section members ... 172
### 6.4  The Overall Stability of the Beam ... 175
#### 6.4.1  Several concepts ... 175
#### 6.4.2  Calculation of critical bending moment of beams ... 176
#### 6.4.3  Main factors affecting the overall stability of steel beams ... 179
#### 6.4.4  The overall stability factor of the beam ... 180
#### 6.4.5  Measures to ensure the overall stability of beams and calculation methods ... 181
### 6.5  Local Stability of Steel Plate Composite Beam ... 183
#### 6.5.1  Local stability of the compression flange ... 183
#### 6.5.2  Local stability of the web ... 186
#### 6.5.3  The construction details and calculation of stiffener of the beam web ... 197
### 6.6  Design Considering the Post-Buckling Strength of Beam Web ... 203
#### 6.6.1  Working performance of beam web after buckling ... 203
#### 6.6.2  Design features after web buckling ... 203
### 6.7  The Design of the Beam ... 207
#### 6.7.1  Section selection of beams ... 207
#### 6.7.2  Checking the beam ... 210
### 6.8  Working Examples ... 212
### 6.9  Chapter Summary ... 215
### Questions ... 217
### Exercises ... 218

## Chapter 7  TENSION-BENDING MEMBERS AND COMPRESSION-BENDING MEMBERS ... 220
### 7.1  Introduction ... 220
### 7.2  Strength and Stiffness of Tension-Bending and Compression-bending Members ... 222
#### 7.2.1  The ultimate state of bearing capacity calculated based on the edge fiber yield ... 224

|  |  |  |
|---|---|---|
| 7.2.2 | Taking the full plastic limit state of the section as the limit state of bearing capacity | 224 |
| 7.2.3 | Taking the plastic development of the section as the strength calculation criterion | 226 |
| 7.2.4 | Stiffness | 227 |

7.3 Global Stability of Solid Web Compression-Bending Members ............ 227

- 7.3.1 Stability calculation of solid web unidirectional compression-bending members in the plane of bending moment ............ 227
- 7.3.2 Stability calculation of solid web unidirectional compression-bending members out-plane of bending moments ............ 234
- 7.3.3 Stability calculation of solid web bidirectional compression-bending members ............ 240

7.4 Local Stability of Solid Web Compression-Bending Members ............ 241

- 7.4.1 The width-to-thickness ratio of the compression-bending members ............ 241
- 7.4.2 The post-buckling strength of the compression-bending members ............ 242

7.5 The Effective Length of the Compression-Bending Members ............ 243

- 7.5.1 The effective length of the single bending members ............ 243
- 7.5.2 The effective length of the frame column ............ 243
- 7.5.3 The effective length of single-story frame with equal section column in the frame plane ............ 244
- 7.5.4 The effective length of multi-story and multi-bay frame with equal section column in the frame plane ............ 246
- 7.5.5 The effective length of single-story frame stepped column in the frame plane ............ 248
- 7.5.6 The effective length of the frame column out of the frame plane ............ 250

7.6 Design of Solid Web Compression-Bending Members ............ 251

- 7.6.1 Design requirements ............ 251
- 7.6.2 Construction requirements ............ 252
- 7.6.3 The selection of cross-section of solid web compression-bending members ............ 252

7.7 Design of Laced Compression-Bending Members ............ 254

- 7.7.1 Stability calculation of laced compression-bending members ............ 255
- 7.7.2 Lacing calculation and construction requirements ............ 259

7.8 Beam-to-Column Connections of the Frame ............ 259

- 7.8.1 Flexible connection ............ 260
- 7.8.2 Rigid connection ............ 260
- 7.8.3 Semi-rigid connection ............ 264

7.9 Working Examples ............ 264

7.10 Chapter Summary ............ 277

Questions ............ 279

Exercises ............ 279

| | | |
|---|---|---|
| APPENDIX A | CHEMICAL COMPOSITION OF STEEL | 282 |
| APPENDIX B | DESIGN STRENGTH OF STEEL | 284 |
| APPENDIX C | GLOBAL STABILITY FACTORS OF BEAM | 286 |
| APPENDIX D | GLOBAL STABILITY FACTORS OF AXIALLY LOADED COMPRESSION MEMBERS | 291 |
| APPENDIX E | EFFECTIVE LENGTH FACTOR OF COLUMNS | 296 |
| APPENDIX F | TYPICAL STEEL SHAPES | 308 |
| REFERENCES | | 330 |

# Chapter 1　INTRODUCTION

**GUIDE TO THIS CHAPTER:**

➤ *Knowledge points*

  *The main characteristics of steel structures; development history of steel structures; application scope of steel structures; classification of main components of steel structures.*

➤ *Emphasis*

  *The main characteristics of steel structures.*

➤ *Nodus*

  *Appropriate application of steel structure according to the advantages and disadvantages of structural steel.*

## 1.1　Characteristics of Steel Structures

With the advancement of human science and technology, construction materials and construction technologies have made qualitative leaps. Building materials have also evolved from traditional timber and earth-rock to modern materials such as steel and reinforced concrete. For example, the world's first furnace steel was born in 1856. Nowadays, steel has been widely used in the following fields: high-rise buildings, super high-rise buildings, large-span bridges, spatial structures, modern single-story factory buildings, heavy-duty workshops, pressure vessel equipment, and pipelines in the petrochemical industry, marine drilling platforms, and other civil constructions.

### 1.1.1　Advantages

Why can steel structures be widely used in the types of buildings and structures mentioned above? Because steel has many advantages compared to other construction materials, such as timber and reinforced concrete.

(1) Steel has high strength and low weight. There is a critical factor $\alpha$ which is the ratio of material density to its strength. The lower $\alpha$ means the lower mass of the structure constructed under the same load. The $\alpha$ of steel is about 1.7-3.7, the $\alpha$ of timber is about 5.4, and the $\alpha$ of concrete is about 18, the unit is $10^{-4}/m$.

(2) Steel is a uniform material, and its properties have better predictability. In addition, steel is an isotropic material, and some basic formulas of material mechanics can be applied directly.

(3) Steel has good plasticity and toughness, which means steel structures can be used directly under dynamic load and have good seismic performances under earthquakes.

(4) Steel is good airtight and water-tight material. Therefore, after the steel structure is connected by welding, it can meet the airtight and water-tight requirements of some high-pressure vessels and pipelines.

(5) Steel structures are easy to manufacture and construct. Steel structures are convenient for prefabrication in the factory. Steel structures are easy to install on the construction site. So, steel structures are natural prefabricated structures and can help to reduce the construction cost.

(6) The seismic performance of the steel structure is good. Under appropriate design conditions, the steel structure is the least damaged in previous world-renowned earthquakes. As a result, it has been recognized as a structure suitable for use in earthquake-resistant areas.

### 1.1.2 Disadvantages

Every coin has two sides, and steel structures also have shortcomings that cannot be overcome by themselves. These shortcomings sometimes affect the application of steel structures.

(1) The first is weak corrosion resistance. Steel structures need to be repainted with anti-corrosion paint regularly, causing higher maintenance costs.

(2) The second is weak fire resistance. As a result, the steel structure's surface needs to be painted with fireproof paint or wrapped with fireproof materials, causing higher construction costs.

(3) The structural cost is relatively high. Because of the weak corrosion resistance, fire resistance, and high material price, it is challenging to lower the construction costs and maintenance costs. However, high costs can be offset by a larger usable floor area and a low loan period.

(4) Under low temperature and other conditions, such as fatigue state, a somewhat brittle fracture will occur.

## 1.2 Development History of Steel Structures

Before the Second World War, China was the pioneer of using irons as load-carrying members, primarily used in two areas. One was the religious buildings, shown in Figure 1.1. *Yuquan Temple* is one of the cultural relics (ancient buildings) proposed to be protected by the Chinese government. And the other was chain bridges. There are many chain bridges in China, the oldest and most famous of which is *Luding Bridge*, constructed in 1705, shown in Figure 1.2.

## 1.2 Development History of Steel Structures

Figure 1.1  Iron Tower of Yuquan Temple (CN, 1061)

Figure 1.2  Luding Bridge (CN, 1705)

America is the pioneer of steel structures. The age of steel began when it was first manufactured in 1856, and the representative is Eads Bridge, shown in Figure 1.3. *Home Insurance Company Building* was the first high-rise steel-framed building globally and was constructed in 1884 (Figure 1.4). After the *Home Insurance Company Building*, several famous high-rise buildings were built in America, such as *Park Row Building* (Figure 1.5) and *Empire State Building* (Figure 1.6). However, during the Second World War, the development of the steel structure was at a standstill. After the Second

Figure 1.3  Eads Bridge (USA, 1868—1874)

World War, steel structures entered the recovery and development stage. In the 1950s to the 1970s, Europe and Japan entered a rapid development phase. As a result, 81% of the tallest completed buildings were steel structures before 1991. However, after the *September 11 Attacks*, the proportion of steel structures in high-rise had dropped to 8%.

Figure 1.4　Home Insurance Company Building
(USA, 1884)

Figure 1.5　Park Row Building
(USA, 1899)

Figure 1.6　Empire State Building
(USA, 1931)

Since the People's Republic of China was founded in 1949, with its steel production in different periods, steel structure development has experienced four stages: tried not to use steel structures, restricted steel structures, rational use of steel structures, and encouraging the use of steel structures. At present, China's annual steel output has been ranked first globally for many consecutive years, and vigorously developing steel structures will also become a basic national policy to revitalize the economy.

## 1.3 Applications of Steel Structures

As stated in section 1.1, today, steel has been widely used in high-rise buildings, super-high-rise buildings, large-span and spatial structures, modern workshops, pressure vessel equipment, etc. According to Chinese practical experience, within the industrial and civil constructions, the application scope of steel structures is roughly as follows:

(1) Multi-story and high-rise buildings

For the multi-story buildings, the moment resistance steel frame (Figure 1.7) and steel frame with brace-system (Figure 1.8) are the most characteristic buildings. At present, in the field of high-rise buildings, the proportion of high-rise buildings with a simple steel structure is tiny, Only about 8%, but the composite of steel and concrete is used widely. A representative example of a high-rise building is shown in Figure 1.9.

Figure 1.7  Moment resistance steel frame (CN, unknown)

Figure 1.8  Steel frame with brace-system (CN, unknown)

(2) Low-rise and single-story buildings

The single-story buildings mainly include a lightweight portal frame(Figure 1.10) and a heavy-duty industrial workshop (Figure 1.11). A lightweight portal frame is a popular industrial building in China; in some cases, it only uses bridge cranes no more than 20 tons. Heavy-duty industrial workshops are widely used in metallurgical enterprises, thermal power plants, and heavy machinery manufacturing plants, etc. Moment resistance steel frames and cold-formed steel buildings (Figure 1.12) are mainly used in fields of low-rise steel buildings.

Figure 1.9  Willis Tower Chicago (USA, 1974)

Figure 1.10  Lightweight portal frame (CN, unknown)

Figure 1.11  Heavy-duty industrial workshop (CN, unknown)

Figure 1.12  Cold-formed steel building

### (3) Towering structure

The towering structure is mainly used for radio and television signal towers, communication towers, watchtowers, etc. Many towers in the world are already sightseeing attractions or landmarks, such as *Eiffel Tower* in France (Figure 1.13) and *Canton Tower* in China (Figure 1.14).

Figure 1.13　Eiffel Tower (FR, 1889)　　　Figure 1.14　Canton Tower (CN, 2009)

### (4) Spatial structure

With the development of the social economy, more and more steel structures began to use in spatial structures. A plate-like space truss with a simple structure has been applied earlier in China (Figure 1.15). The reticular shell structure has been used in many famous buildings. For example, the *National Centre for the Performing Arts of China* (Figure 1.16) is the palace of essential cultural exchange in China. The special-shaped space frame has been used in

Figure 1.15　Beijing Airport Hangar　　　Figure 1.16　National Centre for the Performing Arts of
(CN, unknown)　　　　　　　　　　　　China(CN, 2007)

some unique stadium buildings. For example, the *National Stadium* (Figure 1.17) was the main stadium for the 2008 Olympic Games. Cable-suspended structures have many applications in airport buildings, and one example is *Washington Dulles International Airport* (Figure 1.18). Architects widely love cable-dome structures for their beautiful appearance, for instance, the *Millennium Dome* (Figure 1.19).

Figure 1.17  National Stadium (CN, 2008)

Figure 1.18  Washington Dulles International Airport (USA, unknown)

Figure 1.19  Millennium Dome (UK, 1999)

## 1.3 Applications of Steel Structures

(5) Bridge

The use of steel in bridges was earlier than that in buildings. Some famous bridges include the *Zhongshan Bridge* (Figure 1.20) on the *Yellow River* in China and the *Golden Gate Bridge* in the United States (Figure 1.21).

Figure 1.20  Zhongshan Bridge (CN, 1909)

Figure 1.21  Golden Gate Bridge (USA, 1937)

(6) Pressure vessels and pipelines

Because the steel has good airtightness and water tightness, the pressure vessels and pipelines are almost steel structures. Figure 1.22 shows an oil refinery, including the pressure vessels and pipelines.

Figure 1.22  An oil refinery (CN, unknown)

## 1.4  Classification of Structural Steel Members

There are five types of basic members in the steel structure according to the characteristics of the loading.

(1) The tension member

The tension member is a structural member subjected to direct axial forces that tend to elongate the member, shown in Figure 1.23.

(2) The axially loaded compression member

When the force's direction in Figure 1.23 is exactly the opposite, the axial tension member becomes an axially loaded compression member. One of the axially loaded compression members is the column member, shown in Figure 1.24. In a steel truss bridge

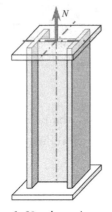   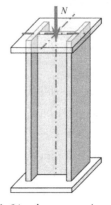

Figure 1.23  A tension member           Figure 1.24  A compression member

shown in Figure 1.25, the tension members are the truss's low chords or some diagonals. The axially loaded compression members are top chords of the truss and some diagonals.

Figure 1.25　A steel truss bridge

(3) Beam

Beams are structural members that carry loads applied at right angles to the member's longitudinal axis, and these loads cause the member to bend. Figure 1.26 shows the primary beams and secondary beams in a steel frame.

Figure 1.26　Primary beams and secondary beams of a steel frame

(4) Members under combined axial forces and bending moments

There are two members under combined axial forces and bending moments, shown in Figure 1.27. For the members subject to compressive forces and bending moments, we call them compressive-bending members in China. In fact, its global name is beam-column. In steel structures, the beam-column is the most crucial member. Prof. W. F. Chen, one of the most famous scholars in the steel structures' research field, has published a book named *Theory of Beam-Column*.

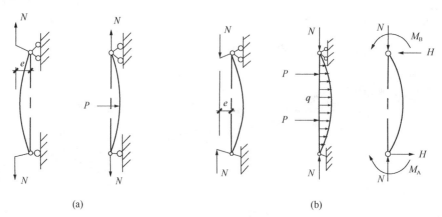

Figure 1.27 Members under combined axial forces and bending moments
(a) tensile forces and bending moments; (b) compressive forces and bending moments (beam-columns)

To accurately analyze and calculate the mechanical performance of various structures of steel structures, it is necessary to grasp the basic knowledge of the construction of steel structures and the basic theories of analysis—the main purpose of the Basic Principles of Steel Structures. The following chapters elaborate on steel structure materials, beams, tension members, axially loaded compression members, tension-bending and compression-bending members, and the steel structure's connection methods (welding and bolting). In addition, the basic principles of their analysis are explained.

## 1.5 The Basic Design Method of Steel Structures

### 1.5.1 The purpose of steel structure design

The purpose of structural design is to ensure that the designed structure and structural components can meet the expected safety and usability requirements during the construction and the service life. Therefore, the structural design criteria could be defined: the effects (internal force and deformation) of the structure produced by various loads are not greater than the resistance or specified limits of the structure (including connections) determined by the material properties and geometric factors. The steel structure designed according to the above design criteria should meet the following structural-functional requirements.

(1) Safety

The structure can maintain the necessary overall structure and component stability under various loads and actions considered in the design.

(2) Applicability

It can meet normal use requirements under normal loads, such as no excessive deformation or vibration that affects normal use.

(3) Durability

Under normal maintenance, the structure has sufficient durability. For example, no abnormal rust will occur and affect the service life of the structure.

**1.5.2 Probabilistic limit state design method**

The probabilistic limit state design method is the basic engineering structure design method in China, including the steel structure design. When designing a structure according to the limit state, the limit state concept should be clarified first. When the structure or its components exceed a specific state and cannot meet a certain functional requirement specified by the design, this specific state is called the function's limit state. The limit state of the structure can be divided into the following two categories.

(1) Ultimate state of bearing capacity

The state that the structure or structural member reached the maximum bearing capacity or occurred deformation that is not suitable for continued bearing, including overturning, strength failure, fatigue failure, loss of stability, the structure which has become a mobile system, or excessive plastic deformation, etc.

(2) Normal service limit state

It corresponds to a certain specified limit of normal use or durability performance of a structure or structural member, including deformation that affects normal use or appearance, local damage that affects normal use or durability, and vibration that affects normal use.

Let the structure or member's resistance be denoted by $R$, the effect of the action on the structure or component is denoted by $S$, and $Z = R - S = 0$ represents the limit state. $R$ and $S$ are affected by many random factors and have uncertainties. Therefore, they are random variables and should be analyzed by probability theory.

The fixed value design method considers that $R$ and $S$ are both deterministic. As long as the structure is designed as $Z \geqslant 0$ and given a certain safety factor, it is safe. But it is the fact that structural failure instances are still heard from time to time. That is due to the basic variables' uncertainty, indicating that the structure's load may have a high value. The material properties may also have a low value, even if the designer adopts a fairly conservative design plan. But after the structure is put into use, no one can guarantee that it is reliable, so the function of the designed structure can only be guaranteed with a certain probability. As long as the probability of failure is small to an acceptable level, the designed structure can be considered safe.

### 1.5.3 Design expression

The structural failure probability analysis or structural reliability analysis is more complex. The probabilistic limit state design method is transformed into a series of design expressions with the basic variable's standard values and partial coefficients. The design expression can adopt the formula stipulated in the latest version of the *Load Code for the design of Building Structure* GB 50009. The partial load coefficient can adopt the values specified in the latest version of the *Unified Standard for Reliability Design of Building Structures* GB 50068. The partial coefficients of steel resistance are adopted following the latest *Standard for Design of Steel Structures* GB 50017.

The ultimate state design of the bearing capacity of the structure or structural member damage or excessive deformation shall meet the following equation requirements.

$$\gamma_0 S_d \leqslant R_d \tag{1-1}$$

where $\gamma_0 =$ the structural importance coefficient, and its value is adopted following the *Unified Standard for Reliability Design of Building Structures* GB 50068;

$S_d =$ the design value of the effect of the combination of actions, such as the design value of the axial force, the design value of the moment or the design value of several axial forces, the design value of the moment vector;

$R_d =$ the design value of the resistance of the structure or structural member.

For permanent design conditions and short-term design conditions, the basic combination of functions should be used. The effect design value of the basic combination should be determined according to the most unfavorable value in Equation (1-2).

$$S_d = S\left(\sum_{i \geqslant 1}\gamma_{Gi}G_{ik} + \gamma_P P + \gamma_{Q1}\gamma_{L1}Q_{1k} + \sum_{j>1}\gamma_{Qj}\psi_{cj}\gamma_{Lj}Q_{jk}\right) \tag{1-2}$$

where $S(\cdot) =$ effect function;

$P =$ relevant representative value of prestress;

$Q_{1k} =$ standard value of the first variable effect;

$Q_{jk} =$ standard value of the $j$-th variable effect;

$\gamma_{Gi} =$ the permanent effect sub-factor should be adopted according to Table 1.1;

$\gamma_P =$ partial coefficient of prestressing should be adopted according to Table 1.1;

$\gamma_{Q1} =$ the first variable effect sub-factor should be adopted according to Table 1.1;

$\gamma_{Qj} =$ the $j$-th variable effect sub-factor should be adopted according to Table 1.1;

$\gamma_{L1} =$ the first load adjustment factors considering the structural design service life should be adopted according to Table 1.2;

$\gamma_{Lj} =$ the $j$-th load adjustment factors considering the structural design service life should be adopted according to Table 1.2;

$\psi_{cj}$ = the combined value coefficient of the $j$-th variable action shall be adopted following the relevant design specifications.

**Partial coefficient of building structures effect**      Table 1.1

| Partial coefficient \ Applicable situation | When the action effect is unfavorable to the bearing capacity | When the action effect is favorable to the bearing capacity |
| --- | --- | --- |
| $\gamma_G$ | 1.3 | $\leqslant$1.0 |
| $\gamma_P$ | 1.3 | 1.0 |
| $\gamma_Q$ | 1.5 | 0 |

**Load adjustment factor $\gamma_L$ considering the design service life of the building structures**      Table 1.2

| The design service life of the building structures/year | $\gamma_L$ |
| --- | --- |
| 5 | 0.9 |
| 50 | 1.0 |
| 100 | 1.1 |

The normal service limit state must meet the requirements of the *Unified Standard for Reliability Design of Building Structures* GB 50068. The standard combination of load, frequent combination, and the quasi-permanent combination is used for the design, and the deformation and other designs should not exceed the corresponding specified limit.

The steel structure only considers the standard combination of loads, and its design formula is:

$$v_{Gk} + v_{Q1k} + \sum_{i=2}^{n} \psi_{ci} v_{Gik} \leqslant [v] \tag{1-3}$$

where   $v_{Gk}$ = the standard value of the permanent load is the deformation value produced in the structure or structural member;

$v_{Q1k}$ = the deformation value produced by the standard value of the first variable load in the structure or structural member;

$v_{Qik}$ = the deformation value produced by the standard value of the i-th variable load in the structure or structural member;

$[v]$ = allowable deformation value of the structure or structural member. For tensile members, compression members, and beam-columns, the deformation refers to the slenderness ratio; for beams and trusses, the deformation refers to deflection.

## 1.6 Chapter Summary

The steel structure has a long history, and humanity has a long history of theoretical research and practical application of steel structure. Compared to reinforced concrete structures and wooden structures, steel structures have the advantages of high industrialization, high strength, good toughness, excellent seismic performance, and environmental protection. This chapter briefly introduces the characteristics, development history, application scope, classification of steel components, and basic design methods of steel structures. In order to study this course smoothly in the future, it is necessary to review the relevant knowledge of Material Mechanics and Structural Mechanics, actively formulate a scientific study plan and learn the relevant content of this course effectively.

## Questions

1.1  Give examples to illustrate the advantages and disadvantages of steel structures.

1.2  What are the common structural forms of steel structures?

1.3  What are the common member forms of steel structure?

1.4  What is the application trend of steel structures in super high-rise buildings?

1.5  Regarding the current design specifications, write out the steel structure's design expressions corresponding to the ultimate state of bearing capacity and give each symbol's meaning and value.

# Chapter 2  STRUCTURAL STEEL

**GUIDE TO THIS CHAPTER:**

➢ *Knowledge points*

   *The working performance of structural steel; the steel structure requirements for material performance; the main factors affecting steel performance; the type, specification, and selection principle of structural steel.*

➢ *Emphasis*

   *The requirements of steel structure for material performance; the main factors affecting the performance of steel.*

➢ *Nodus*

   *Appropriate selection of structural steel.*

## 2.1 Introduction

The steel material properties strongly influence a structure's behavior under load in a steel structure's construction. There are many types of steel in China, and material properties are also different. Among hundreds of carbon steel and alloy steel, the *Standard for Design of Steel Structures* GB 50017 only recommends one carbon structural steel Q235, and five low-alloy high-strength structural steels are Q345, Q390, Q420, Q460, and Q345GJ. The steel used for the steel structure must meet the following requirements.

(1) High tensile strength and yield strength

Tensile strength measures the strength of steel after a large deformation, and it directly reflects the quality of the steel's internal structure. At the same time, the high tensile strength can increase the safety of the structure. The yield strength is an index to measure the bearing capacity of the structure. Therefore, high yield strength can reduce the structural weight, steel consumption, and structural costs. In this book, the tensile strength is expressed as $f_u$, and the yield strength is expressed as $f_y$.

(2) Good plasticity and toughness to reduce the tendency to brittle failure

Steel structures require sufficient strain capacity under static and dynamic loads, but steel structures tend to undergo brittle failure under certain design conditions or use environments. Therefore, the good plastic properties and toughness of steel enable structural members to adjust the local stress through larger plastic deformation, and at the same time, have better resistance to repeated loads.

## Chapter 2 STRUCTURAL STEEL

### (3) Excellent process performance

Excellent process performance means it does not significantly affect the structure's mechanical performance. Excellent process performance refers to cold working, hot working, and welding performance.

According to the above requirements, the *Standard for Design of Steel Structures* GB 50017 specifically stipulates that the load-bearing structure's steel should have tensile strength, elongation, yield strength, and qualified guarantee sulfur and phosphorus content; and the qualified guarantee of carbon content for welded structure. The steel materials used in welded load-bearing structures and important non-welded load-bearing structures shall also be qualified for the cold bending test. For structural steel that need to check fatigue strength, according to the specific circumstances, it shall have the qualified guarantee of normal temperature or negative temperature impact toughness.

## 2.2 Several Important Performance Indexes of Structural Steel

### 2.2.1 Strength performance

The unidirectional tension test is the most basic mechanical property test of steel. A typical stress-strain relationship for low carbon or low alloy steel is shown in Figure 2.1. There are several important data points on this curve that need special attention. The typical stress-strain diagram reflects that the steel will go through four ranges, from the beginning of the test to the breaking.

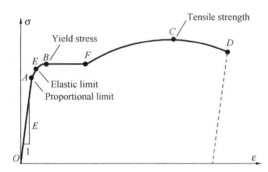

Figure 2.1 Stress-strain relationship for low carbon steel or low alloy steel

(1) Elastic range ($O$-$E$ segment)

In Figure 2.1, the $O$-$A$ segment is a straight line that indicates that the steel has full elastic properties. At this time, the stress can be defined as $\sigma = E\varepsilon$, $E$ is Young's elastic modulus. The stress corresponding to point $A$ is called the proportional limit $f_p$. The curve's $A$-$E$ segment is still elastic, but it is nonlinear in the nonlinear elastic range. The modulus is called the tangent modulus $E_t = d\sigma/d\varepsilon$, and the stress $f_e$ at the upper limit $E$ point of this segment is called the elastic limit. The elastic limit and the proportional limit are very close. However, it isn't easy to distinguish, so there is usually only a proportional limit.

(2) Yielded range ($E$-$B$-$F$ segment)

As the load increases, the curve reaches the $E$-$B$ segment. At this time, it is inelastic. That is, the unloading curve will leave permanent residual deformation. The stress $f_y$

## 2.2 Several Important Performance Indexes of Structural Steel

at the upper limit point $B$ of this segment is called the yield stress. For low carbon steel, there is often an obvious yield step $B$-$F$ section. That is, the strain continues to increase under the condition that the stress maintains little fluctuation.

(3) Strengthening range ($F$-$C$ segment)

After point $F$ of the yielding platform is exceeded, the material undergoes strain hardening, and the curve rises to point $C$ at the highest point of the curve. The stress $f_u$ at this point is called the tensile strength or ultimate strength. When the stress $f_y$ at the yield point is used as the strength limit, the tensile strength $f_u$ becomes the material's strength reserve.

(4) Neckingrange ($C$-$D$ segment)

When the stress reaches point $C$, the carbon steel specimen shrinks and finally breaks at point $D$.

Instead of the real model of Figure 2.1, the idealized model of Figure 2.2 is used to control the engineering design. In China, to simplify the calculation, it is usually assumed that the steel is completely elastic before the yield point. Generally, only the first two ranges are taken, and the material is assumed to be an ideal elastic-plastic model. Thus, the yield strength $f_y$ and tensile strength $f_u$ are the strength performance indexes of structural steel.

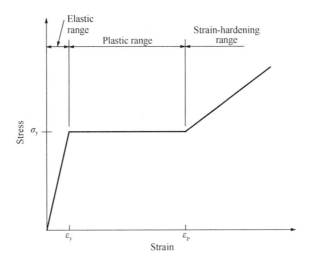

Figure 2.2 Idealized stress-strain diagram for low carbon steel or low alloy steel

High-strength steel has no obvious yield point and yields platform. Therefore, this type of steel's yield conditions is artificially specified based on the test analysis results, conditional yield point, or yield strength. For example, the conditional yield point is denoted by $f_{0.2}$, which corresponds to 0.2% residual strain after unloading, as shown in Figure 2.3. Since this type of steel does not have an obvious plastic platform, its plasticity should not be used in the design.

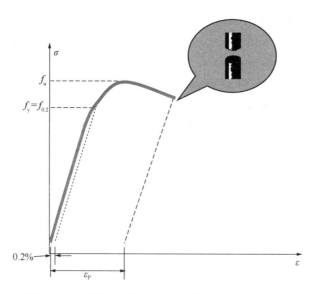

Figure 2.3 The yield strength of high-strength steel

### 2.2.2 Plastic performance

The percentage of the specimen's absolute deformation value to the original gauge length when the specimen is broken is called the elongation $\delta$. Elongation $\delta$ represents the plastic strain capacity of the material when stretched in one direction. The area shrinkage rate $\psi$ refers to the percentage of the reduced value of the cross-sectional area of the necked area to the original cross-sectional area after the specimen is broken. The area shrinkage rate $\psi$ is the maximum plastic deformation that the steel can produce under the necking zone's stress state. The area shrinkage rate $\psi$ is a more realistic and stable index to measure steel's plasticity, but it isn't easy to measure accurately. Therefore, the plasticity index of steel is still guaranteed by elongation $\delta$. Yield strength, tensile strength, and elongation are the three most important mechanical properties of steels.

### 2.2.3 Steel physical performance index

The steel's force performance under unidirectional compression (non-buckling conditions) is the same as it under the unidirectional tension state. Likewise, the performance under shear is similar, but the shear modulus $G$ is also lower than the elastic modulus $E$.

The elastic modulus $E$, shear modulus $G$, linear expansion coefficient $\alpha$, and mass density $\rho$ of structural steel are shown in Table 2.1.

Physical performance indicators of steel and steel castings  Table 2.1

| Elastic modulus $E$ (N/mm²) | Shear modulus $G$ (N/mm²) | Linear expansion coefficient $\alpha$ (/℃) | Mass density $\rho$ (kg/m³) |
|---|---|---|---|
| $2.06 \times 10^5$ | $7.9 \times 10^4$ | $1.2 \times 10^{-5}$ | 7850 |

## 2.2.4 Cold bending performance

The cold bending performance is measured by the cold bending test (Figure 2.4). During the test, press the punch on the testing machine according to the specified bending core diameter to make the specimen bend 180°. If there is no crack or delamination on the specimen's surface, it is qualified. The cold bending test can directly test the steel's bending deformation ability or plastic properties and expose the steel's metallurgical defects, such as the segregation of sulfur and phosphorus and sulfide doping and oxide. These defects will reduce the cold bending performance of steels. Therefore, the cold bending performance is a comprehensive index to identify the steel's plastic strain capacity in the bending state and the steel quality.

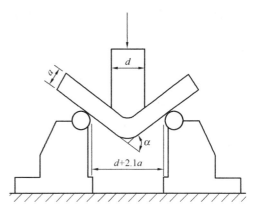

Figure 2.4 Test set-up for the cold bending test

## 2.2.5 Impact toughness

The steel properties, such as strength and plasticity, shown in the tensile test, are static and cannot reflect the steel's ability to prevent brittle fracture. Moreover, toughness is the ability of steel to resist impact loads. It is measured by the total energy absorbed by the material when it is broken, including elastic energy and inelastic energy. Thus, toughness is a comprehensive index of steel strength and plasticity. Generally, the increase in steel strength and the decrease in toughness indicate that the steel tends to be brittle.

In China, the Charpy specimen with V-notch is used to check impact toughness on a special Charpy test machine, shown in Figure 2.5. The Charpy specimen's size is 10mm×10mm×55mm. The specimen will rupture when the hammer drops at a certain height. The impact of work done by the hammer is the impact toughness, and the unit is Joule.

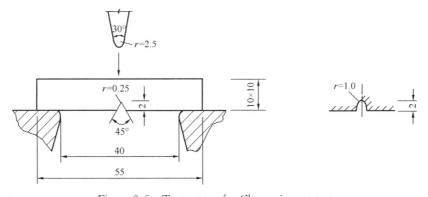

Figure 2.5 Test set-up for Charpy impact test

Because low temperature has a significant influence on the brittle failure of steel, structures built in cold regions require not only the impact of toughness at room temperature (20℃) but also the impact toughnesses at 0℃, −20℃, and −40℃, to ensure that the structure has sufficient resistance to brittle failure.

### 2.2.6 Structural steel weldability

In steel structures, welding is the most commonly used connection method, so structural steel weldability must meet specific requirements. First, there are no welding cracks in the weld and the adjacent metal. Second, the weld plasticity and mechanical properties and the nearby metal are not lower than the base material.

Chemical elements significantly influence the weldability of low-carbon steel, and the specific content will introduce it in the next section. The control of weldability is achieved for low alloy steels by controlling the steel's carbon equivalent, calculated according to Equation (2-1). Carbon equivalent $C_E$ is less than 0.38%.

$$C_E = C + \frac{Mn}{6} + \frac{(Cr + Mo + V)}{5} + \frac{(N_i + Cu)}{15} \tag{2-1}$$

It should be pointed out that the method of using carbon equivalent to determine the weldability of low-alloy steel is not completely reliable and cannot completely guarantee the safety and reliability of steel welding. Therefore, the welding method, the electrode, and the welding temperature must be considered comprehensively. Steels with poor weldability (high carbon equivalent) require strict welding processes.

## 2.3 Stress-Strain Relationship of Structural Steel under Different Loadings

Figure 2.1 shows a typical stress-strain relationship for low carbon or low alloy steel under unidirectional tension loading. The steel performance under repeated uniaxial stress can be expressed by the stress-strain relationship curve shown in Figure 2.6. When the component stress is $|\sigma| < \sigma_y$, the steel is in the elastic stage, the elastic deformation is recoverable, and there is no residual deformation. At this time, the material properties of the steel do not change under repeated stress. When the stress of steel is $|\sigma| \geqslant \sigma_y$, the steel is in the elastoplastic stage, and repeated stress will cause the growth of plastic deformation. When $\sigma$ exceeds the yield strength $\sigma_y$ of the steel, reloading following unloading immediately does not change the stress-strain relationship, as shown in Figure 2.6(a). If there is a certain intermittent time before reloading, for example, more than five days at room temperature, the curve after reloading is shown in Figure 2.6(b). The steel's yield strength is improved, the plasticity is reduced, and the tensile strength is slightly improved. This phenomenon is called the aging effect of steel. Figure 2.6(c) shows that

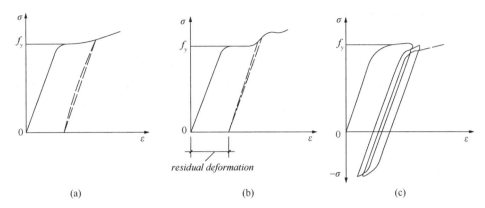

Figure 2.6  σ-ε curve of steel under repeated or cyclic loading

steel's compressive performance has deteriorated after being tensioned, called the *Bauschinger Effect*.

The stress-strain relationship curve forms a hysteresis loop (hysteresis curve). The area enclosed by the hysteresis curve represents the energy absorbed by a unit volume of steel during a load cycle.

Under repeated loads, the structure's resistance and performance will undergo important changes, and even fatigue failure will occur. Fatigue failure is manifested as a brittle fracture that occurs suddenly. The experiments have proved that steels with low component stress levels or few repetitions generally do not undergo fatigue failure. Therefore, the effect of fatigue does not need to be considered in the calculation. However, structural members and their connections subjected to frequently repeated loads for a long time, such as crane beams of heavy-duty cranes, must be designed to resolve structural fatigue problems. Regarding the fatigue failure of steel, we will introduce it in more detail in the next chapter.

## 2.4 Performance of Steel under Complex Stress

In the uniaxial tensile test of steel, when the uniaxial stress reaches the yield point $\sigma_y$ of the material, the steel enters the plastic state. However, under complex stress, such as a plane or three-dimensional stress state (Figure 2.7), the equivalent stress $\sigma_{zs}$ is usually used to determine the real stress situation. Equation (2-2) expresses the calculation method of the equivalent stress $\sigma_{zs}$ according to the energy intensity theory (or the fourth intensity theory). Using equivalent stress and the material's yield strength, we can judge the steel's stress state under a complex stress state. If the equivalent stress is larger than the yield strength $\sigma_y$, the steel is in the

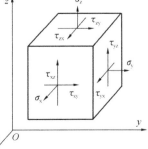

Figure 2.7  Three-dimensional stress diagram

plastic state. Otherwise, the steel is in an elastic state.

$$\sigma_{zs} = \sqrt{\sigma_x^2 + \sigma_y^2 + \sigma_z^2 - (\sigma_x\sigma_y + \sigma_y\sigma_z + \sigma_z\sigma_x) + 3(\tau_{xy}^2 + \tau_{yz}^2 + \tau_{zx}^2)} \quad (2\text{-}2)$$

If one of the stresses in three-dimension is very small, such as the thickness of the steel plate is small, the stress in the thickness direction is negligible (or zero), it belongs to the plane stress state, and Equation (2-2) is rewritten as Equation (2-3).

$$\sigma_{zs} = \sqrt{\sigma_x^2 + \sigma_y^2 - \sigma_x\sigma_y + 3\tau_{xy}^2} \quad (2\text{-}3)$$

In a general beam, there are only normal stress and shear stress, then,

$$\sigma_{zs} = \sqrt{\sigma^2 + 3\tau^2} \quad (2\text{-}4)$$

For the pure shear state, let $\sigma=0$ in Equation (2-4), then there is, $\sigma_{zs} = \sqrt{3\tau^2} = \sqrt{3}\tau = f_y$
The shear yield strength of steel can be expressed as,

$$\tau = f_y/\sqrt{3} = 0.58 f_y \quad (2\text{-}5)$$

## 2.5 Main Factors Affecting the Performance of Structural Steel

### 2.5.1 Chemical compositions of steel

Steel is composed of various chemical components, and the chemical composition and its content have an important influence on the properties of steel, especially the mechanical properties. Ferrum is the fundamental element, accounting for about 99% of carbon structural steel. Carbon and other elements account for 1%. Still, they have a decisive influence on the mechanical properties of steel. Other elements include silicon, manganese, sulfur, phosphorus, nitrogen, oxygen, etc. In low alloy steel, the total alloying elements are less than 5% content, including copper, vanadium, titanium, niobium, chromium, etc.

(1) Fundamental elements

The ferrum and corban are the fundamental elements in structural steel. For carbon structural steel, the finer the ferrite grains are, the better the steel performances. In carbon structural steel, carbon is another fundamental element which directly affects strength, plasticity, toughness, and steel weldability. As the carbon content increases, the strength of the steel increases, while the plasticity, toughness, and fatigue strength decrease, and at the same time deteriorate the welding performance and corrosion resistance of the steel. Therefore, although carbon is the main element to obtain sufficient steel strength, the carbon content of carbon structural steel used in steel structures should be limited. Generally, it should not exceed 0.22%. In welded structural steel, it should be less than 0.20%. When low-alloy, high-strength structural steel is used for the steel structure, its carbon equivalent (see Section 2.2.6) is generally used to evaluate the steel weldability.

### (2) Harmful elements

Sulfur, phosphorus, nitrogen, and oxygen are harmful elements. Increasing sulfur content will cause hot brittleness of steel. However, increasing sulphur content can reduce plasticity, toughness, weldability, fatigue performance, and rust resistance. Increasing phosphorus content can increase steel strength and rust resistance, but it will cause cold brittleness.

Moreover, increasing phosphorus content reduces plasticity, toughness, and cold-bend performance. Oxygen and sulfur are elements of the same family. So the effects of oxygen and sulfur are similar, and the results are severer than sulfur. Nitrogen and phosphorus are elements of the same family. So the effects of nitrogen and phosphorus are similar. Sulfur and phosphorus content is less than 0.045% in carbon structural steel and is less than 0.035%, respectively.

Copper is an impurity component in carbon steel, improving corrosion resistance and strength, reducing weldability.

### (3) Beneficial elements

Silicon and manganese are both deoxidizers for steel making. Manganese is a beneficial element. Increasing manganese content can increase steel strength and can reduce weldability. In carbon structural steel, manganese content is less than 0.7%. Silicon is another beneficial element. Increasing silicon content can increase steel strength and can reduce weldability and rust resistance. In carbon structural steel, the silicon content is less than 0.4%. Besides, vanadium and titanium are alloy elements in steel, improving steel strength and corrosion resistance without significantly reducing steel plasticity.

### 2.5.2 Production process

The composition and proportion of trace elements in steel can be adjusted through smelting. But the smelting process will also bring some defects to the steel, including inconsistency, evenness of chemical composition, and non-metallic inclusions. The casting process determines whether the steel is boiling steel or semi-killed steel, or killed steel. However, casting defects are challenging to avoid. Steel rolling can improve the internal structure, eliminate defects, and improve mechanical properties. Steel rolling generally divides into hot rolling and cold rolling.

After the steel is hot rolled, residual stress will be generated due to uneven cooling. Generally, tensile stress will be generated where cooling is slow, and compressive stress will be generated where cooling is fast. Figure 2.8 shows the residual stress of hot-rolled steel. Residual

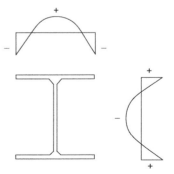

Figure 2.8 Residual stress distribution of hot-rolled H-shaped steel

stress is internal self-balanced stress.

### 2.5.3 Steel hardening

Cold working such as cold drawing, cold bending, punching, mechanical shearing, etc., can cause large plastic deformation of the steel, increasing the steel's yield point and reducing the steel's plasticity and toughness. This phenomenon is called cold work hardening or strain hardening. Nitrogen and carbon are gradually precipitated from pure iron to form free carbides and nitrides with time growth, restraining a pure iron body's plastic deformation. This phenomenon is called age hardening, commonly known as aging. The process of age hardening is generally very long.

However, if the material is heated after plastic deformation, the age hardening can develop particularly rapidly. This method is called artificial aging. Besides, strain aging is strain hardening, and aging hardening is added after cold work hardening. In general, the strength increased by hardening is not used. However, some important structures require the steel to be artificially aged to test its impact toughness to ensure that the structure has sufficient resistance to brittle failure. Also, the locally hardened part should be eliminated by edge planning or drilling.

### 2.5.4 Temperature

The tendency of steel properties to change with temperature changes is that when the temperature increases, the steel strength decreases, and the strain increases. On the contrary, when the temperature decreases, the steel strength increases slightly, but the plasticity and toughness decrease and become brittle, as shown in Figure 2.9. When $T \leqslant 200$ ℃, there is no significant change. When the temperature is between 430 ℃ and 540 ℃, the steel strength drops sharply. At 600 ℃, the steel is shallow in strength and cannot bear the

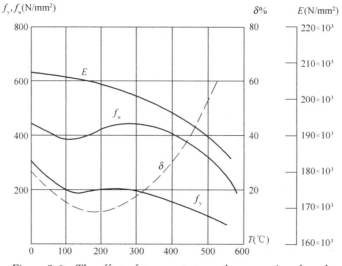

Figure 2.9 The effect of temperature on the properties of steel

load. But at around 250 ℃, the strength of steel is slightly increased, while plasticity and toughness are reduced. As a result, the material tends to become brittle. The oxide film on the steel surface appears blue, which is called blue brittleness. Steel should be avoided in the blue brittleness temperature range in hot processing. When the temperature is between 430 ℃ and 540 ℃, the steel will continue to deform at a very slow speed under the condition of constant stress. This phenomenon is called the creep phenomenon.

### 2.5.5 Stress concentration

The working performance and mechanical performance indicators of steel are based on the uniform distribution of stress in the axial tension member along the section. However, there are often holes, notches, concave angles, sudden changes in section and steel internal defects, etc. At this time, the stress distribution in the component will no longer remain uniform. Still, local peak stress will be generated in some areas, and the stress will be reduced in other areas, forming the so-called stress concentration phenomenon. Three leading reasons cause stress concentration. The first is the cross-section changing of members. The second is the holes, notches, and concave corners of members. The third is the internal defects in the steel of members. Figure 2.10 shows the schematic diagram of stress concentration. It can be seen that the peak stress appears near the hole.

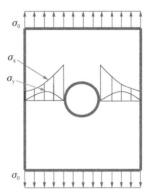

Figure 2.10 Schematic diagram of stress concentration

Stress concentration will put the structure in a complex stress state, which is the source of brittle fracture of the structure working under negative temperature or dynamic load. Therefore, measures should be taken to avoid or reduce stress concentration, and high-quality steel should be selected.

## 2.6 Types and Specifications of Structural Steel

### 2.6.1 Classification of steel in China

In China, carbon structural steel, low alloy structural steel, high-quality carbon structural steel, weathering steel, and fire-resistant steel are the main kinds of steel used in steel structures. However, the classification of steel is various in China.

According to the deoxygenation method, steels can be divided into boiling steel, killed steel, semi-killed steel, and special killed steel. The code for them is F, Z, b, and TZ, respectively. Boiling steel has gradually withdrawn from the market, and killed steel is the most popular steel product in the market. However, it should be noted that F and b must be used with steel grades, while Z and TZ can be omitted when used with steel grades. As a result, the market share of semi-killed steel and special killed steel are both minimal.

According to the molding method, steels are divided into rolled steel, forged steel, and cast steel. The rolled steels are divided into hot-rolled steel and cold-rolled steel. Hot-rolled steel is commonly used in civil construction, and forged steel is rarely used in civil construction. The price of cast steel is generally very high, and it is always used in complex parts of the steel structure, such as complex joints or connections.

According to carbon content and chemical composition, structural steels are divided into carbon structural steel and low alloy structural steel.

(1) Carbon structural steel

There are five grades of carbon structural steel in China according to *Carbon Structural Steels* GB 700, which are Q195, Q215, Q235, Q255, and Q275. Only Q235 can be used in the building structure as structural steel. Therefore, Q represents the steel's yield strength, and the following number represents the value of the yield strength, and the unit is $N/mm^2$.

To ensure that the steel's mechanical properties meet the standard requirements, the carbon, manganese, and silicon content of level A steel cannot be used as delivery conditions. Still, its content should be indicated in the quality certificate. Likewise, levels B, C, and D require all the mechanical properties such as yield strength, tensile strength, elongation, cold bending, and impact toughness should be guaranteed.

There are four quality levels of carbon structural steel, A, B, C, and D. Level A does not require impact toughness. Therefore, the cold-bend test only carries out when required by the buyer. Levels B, C, and D require impact toughness at 20 ℃, 0 ℃, and −20 ℃, respectively, and all need the cold-bend test. For Q235 steel, level A and level B are boiling steel or killed steel, or semi-killed steel. However, level C must be killed steel, and level D must be the special killed steel.

(2) Low alloy structural steel

Low alloy steel is the addition of several small amounts of alloying elements to ordinary carbon steel. As the total amount is less than 5%, it is called low alloy steel. According to *High Strength Low Alloy Structural Steels* GB/T 1591, there are five low alloy structural steel grades which are Q295, Q355, Q390, Q420, and Q460. The last four steel grades can be used in the building structure as structural steel following the latest *Standard for Design of Steel Structures* GB 50017. Same as carbon structural steel, the number represents the yield strength, and the unit is $N/mm^2$. It is worth pointing out that Q355 is the Q345 steel in the *Standard for Design of Steel Structures* GB 50017.

There are five quality levels of low alloy structural steel, A, B, C, D, and E. For levels A, B, C, and D, low alloy structural steel requirements are the same as carbon structural steel. The level E of low alloy structural steel requires −40℃ impact toughness and a cold-bend test.

All grades of low alloy structural steels are killed steel or special killed steel, of which C, D, E grades should be special killed steels.

Low-alloy structural steel has high yield strength and tensile strength. It also has good plasticity and impact toughness, especially low-temperature impact toughness, corrosion resistance, and low-temperature resistance. The main purpose of using low-alloy structural steel is to reduce structural weight, steel consumption and extend the service life.

(3) High-quality carbon structural steel

From the manufacturing method, high-quality carbon structural steel is carbon steel after heat treatment. The main difference between high-quality carbon structural steel and carbon structural steel is that the steel contains fewer impurity elements. For example, sulfur and phosphorus content is no more than 0.035%, and other defects are strictly restricted. High-quality carbon structural steel is mainly used in certain joints of steel structures or as connectors. For example, No. 45 high-quality carbon structural steel is used to manufacture high-strength bolts requiring heat treatment, which has high strength, while plasticity and toughness are not significantly affected.

(4) Steel plate for building structure

For a high-rise building structure, there is an unique steel plate named high-performance building structural steel. Its code is capital GJ. There are five grades and four quality levels of GJ steel. Q235GJ, Q345GJ, Q390GJ, Q420GJ, and Q460GJ. The quality levels of Q235GJ and Q345GJ are B, C, D, and E. The quality levels of Q390GJ, Q420GJ, and Q460GJ are C, D, and E. However, the design standard GB 50017 only recommends Q345GJ.

Compared with low-alloy structural steels of the same grade, the content of harmful elements such as sulfur and phosphorus in GJ series steels is limited. As a result, the content of trace alloying elements is well controlled, the range of yield strength changes is small, and the plasticity is better.

(5) Weathering steel and fire-resistant steel

Weathering steel improves resistance to atmospheric corrosion. A small number of specific alloy elements are added to form a protective layer on the metal substrate surface during the steel smelting process. Its code is capital NH. The design standard GB 50017 recommends three kinds of NH steel, which are Q235NH, Q355NH, and Q415NH.

Fire-resistant steel adds a small number of precious metals to the steel to make it have high-temperature resistance. However, the high cost of fire-resistant steel makes it rarely used in buildings.

## 2.6.2 The choice principles of structural steel

The choice of structural steel is the most important part of the design of steel structures. The purpose of the choice is to ensure safety, reliability, and economy. Therefore, the following factors should be considered when choosing steels.

(1) The importance of structure

High-quality steel should be considered for important structures such as heavy industrial building structures, large-span structures, high-rise or super high-rise civil building structures, or special structures. However, ordinary quality steel can be selected for general industrial and civil building structures according to the nature of the work.

(2) Load situation

Loads can be divided into static loads and dynamic loads. Structures directly bearing dynamic loads and structures in strong earthquake areas should use steel with good overall performance. Generally, lower-priced steel can be used for structures bearing static loads.

(3) Connection method

There are two ways to connect steel structures: welding and non-welding. The welding process, welding deformation, welding stress, and other welding defects such as undercuts, pores, cracks, slag inclusions, will occur. As a result, there is a risk of cracking or brittle fracture in the structure. Therefore, the material requirements of the welded structure should be stricter.

(4) The temperature and environment

Steel is easy to be cold and brittle when it is at a low temperature. Therefore, the structure that works under low-temperature conditions, especially the welded structure, should be used to kill steel with good resistance to low-temperature brittle fracture. Besides, the open-air structure's steel is prone to aging, and the steel of the harmful medium is prone to corrosion, fatigue, and fracture. Therefore, different materials should also be selected differently.

(5) The thickness of structural steel

Thin steel has more rolling times, the compression ratio of rolling is large, and the compression ratio of thick steel is small. Therefore, the thick steel has lower strength and poor plasticity, impact toughness, and welding performance. This is why the thicker welded structure should be made of high-quality steel.

## 2.6.3 Steel selection recommendations

The steel used in the load-bearing structure should have yield strength, tensile strength, elongation after fracture, and the qualified guarantee of sulfur and phosphorus content. The welded structure should still have a carbon limit content guarantee or a carbon equivalent qualification guarantee. The steel used for welded load-bearing structures and important non-welded load-bearing structures shall have the qualification guarantee by a cold bending test. The steel used for the components directly bearing dynamic load or requiring fatigue calculation shall have a qualification guarantee of impact toughness. The selection of steel quality Level should meet certain requirements.

Level A steel can only be used for structures that do not require a fatigue check if the structural working temperature is higher than 0 ℃, and Q235A steel should not be used for welded structures.

When the structural working temperature is higher than 0 ℃, the steel should not be lower than level B for welded structures that require a fatigue check. When the structural working temperature is below 0 ℃ but higher than $-20$ ℃, the quality grade of Q235 steel and Q355 steel should not be lower than level C. Likewise, the quality grade of Q390 steel, Q420 steel, and Q460 steel should not be lower than level D. When the structural working temperature is not higher than $-20$ ℃, the quality grade of Q235 steel and Q355 steel should not be lower than level D. The quality level of Q390 steel, Q420 steel, and Q460 steel should be level E.

For non-welded structures that require fatigue calculations, the steel quality level requirements can be lower than the above-mentioned welded structures by one level. Still, they should not be lower than level B. Intermediate working crane beams with a crane lifting capacity of not less than 50 t, and their steel quality level requirements need to be checked. That is because the fatigue components are the same.

## 2.7 Structural Steel Products in China

### 2.7.1 Hot-rolled steel products in China

Common steel structure buildings use hot-rolled steel plates, hot-rolled shapes, etc.

(1) Hot-rolled plates (PL) and flat bars (FLT)

The first type of hot-rolled steel product is the plates (Figure 2.11) and flat bars (Figure 2.12). Plates and flat bars, all rectangular in cross-section, come in many widths and thicknesses. A flat shape has historically been classified as a bar if its width less than

Figure 2.11  Hot-rolled steel plates

Figure 2.12  Hot-rolled flat steel

and equal to 200 mm. Flat bars are rolled between horizontal and vertical rollers and trimmed to length by shearing or flame cutting on the ends. Plates are produced from billets by squeezing the hot metal between smooth cylindrical rolls adjusted to form the required width and thickness.

The thickness of steel plates from 4.5 mm to 60 mm is called thick steel plate. The thickness of steel plates from 0.35 mm to 4 mm is called thin steel plate. The width of hot-rolled plates is from 700 mm to 3,000 mm, and the width of flat bars is from 12 mm to 200 mm. Representation method: —width×thickness×length, unit: mm.

(2) Hot-rolled angle shapes

The second type of hot-rolled steel product is the *angle shape*. *Angle shapes* are sections whose cross-section is composed of two rectangular elements called legs perpendicular to one another. The angles having legs of equal lengths are called *equal leg angles* (Figure 2.13), while those with unequal legs are known as *unequal leg angles* (Figure 2.14). The inner and outer surfaces of each leg are parallel, and the thickness of both the legs is the same no matter whether the angle has equal or unequal legs. Angles are designated by the alphabetical letter symbol L, followed by leg lengths (always the longer leg first) and thickness in millimeters. For instance, *unequal legs angles*: L long-leg length × short-leg length × thickness; *equal legs angles*: L leg length × thickness, and the unit is mm. Thus, the designation L100×60×6 stands for an unequal leg angle whose long leg is 100 mm long and 6 mm thick, and whose short leg is 60 mm long and 6 mm thick. Similarly, the designation L100×6 stands for an equal leg angle whose legs are 100 mm long and 6 mm thick.

Figure 2.13 Equal leg angles

Figure 2.14 Unequal leg angles

(3) Hot-rolled I-shapes

The third type of hot-rolled steel product is the *I-shapes*, divided into *universal I-shapes* (Figure 2.15) and *lightweight I-shapes* (Figure 2.16). An *I-shape* has two horizontal rectangular elements, called flange, and a vertical rectangular element called web, connected by fillets as shown in Figures 2.15 and 2.16. The cross section has two axes of symmetry. These shapes have a depth much greater than the flange width, and the flange thickness is generally greater than the web thickness. The inner and outer surfaces of the top and bottom flange are not parallel because the thickness of the flange near the web is greater than the thickness near the edge. The *universal I-shape* is identified by the alphabetical symbol I, followed by the normal depth in centimeters. Thus, the designation I20 represents an *I-shape* Whose depth is 20 cm.

Figure 2.15   Universal I-shapes        Figure 2.16   Lightweight I-shapes

Similarly, the *lightweight I-shape* is identified by the alphabetical symbol QI, followed by the normal depth in centimeters. Thus, the designation QI16 represents a *lightweight I-shape* whose depth is 16 cm.

For I-shapes over I20, there are three web thicknesses for the same number: a, b, and c, for example, I30a, I30b, and I30c.

(4) Hot-rolled channels

The fourth type of hot-rolled steel product is the *channels*, divided into *universal channels* (Figure 2.17) and *lightweight channels* (Figure 2.18). Like the standard I-shape, a *channel shape* has a web and two parallel flanges, with the inner surfaces of both flanges having a lope. Compared to the *universal channels*, and the *lightweight channels* have thinner webs and flanges than *universal channels*. Also, for these shapes, the actual

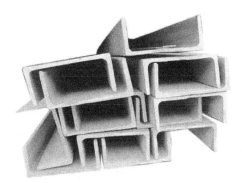

Figure 2.17   Universal channels        Figure 2.18   Lightweight channels

depth equals the nominal depth. A channel shape has only one axis of symmetry. The *universal channel* is identified by the symbol [, followed by the normal depth in centimeters. Thus, the designation [20 represents a *channel shape* whose depth is 20 cm.

Similarly, the *lightweight channel* is identified by the letter and symbol Q [, followed by the normal depth in centimeters. Thus, the designation Q [18 represents a lightweight I-shape whose depth is 18 cm.

For channels over [20, there are two or three web thicknesses for the same number: a, b, and c, for example, [25a, [25b, and [25c.

Figure 2.19 H-shapes

(5) Hot-rolled H-shapes

The fifth type of hot-rolled steel product is the *H-shapes* (Figure 2.19). There are three types of H-shapes products in China: *wide flange H-shapes*, *middle flange H-shapes*, and *narrow flange H-shapes*.

A wide flange H-shape has two horizontal rectangular elements, called flange, and a vertical rectangular element called a web, connected by fillets, as shown in Figures 2.19. The cross section has two axes of symmetry. These shapes have a depth equal to or less than the flange width. The inner and outer surfaces are parallel. The wide flange H-shape is identified by letters HW, followed by depth, flange width, web thickness, and flange thickness in millimeters. For instance, the designation HW $200 \times 200 \times 8 \times 12$ represents a wide flange H-shape section, 200 mm in depth, 200 mm in flange width, 8 mm in web thickness, and 12 mm in flange thickness.

A middle flange H-shape also has two horizontal flanges and a web and connected by fillets. These shapes have a flange width approximately equals to 2/3 depth. The middle flange H-shape is identified by letters HM, followed by depth, flange width, web thickness, and flange thickness in millimeters. For instance, the designation HM $550 \times 300 \times 11 \times 18$ represents a middle flange H-shape section that is 550 mm in depth, 300 mm in flange width, 11 mm in web thickness, and 18 mm in flange thickness.

Similarly, a narrow flange H-shape also has two horizontal flanges and a web and connected by fillets. These shapes have a flange width approximately equals to $1/3 \sim 1/2$ depth. The narrow flange H-shape is identified by letters HN, followed by depth, flange width, web thickness, and flange thickness in millimeters. For instance, the designation HN $400 \times 200 \times 8 \times 13$ represents a narrow flange H-shape section that is 400 mm in depth, 200 mm in flange width, 8 mm in web thickness, and 13 mm in flange thickness.

(6) Structural Tees

The sixth type of hot-rolled steel product is *structural Tees* (Figure 2.20). The cross section of a structural Tee resembles the letter T. The top part of a T-section is called its flange, and the lower part (vertical) is called its stem (more often called the web in China). T-sections are manufactured at the mill in China by cutting an HW- shape, HM- shape, or HN-shape in half along the web center using a rotary shear, resulting in two TW-shapes, TM-shapes, or TN-shapes, respectively. Thus, TW100×200×8×12 designates a structural Tee whose depth (tip of the stem to outside flange surface) is nominally 200 mm, flange width 200 mm, web thickness 8 mm, and flange thickness 12 mm.

Figure 2.20  Structural Tees

(7) Structural steel tubings

The seventh type of hot-rolled steel product is structural steel tubings, including square tubes (Figure 2.21) and round pipes (Figure 2.22). There are two types of square steel tubes commonly used in construction engineering. One is the hot-rolled seamless square tube, and the other is the welded square steel tube made of hot-rolled steel plates. Square steel tubes representation method is □ section width×section height×thickness, unit: mm. Like the square steel tubes, two kinds of steel pipes, one is the hot-rolled seamless steel pipe, and the other is the welded steel pipe made of hot-rolled steel plates. Steel pipes representation: $\phi$ outer diameter×thickness, unit: mm.

Figure 2.21  Square tube          Figure 2.22  Round steel pipes

All of the hot-rolled shapes are manufactured to certain tolerance with respect to variations in cross-sectional dimensions, lengths, squareness, flatness, weights, sweep, camber, and so on. The tolerances are contained in some production standards in China.

### 2.7.2 Cold-formed steel shapes in China

Cold-formed steel(CFS) shapes are often made of cold-rolled thin steel plates, generally Q215, Q235, or Q355. Cold-formed structural steel shapes are produced by passing sheet or strip steel at room temperature through rolls or press brakes and then bending the steel into desired shapes. Cold-formed members may be distinguished from hot-rolled shapes in that their profiles contain rounded corners and slender flat elements and that all these elements have the same thickness. Cold-formed steel shapes can be divided into two types: *framing members* and *surface members*.

(1) Framing members

Framing members have the general outline of well-established hot-rolled shapes, such as angles, channels, Zees, round, square, rectangular, and hat shapes. Figure 2.23 shows the typical section forms of CFS framing members. The wall thickness of CFS framing members is usually 1.5 mm to 25 mm. These shapes and dimensions frequently result in slender plate elements (high flat width-to-thickness ratios) requiring lips or other edge stiffeners along free edges to prevent premature plate local buckling.

Figure 2.23 The typical section forms of CFS framing members

(2) Surface members

Surface members are load-resisting shapes that also provide useful surfaces. They are extensively used in roof, floor, wall, and partition construction due to their favorable weight-strength and weight-stiffness ratios. The surface members used in roofs and walls are sometimes called color profiled steel sheeting (Figure 2.24) in China. Those used in the floor are called profiled steel sheeting (Figure 2.25) or deck. The usual surface members are made of cold-rolled

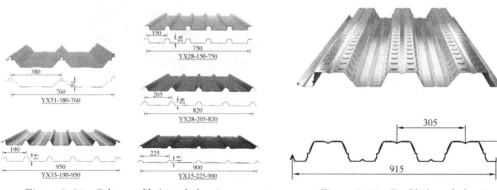

Figure 2.24  Color profiled steel sheetings       Figure 2.25  Profiled steel sheeting

thin steel plates, generally Q215 and Q235. The substrate thickness of profiled steel sheeting is usually from 0.75 mm to 2.0 mm, and the substrate thickness of color profiled steel sheeting is generally from 0.4 mm to 1.6 mm. The profiled steel sheeting is often used to make composite slabs with concrete. In addition, the color profiled steel sheeting is often used to fabricate roof plates or wall plates.

## 2.8 Chapter Summary

(1) Although there are many types of steel in China, the Standard for Design of Steel Structures GB 50017 only recommends one carbon structural steel and five low-alloy high-strength structural steel. That means steel must have high tensile strength and yield strength, good plasticity and toughness to reduce the tendency to brittle failure, and excellent process performance, which refers to cold working, hot working, and welding performance.

(2) The four working stages of steel in unidirectional uniform tension and the corresponding yield strength, tensile strength, deformation, and elongation describe the most basic steel properties and are of great significance. In addition, the simplified ideal elastoplastic model is very important for structural analysis and design. Therefore, these concepts will be used frequently in the following chapters and need to be mastered and understood.

(3) For the performance indicators of low-carbon steel, beginners should be proficientin choosing the steel skillfully and correctly. These indicators include strength, deformation, plasticity (including elongation and reduction of cross-sectional area), toughness (impact toughness at room temperature and impact toughness at negative temperature), cold bending performance, weldability, etc.

(4) Understand the failure behavior of steel under complex stress, especially the different effects on the strength and plasticity of steel under the three-dimensional compressive stress field and the three-dimensional tensile stress field.

(5) Factors such as chemical elements, metallurgical defects, steel hardening, temperature, stress concentration, and others impact steel properties. Among them, chemical elements and metallurgical defects are innate factors that need to be resolved at the source of production. In addition, the hardening of steel, temperature, and stress concentration is acquired factors alleviated by appropriate design and technical means such as changing the usage environment.

(6) In order to be proficient in knowing the types and expression methods of hot-rolled steel shapes, cold-formed steel members and profiled sheetings commonly used in engineering, it is very important to understand Chinese structural steel composition.

Chapter 2 STRUCTURAL STEEL

## Questions

2.1 What are the material requirements for steel structures?

2.2 What are the main mechanical properties of steel? Which methods are used to determine? What are the common structural forms of steel structures?

2.3 What are the main chemical components that affect the performance of steel? What are their respective influences?

2.4 What are the factors that affect the welding performance of steel?

2.5 What effect does temperature have on steel properties?

2.6 How to choose structural steel reasonably?

# Chapter 3  POSSIBLE FAILURE OF STEEL STRUCTURES

**GUIDE TO THIS CHAPTER:**

➤ *Knowledge points*

  The strength failure; overall instability and local instability of members and structures; section classification determined by the width-to-thickness ratio of the plate; fatigue failure of steel structures; deformation failure of steel structures; fracture failure of steel structures.

➤ *Emphasis*

  Possible failure of steel structures or steel members and effective preventive measures.

➤ *Nodus*

  The difference between overall instability and local instability; the difference between brittle fracture failure and instability failure.

## 3.1  Introduction

The design principles of steel structures are the adoption of advanced technologies that are economical and appropriate, safe and applicable, and ensure the quality to achieve the design goal, enabling the structure to complete the predetermined functions of safety application and durability. However, the above requirements and goals must be achieved without any damage to the steel structure. Therefore, practitioners must understand the possible failures of steel structures and master the technical measures to prevent them from happening.

One of the main reasons for the failure of steel structures is caused by the destruction of materials. The main forms of steel failure are plastic failure, brittle failure, and fatigue. The plastic failure of steel refers to the failure under the action of external force that the stress exceeds the yield strength and reaches the tensile strength, and there is an obvious sign (large deformation) before the failure. The brittle failure of steel refers to the fracture failure in which the stress of the steel is at a relatively low level under the action of external force, and there is no obvious warning of deformation. Its salient feature is that it has no obvious signs of deformation and occurs suddenly, which is a huge hazard to people's lives and property. The fatigue of steel is also a fracture failure that occurs suddenly when the stress is low. Still, the prerequisite for its occurrence is that the steel must undergo the long-term action of continuously repeated loads. The fatigue of steel is a special form of brittle failure of steel. Generally, it can be prevented by presetting the fatigue life of steel and monitoring the number of stress cycles.

The steel structure damages mentioned above are all on the material level, and the other reason for the steel structure damage is the structure itself. For example, the strength failure of the member or joint material; the instability failure of the member or structure (global and local); the brittle fracture of the structure and the weld connection; the deformation failure of the structure or the component.

## 3.2　The Strength Failure of Steel Structures

When the steel structure's overall stability and local stability (member) are guaranteed, the member section's stress reaches the material's tensile strength as the load increases. As a result, the member will undergo strength failure. In the trussing steel structure, two types of members may undergo strength failure: tension members and bending members with guaranteed stability.

### 3.2.1　The strength failure of tension members

Under the static load effect and working under normal temperature conditions, if there are no serious stress concentration, strain hardening, and time hardening defects, as the load increases, the member section's stress will reach the yield strength $f_y$ of the steel. Entering the plastic deformation stage (segment B-F in Figure 2.1), the member appears obvious elongation. Then the steel enters the strengthening stage (segment F-C in Figure 2.1), and the tensile stress of the member continues to increase. Finally, when the tensile stress reaches the tensile strength $f_u$, the member is broken. This type of damage is a ductile failure. However, the member has obvious elongation deformation before it is broken, and it is easy to find, and preventive measures can be taken.

However, due to improper design, manufacturing, and usage during the service life, tension members may also experience strength failures without significant deformation, named brittle fracture failures. For example, a sudden change in the cross-section of the member causes a severe stress concentration; the member bearing a large dynamic loading; random welding on the member causing the material to become brittle; low temperature, strain hardening, and failure hardening seriously reduce the toughness of the steel.

### 3.2.2　The strength failure of bending members

Bending members with guaranteed overall stability and local stability will go through the elastic working phase, the elastoplastic working phase, and the plastic working phase formed by the plastic hinge as the external load gradually increases. A plastic hinge occurring on the beam means that the structure is transformed into a mechanism and destroyed for a simple-supporting beam. Still, there will be a large deformation before the failure as a warning sign. However, for bending members in statically indeterminate structures such as steel frames, the appearance of a plastic hinge on a beam will not cause any damage to the structure but will cause a redistribution of internal structural forces. With the continuous appearances of beam plastic hinges and the constant change of internal force redistribution, the structure will eventually fail.

For the statically indeterminate steel beam, if it can be ensured that only strength and ductility failures occur in the structure, the steel plastic properties can be used to design the structure and reasonably control the structural steel consumption.

The strength failure of steel members is likely to lead to the failure of the overall structure. Still, in real steel structures, pure strength failure rarely occurs because the buckling failure of some compressive members or compressive-bending members in the structure may occur earlier before the tension or bending members' strength failure.

## 3.3 Overall Instability Failure of Steel Structures

The first chapter mentioned that the strength of steel is much higher than that of concrete and timber. Therefore, under the same design conditions, the section of steel members is much smaller, and the slenderness ratio of steel members is larger. For this reason, the design of compressive members in steel structures is usually not controlled by strength but by stability or stiffness (slenderness ratio).

The stability analysis of steel structures and members is different from the general strength analysis.

### 3.3.1 Aspects that must be considered

(1) Geometric nonlinearity

Three types of geometric nonlinearities need to be considered in the stability analysis of steel structures and steel members. The first type is small displacement and small rotation angles that consider structural deformation influencing external force effects. The analysis method that considers this geometric nonlinearity is called second-order analysis (compared to the first-order analysis of the *Structural Mechanics*). This method is used in the overall stability analysis of steel members, steel frames, and steel arches. The second type is a large displacement and a small rotation angle. This method is adopted when the steel frame considers both the members' and the structures' overall instability. The third is a large displacement and a large rotation angle. This method uses structural members' analysis considering post-buckling strength and when the structural members consider the global and local relevant buckling.

(2) Material nonlinearity

When a steel structure or steel member fails, it generally enters the elastoplastic stage. Therefore, the stability analysis must consider the dual effects of material nonlinearity and geometric nonlinearity simultaneously to obtain the steel structures or members' ultimate buckling load.

(3) The initial defects of the structure and members

Studies have shown that the initial defects of structures and members have a significant impact on stability. Common initial defects include initial bending of members, the initial eccentricity of load, residual stress, etc.

### 3.3.2 Overall instability of steel structures

The structure's overall instability failure followed external load gradually increasing, and the external load on the structure has not reached the failure load calculated by the strength. As a result, the structure cannot bear the load, and large deformation occurs, and the entire structure deviates from the original balanced position. The value of the external load that the structure occurs overall instability is called the critical load.

When the external load reaches the critical load and continues to increase, the structure is unstable. In this process, the structure's deformation increases rapidly and continuously, and the structure will lose its bearing capacity and be destroyed in a short time. Therefore, stability is a problem of the ultimate state of bearing capacity. Generally, a certain member or joint failure often induces the overall instability failure of steel structures.

For instance, in August 1907, the *Quebec Bridge* collapsed suddenly. Seventy-five workers were killed in this accident, and there were only eleven survivors from the workers on the span. A distinguished panel was assembled to investigate the disaster. The panel's report found that the main cause of the bridge's failure was the improper design of the latticing on the compression chords. The collapse was initiated by the buckling failure of Chord A9L on the anchor arm near the pier, immediately followed by Chord A9R (Figures 3.1, 3.2).

Figure 3.1 Quebec Bridge just before collapse (Modjeski et al. 1919 used by permission of Library and Archives Canada, PA 029229)

### 3.3.3 Overall instability of steel members

Due to the different cross-section forms and different steel members' stress states, overall instability failures are also different.

(1) Global buckling of the axially loaded

## 3.3 Overall Instability Failure of Steel Structures

Figure 3.2 Wreckage of the Quebec Bridge
(Modjeski et al. 1919 used by permission of Library and Archives Canada, C009766 and PA020614)

compression members

The overall instability (global buckling) of the axially loaded compression steel members includes flexural instability (buckling), torsional instability (buckling), or flexural-torsional instability (buckling). The axial compression member's overall instability (global buckling) form with biaxial symmetric I-section or box-section is flexural buckling (Figure 3.3). Axially loaded compression members with cruciform sections generally suffer from torsional buckling (Figure 3.4), and members with large slenderness ratios may also suffer from flexural buckling. The axial compression member with the uniaxial symmetrical section is flexural buckling when it loses stability around the asymmetric axis, flexural-torsional buckling when it loses stability on the symmetry axis (Figure 3.5).

Figure 3.3 Flexural buckling    Figure 3.4 Torsional buckling    Figure 3.5 Flexural-torsional buckling

(2) Global buckling of the bending members

The overall instability (global buckling) form of the bending member, sometimes refers to as the beam, is lateral-torsional buckling.

(3) Global buckling of the beam-column

The overall instability form of solid web unidirectional beam-column is as follows.

① Biaxial symmetrical section, the overall instability in-plane of the moment is flexural buckling; the instability out-plane of the moment is lateral-torsional buckling.

② Uniaxial symmetrical section, when the moment acts on the asymmetrical axis plane (around the symmetrical axis) and acts out-plane or around two axes, overall instability is lateral-torsional buckling.

③ Uniaxial symmetrical section, when the moment acts in-plane of symmetry axis (around the asymmetric axis), the overall instability (global buckling) in-plane of the moment (around the asymmetric axis) is flexural buckling. Out-plane of the moment (around the symmetric axis), the overall instability (global buckling) is lateral-torsional buckling.

(4) The overall instability (global buckling) form of solid web bidirectional beam-columns is lateral-torsional buckling.

(5) For the frame and arch, flexural instability occurs in the plane, and lateral-torsional instability occurs out of the plane.

## 3.4 Local Instability Failure of Steel Structures

### 3.4.1 Local instability of steel structures

The local instability failure of the structure means that in the process of the external load gradually increasing, under the condition that the structure maintains the overall stability and without strength failure, the structure's local members can no longer withstand the external load and lose stability. For example, in steel Moment-Resisting Frames (MRFs), the local member that is unstable may be a beam-column or a beam.

Commonly, steel structures are mostly statically indeterminate structures. Therefore, the instability failure of a certain local member will not immediately lose the overall structure's load-bearing capacity. Still, it will continuously deteriorate the structure's overall load-bearing capacity, which may induce the overall destruction of the structure in advance. For example, the local buckling range gradually spreads from a member (or joint) to the surroundings in the reticular shell structure. Finally, a large depression is formed on the reticular shell (Figure 3.6).

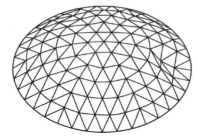

Figure 3.6 Local buckling of a reticular shell

### 3.4.2 Local instability of steel members

The local instability failure of a member refers to the fact that in increasing external load, under the condition that the member maintains overall stability and does not undergo strength failure, the member's local plates can no longer withstand the external load and lose stability. For example, the local buckling of the web or the flange of an I-section beam-column is shown in Figures 3.7 and 3.8.

Figure 3.7　Local buckling of a web　　　Figure 3.8　Local buckling of flanges

The steel member's local instability will deteriorate its working condition and may induce its overall instability or strength failure in advance. Therefore, the current *Standard for Design of Steel Structures* GB 50017 stipulates that the compression flange's post-buckling strength cannot be used for conventional steel members. It must meet the requirements of the width-to-thickness ratio limit. Under appropriate conditions, the web's post-buckling strength of the beam, axially compressive column, and beam-column can be considered to achieve the purpose of saving steel.

### 3.4.3 Classification of cross-section

Steel structural members of various cross-sections are composed of plates (flat or curved plates). When the plate is under tension, the steel can theoretically reach the yield strength or tensile strength. However, local instability of the plate may occur under pressure. Under the plate's same boundary conditions, the plate's buckling load for local instability is related to its width-to-thickness ratio. The larger the width-to-thickness ratio, the smaller the buckling load. Taking common I-section members as an example, the flange plate's width-to-thickness ratio generally refers to the flange's outer extension width to its thickness. The wed's width-to-thickness ratio is generally expressed by its depth-thickness ratio, which refers to the ratio of its depth to its thickness. The *Standard for Design of Steel Structures* GB 50017 divides the member into five grades on the basis of the plate's width-to-thickness ratio according to different applications and gives the corresponding limit width-to-thickness ratio of the plate, as shown in Table 3.1. According to the seismic grade of steel frames, the *Code for Seismic Design of Buildings* GB 50011 limits frame columns and beams' width-to-thickness ratio.

According to the section's different bearing capacities and plastic rotation deformation capacities, steel member sections are generally divided into four classes. However, considering that China uses section plastic development coefficients in bending members' design, The *Standard for Design of Steel Structures* GB 50017 divided the section into five classes based on the plate's width-to-thickness ratio.

(1) Class S1 section

Up to full section plasticity, the plastic hinge has the rotational capacity required by the plastic design. The bearing capacity does not decrease during the rotation process, called the first-grade plastic section, and can also be called the plastic rotation section. At this time, curve 1, shown in Figure 3.9, can represent the bending moment-curvature relationship. $\Phi_{p2}$ generally requires 8-15 times the curvature $\Phi_p$ obtained by dividing the plastic bending moment $M_p$ by the initial elastic stiffness.

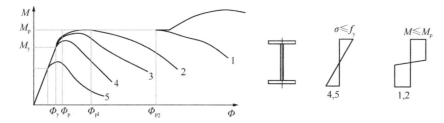

Figure 3.9 Classification of sections and their rotation capacity

(2) Class S2 section

It can reach full section plasticity, but due to local buckling, the plastic hinge rotation capacity is limited, called the secondary plastic section; the bending moment-curvature relationship is shown in curve 2 in Figure 3.9. $\Phi_{p1}$ is approximately 2-3 times $\Phi_p$.

(3) Class S3 section

The flanges yield and the web can develop plasticity that does not exceed 1/4 of the section height, called the elastoplastic section. When used as a beam, its bending moment-curvature relationship is shown in curve 3 in Figure 3.9.

(4) Class S4 section

The edge fiber can reach the yield strength but cannot develop plasticity due to local buckling, called the elastic section. When used as a beam, its bending moment-curvature relationship is shown in curve 4 in Figure 3.9.

(5) Class S5 section

Before the edge fiber reaches the yield stress, the web may be locally buckled, called a thin-wall section. As a beam, its bending moment-curvature relationship is shown in curve 5 in Figure 3.9.

The parameter $\alpha_0$ in Table 3.1 shall be calculated as follows:

$$\alpha_0 = \frac{\sigma_{max} - \sigma_{min}}{\sigma_{max}} \quad (3-1)$$

where   $\sigma_{max}$ = Calculate the maximum compressive stress at the edge of the web (N/mm$^2$);

$\sigma_{min}$ = The corresponding stress on the other edge of the web's calculating height (N/mm$^2$), the compressive stress takes a positive value, and the tensile stress takes a negative value.

$\varepsilon_k = \sqrt{235/f_y}$, it is the steel grade correction coefficient, and $f_y$ is the yield strength of the steel.

The $b$ is the flange extension width of the I-sections or H-shapes, $t$, $h_0$, $t_w$ is the flange thickness, the net height of the web, and the web thickness. The web's net height does not include the arc section at the flange web transition for hot-rolled shaped members. For the box section, $b_0$ and $t$ are the distance between the walls and the wall thickness, respectively. $D$ is the circle pipe outer diameter. $\lambda$ is the slenderness ratio of the member in-plane of the bending moment. According to the H-shapes web, for box-section beams and box-section columns subjected to unidirectional bending, the web limit may be adopted. The width-to-thickness ratio of the web can be reduced by setting stiffeners.

**The classes and limits of the width-to-thickness ratio of the section plate of steel compression and bending members**  Table 3.1

| Member | Grade of width-to-thickness ratio | | S1 | S2 | S3 | | S4 | S5 |
|---|---|---|---|---|---|---|---|---|
| Beam-column (frame column) | H-shape section | Flange $b/t$ | $9\varepsilon_k$ | $11\varepsilon_k$ | $13\varepsilon_k$ | | $15\varepsilon_k$ | 20 |
| | | Web $h_0/t_w$ | $(33+13\alpha_0^{1.3})\varepsilon_k$ | $(38+13\alpha_0^{1.39})\varepsilon_k$ | $0 \leqslant \alpha_0 \leqslant 1.6$ | $(16\alpha_0+0.5\lambda+25)\varepsilon_k$ | $(45+2\alpha_0^{1.66})\varepsilon_k$ | 250 |
| | | | | | $1.6 < \alpha_0 \leqslant 2.0$ | $(48\alpha_0+0.5\lambda-26.2)\varepsilon_k$ | | |
| | Box-shape section | Flange between wall plates (webs) $b_0/t$ | $30\varepsilon_k$ | $35\varepsilon_k$ | $0 \leqslant \alpha_0 \leqslant 1.6$ | $(12.8\alpha_0+0.4\lambda+20)\varepsilon_k$, and $\geqslant 40\varepsilon_k$ | $45\varepsilon_k$ | — |
| | | | | | $1.6 < \alpha_0 \leqslant 2.0$ | $(38.4\alpha_0+0.4\lambda-21)\varepsilon_k$ | | |
| | Round | Diameter-thickness ratio $D/t$ | $50\varepsilon_k^2$ | $70\varepsilon_k^2$ | $90\varepsilon_k^2$ | | $100\varepsilon_k^2$ | — |

continued

| Member | Grade of width-to-thickness ratio | | S1 | S2 | S3 | S4 | S5 |
|---|---|---|---|---|---|---|---|
| Bending member (beam) | I-section | Flange $b/t$ | $9\varepsilon_k$ | $11\varepsilon_k$ | $13\varepsilon_k$ | $15\varepsilon_k$ | 20 |
| | Box-shaped section | Web $h_0/t_w$ | $65\varepsilon_k$ | $72\varepsilon_k$ | $(40.4+0.5\lambda)\varepsilon_k$ | $124\varepsilon_k$ | 250 |
| | | Flange between wall plates (webs) $b_0/t$ | $25\varepsilon_k$ | $32\varepsilon_k$ | $37\varepsilon_k$ | $42\varepsilon_k$ | — |

## 3.5 The Fatigue Failure of Steel Structures

### 3.5.1 The concept of fatigue failure

Fatigue failure refers to the brittle fracture failure when steel or steel members' stress is lower than the ultimate strength or even lower than the yield strength under continuously repeated loads. The fatigue failure of steel occurs after a long period of development. The fatigue failure of steel structures often occurs in steel bridges and heavy-duty steel crane beams. Sometimes, steel structures subjected to repeated earthquakes will also experience fatigue failure. Besides, there will be fatigue damage to offshore marine steel structures. The fatigue of steel structures can be divided into high-cycle fatigue and low-cycle fatigue according to the magnitude of stress before fracture and the number of stress cycles. Crane beams or bridges are high-cycle fatigue, and earthquake action is low-cycle fatigue. Generally, steel structures only consider high-cycle fatigue calculations. The fatigue failure of the steel structure is closely related to the fatigue failure of steel. Therefore, it is necessary to link the two and introduce them together.

The fatigue failure process of steel usually goes through three stages. The formation of cracks, the slow expansion of cracks, the rapid training of cracks. The plastic deformation of steel is extremely small before fatigue failure. Therefore, fatigue failure belongs to brittle fracture without obvious deformation. The fracture is straight. The fracture may penetrate through the base material, penetrate through the connection weld, and penetrate the base material and the weld. The fracture of fatigue failure can generally be divided into two parts. One part is semi-elliptical and smooth, and the rest is rough, as shown in Figure 3.10.

Figure 3.10 Schematic diagram of fracture

The micro-cracks expand with the continuous and repeated action of stress, and the materials on both sides of the crack sometimes squeeze each other and separate, forming a smooth zone. The expansion of the cracks weakened the cross-section. When the residual part of the cross-section is not enough to resist loading, the member suddenly breaks, affecting tearing and forming a rough zone. When the member stress is larger, the smooth zone is smaller; on the contrary, its range is larger.

### 3.5.2 Factors affecting fatigue failure

Fatigue failure is the continuous expansion of microscopic cracks under continuously repeated loads until they reach fracture. The prerequisite is the formation of microscopic cracks and the stress concentration at the crack location. It depends on the stress factor generated by the continuous repeated load, including stress ratio, stress amplitude, and the number of stress cycles.

(1) Micro-cracks and stress concentration

For steel structures, the first stage of the fatigue fracture process, the formation stage of micro-cracks, generally does not exist. However, the steel production and manufacturing process inevitably will have local and minor defects in some parts of the structure, and they play a role similar to micro-cracks. For example, the segregation of the chemical composition of the steel, non-metallic inclusions; cracks, surface irregularities, and delamination formed during rolling; punching, trimming, and nicks during manufacturing; burrs and cracks caused by flame cutting; air bubbles in the weld, slag inclusion, under-welding during welding, etc. These small and local defects are the main parts that may cause cracks. Because when the continuous load is repeated, the stress distribution on the cross-section of these parts will be uneven, resulting in stress concentration, forming a two-way or three-way same sign tensile stress field. The plastic deformation of the material is limited, and microscopic cracks will appear first at the peak stress zone. Then as the crack develops gradually, the effective cross-section gradually decreases accordingly. The stress concentration phenomenon becomes more and more serious, which in turn promotes the cracks to continue. When the repeated load reaches a certain number of cycles, the development of cracks will weaken the section more, resulting in the inability to withstand external forces and eventually brittle fractures and fatigue failures. At the same time, there are stress-concentrated parts, such as the sudden change of cross-section. Due to the peak stress and the influence of repeated loads, even if there are no defects in the place, micro-cracks will occur and cause fatigue failure.

The general stress concentration in the steel structure, under static load, the peak stress is relatively reduced due to the plastic development of the steel. The increase of the stress in the lower stress position makes the cross-section uneven stress uniform, so it does not affect the cross-section's ultimate bearing capacity without considering its influence. However, with the more serious stress concentration, there is always a larger stress field in the peak stress zone, making the plastic deformation of the steel difficult, and brittle fracture occurs. Thus, especially under the action of dynamic load, the structure is more prone to fatigue failure.

(2) Stress ratio and stress amplitude

The repeated changes in the stress under continuously repeated loads are called a cycle. After a certain number of stress cycles, the steel undergoes fatigue failure. The maximum stress in the stress cycle is called fatigue strength. The strength is mainly related to the members' structural details, stress cycles, and the number of cycles. The stress ratio, $\rho = \sigma_{min}/\sigma_{max}$, usually expresses the stress cycle characteristics. Where $\sigma_{min}$ is the peak stress with the smallest absolute value and $\sigma_{max}$ is the peak stress with the largest absolute value, the tensile stress is positive, and the compressive stress is negative. When $\rho = -1$, it is called a fully symmetrical cycle, and the fatigue strength is the smallest. When $\rho = 0$, it is called the pulse cycle. When $\rho = 1$, it is a static load. When $-1 < \rho < 0$, it is a different-sign stress cycle, and the fatigue strength is small. $0 < \rho < 1$ is the same-sign stress cycle, and the fatigue strength is greater.

For the non-welded part of the material or structure, the residual tensile stress is very small, and the fatigue strength is mainly related to the stress ratio $\rho$ and the maximum stress $\sigma_{max}$. But for the welded part, the residual tensile stress in the weld and the base metal is often as high as the yield strength of the material. The nominal stress cycle feature-stress ratio does not represent the fatigue crack's true stress state. The load's stress cycle is superimposed on the material's residual stress as the fatigue crack's actual stress state. Here, the maximum tensile stress starts from the yield strength $f_y$, and $\sigma_{max} = f_y$, which decreases to $\sigma_{min}$. The stress cycle is a completely symmetric cycle or a pulse cycle; the stress amplitude $\Delta\sigma$ can express its stress cycle characteristics. The stress amplitude $\Delta\sigma = \sigma_{max} - \sigma_{min}$. Where $\sigma_{max}$ is the maximum tensile stress in the stress cycle, which takes a positive value; $\sigma_{min}$ is the minimum tensile stress or compressive stress in the stress cycle, the tensile stress takes a positive value, and the compressive stress takes a negative value. The stress amplitude $\Delta\sigma$ is always positive. For welded structures, regardless of the nominal stress ratio under cyclic loading, the fatigue strength is the same as long as the stress amplitude is the same. The fatigue performance of welded connections or welded members is directly related to the stress amplitude and not related to the nominal stress ratio.

(3) Fatigue strength and fatigue life (number of cycles)

The fatigue design should consider the following conditions, the number of stress cycles and the magnitude of the stress amplitude. Studies have shown that fatigue failure of materials or members is related to the magnitude of the stress amplitude $\Delta\sigma$ and the number of stress cycles $n$. At this time, the stress amplitude $\Delta\sigma$ is called the fatigue strength corresponding to the number of stress cycles $n$. Similarly, the number of stress cycles $n$

corresponding to the stress amplitude $\Delta\sigma$ is called the material or member's fatigue life. The *Standard for Design of Steel Structures* GB 50017 stipulates that for structural steel members and their connections that directly bear the repeated action of dynamic loads, fatigue calculations should be performed when the number of cycles $n$ of stress changes is equal to or greater than 50,000.

### 3.5.3 Fatigue calculation and anti-brittle fracture design

The fatigue calculation should adopt the allowable stress amplitude method. The stress should be calculated according to the elastic state. The allowable stress amplitude should be determined according to the member and connection category, the number of stress cycles, and the plate's thickness at the calculation site. For non-welded members and connections, the fatigue strength is not calculated for the parts where tensile stress does not appear in the stress cycle.

Steel structural members that work at low temperatures or are fabricated and installed should be designed to prevent brittle fracture. Therefore, it is necessary to calculate that the steel used for fatigue members should have a qualified guarantee of impact toughness. The details of fatigue calculation and anti-brittle fracture design refer to Chapter 16 of the *Standard for Design of Steel Structures* GB 50017.

## 3.6 The Deformation Failure of Steel Structures

The loss of stiffness causes the steel structure deformation failure. The main reasons for the steel structure's deformation failure are improper design, improper manufacture, and improper use. The steel structure has the characteristics of high strength and good plasticity. In particular, with the application of cold-formed steel and the rapid development of light steel structures, structural steel members' cross-section is getting smaller than before, and the steel plate thickness is getting thinner than before. In this case, coupled with the defects in the materials, processing, manufacturing, installation, and inappropriate craft, the steel structure's deformation failure cannot be ignored.

### 3.6.1 The deformation failure of steel members

The steel structure needs to satisfy the design under the normal service limit state and limit deformation that affects normal use or appearance. The member's stiffness does not meet the use requirements, resulting in too large a deformation of the member that cannot normally be used or installed. For example, when the slenderness ratio of horizontally placed members is too large, there may be deflection that affects the use under the action of its weight. Under dynamic load, members with a large slenderness ratio may exhibit excessive vibration. When the beam deflection is too large, it will affect the appearance and cause the members not to be used normally. For example, excessive deflection of the crane beam may cause the crane to pass through difficulty. Due to inappropriate technology and other factors, the members themselves are severely bent, twisted, and local uneven during the steel structure's manufacturing process. These deformations affect the appearance and bring great difficulties to the installation of structural members. During the use of steel structures, arbitrary changes to the structure's purpose will also lead to large deformation and member failure.

### 3.6.2 The deformation failure of steel structures

The deformation of the structure is generally controlled by design. For example, the frame structure's lateral displacement under the earthquake and the vertex lateral displacement under wind loading are limited in the design specifications. There are different limits of lateral displacement for the crane portal frame and the non-crane portal frame. The deformation and damage of the structure are generally caused by improper design and improper construction. If the structural bracing system is not properly arranged in steel structures, it may cause structural deformation and damage. For example, in an industrial plant structure, the bracing system is important to ensure its stiffness. It plays a major role in resisting horizontal loads and earthquakes and plays a key role in ensuring the structure's normal use. However, when the bracing system is not arranged properly or the position is incorrect, excessive structural vibration and shaking may occur during the crane operation, affecting its normal operation. Structural deformation may also be caused by improper construction. For example, there are many bolted connections in the steel structure. Sometimes on-site reaming may occur due to construction errors that cause installation difficulties, and reaming will cause deformation and damage beyond the design expectations in the later use stage. In summary, structural deformation and damage can be avoided by strictly controlling the design and construction quality.

## 3.7 Brittle Fracture of Steel Structures

### 3.7.1 Examples of brittle fracture of steel structures

Failure of steel structures triggered by fracture is a much-feared situation across various engineering disciplines. Beginning with the Liberty Ship fractures in the 1940s (Sheldon 2010) that catalyzed fracture mechanics' initial development as a field of study, catastrophic structural failures are periodic reminders of fracture consequences. Prominent examples include the Alexander Kielland drilling rig fracture in 1980 (Almar-Naess et al. 1984), resulting in 123 deaths and steel building fractures during the 1994 Northridge and 1995 Kobe earthquakes.

## 3.7 Brittle Fracture of Steel Structures

Starting in the 1960s, engineers began to regard welded steel MRFs as among the most ductile systems in the building code. Many engineers believed that MRFs were essentially invulnerable to earthquake-induced structural damage and thought that should such damage occur, it would be limited to ductile yielding of members and connections. Thus, the earthquake-induced collapse was not believed possible. Partly as a result of this belief, many large industrial, commercial, and institutional structures employing steel MRFs were constructed, particularly in the western United States. The Northridge earthquake on January 17, 1994, challenged this paradigm. Following that earthquake, many steel frame buildings were found to have experienced brittle fractures of beam-to-column connections (Figure 3.11). The Northridge and Kobe earthquakes underscored the problems with these empirical approaches, providing the impetus for studying fracture within a context-specific to civil (especially earthquake) engineering. The brittle fracture of the structure is the most dangerous kind of structure. Before brittle fracture failure, the stress is usually less than the steel's yield strength. No significant deformation occurs. The failure occurs suddenly without any warning, and it is too late to remedy, which will cause catastrophic results.

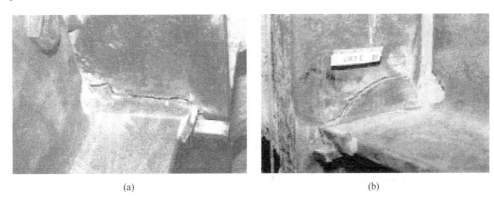

Figure 3.11 Fractures of beam-to-column joints
(a) fracture at fused zone; (b) column flange "Divot" fracture

### 3.7.2 The main cause of brittle fracture

Although the structural steel used in the steel structure has better plasticity and toughness, however, under certain conditions, the brittle fracture may still occur. The main reasons are as follows.

(1) Material defects

When the content of certain elements is too high, the steel's plasticity and toughness will be seriously reduced, and the brittleness will increase accordingly. For example, excessive sulfur and oxygen will cause hot embrittlement. Too much nitrogen and phosphorus will cause cold brittleness. Excessive hydrogen can cause hydrogen embrittlement. Too much carbon will cause the steel to become brittle and deteriorate weldability. In addition, metallurgical defects such as cracks, segregation, non-metallic inclusions, and delamination can also greatly reduce the ability to resist brittle fracture.

(2) The thickness of the steel plate

The thickness of the steel plate has a greater influence on brittle fracture. That is because the toughness of thick steel plates is lower than that of thin steel plates. Therefore, generally, the thicker the steel plate, the greater the possibility of brittle fracture failure. For example, interlaminar tearing between thick steel plates and extra-thick steel plates is also one of the main causes of brittle fracture.

(3) Stress concentration and residual stress

The stress concentration on the member will cause the plane or three-dimensional tensile stress field to appear, limiting the steel's plastic deformation ability and cause the steel too brittle. If the members are accompanied by large residual stress, the situation will further deteriorate. The stress concentration and residual stress are related to the member's structural details, the welding seam's location, and the construction process. For example, the welding seam is too concentrated, and the three-way welding seam intersects, the sudden change of the member section, the incorrect welding process, and welding sequence, etc.

(4) Working temperature

The impact toughness of steel is closely related to temperature. When the working temperature drops to a certain temperature range, the toughness value of steel will drop sharply, and low-temperature cold brittleness will appear. At this time, if the member is subjected to a large dynamic load, it is prone to brittle fracture.

### 3.7.3 Measures to prevent brittle fracture

Brittle fractures are fatal to structural safety and threaten people's lives and property. Therefore, structural engineers must take technical measures to prevent brittle fracture of the structure. The measures to prevent the fracture of the steel structure mainly include the following aspects.

(1) Appropriate use of steel

According to the importance of the structure, load feature, connection method, and the working environment temperature, reasonably select structural steel. For example, welded structures are subjected to greater dynamic loads under low-temperature conditions. Therefore, steel with higher toughness should be selected.

(2) Appropriate design

The structural form should be selected reasonably after considering factors such as fracture toughness of the steel, working environment temperature, load characteristics, stress concentration, and other factors. Especially choose appropriate construction details.

Strive to ensure the geometric continuity and rigidity of the structure. For example, in the design, we should avoid excessive concentration of welds and the appearance of three-way welds; choose appropriate connection methods to avoid stress concentration.

(3) Appropriate processing and installation

Due to sulfur, phosphorus, nitrogen, and oxygen, hot and cold processing may make steel brittle in the production process. In addition, welding is easy to causes cracks and residual welding stress. An inappropriate installation process will also produce residual assembly stress and other structural defects. Therefore, before processing and installing the steel structure, an appropriate processing and installation process must be formulated to reduce defects and reduce residual stress.

(4) Appropriate use and maintenance

During the structure's use, the purpose, load, and ambient temperature of the structure cannot be changed at will. Therefore, welding parts should not be added to the main structure at will. During normal use, regular structural inspection and maintenance are carried out. Repair the defects and damages found in time.

## 3.8 Chapter Summary

Strength, stiffness, and stability are the core factors that control the design of steel structures, and they are also the core content of the *Basic Principles of Steel Structures*. Therefore, any improper design will cause damage to the structure or members. This chapter introduces the various possible forms of damage to steel structures or steel members, the main causes of these damages, and the effective measures to prevent them.

(1) The steel members that may undergo strength failure are mainly axial tension components and bending components. However, other steel components are often not designed under the control of strength conditions, so generally, there will be no strength failure. Thus, the steel structure as a whole seldom occurs strength failure.

(2) Instability (buckling) failure is the main failure mode of steel structures or steel members. However, most steel members or steel structures are designed under the control of stable conditions.

(3) Instability (buckling) damage is diverse. Therefore, the difference between the member's overall instability and the structure's overall instability should be correctly distinguished. In addition, the difference between the local instability of the member and the local instability of the structure should also be correctly distinguished.

(4) Excessive deformation belongs to the category of normal Service Limit State. Although the direct consequences of excessive deformation are generally not serious, if the structure or members that have undergone excessive deformation are not repaired in time, the structure may lose its load-bearing capacity. Therefore, the code regards the structure

as being unsuitable for continued bearing due to excessive deformation as a category of the Ultimate Bearing Capacity.

(5) Brittle fracture is the most dangerous failure of the structure. In order to understand the various causes of brittle failure, it is necessary to take effective measures to prevent brittle fracture damage when designing and constructing steel structures.

## Questions

3.1　What are the main types of failure steel structures?

3.2　Why does the simple strength failure rarely occur in actual structures?

3.3　What is the difference between the overall instability of members and structures?

3.4　What are the overall instability forms of axially compression steel members?

3.5　What is the difference between the local instability of members and the structure?

3.6　What is the difference between instability and buckling?

3.7　What are the main causes of excessive deformation of steel structures?

3.8　What are the main factors affecting fatigue strength?

3.9　Which has a greater influence on fatigue strength, stress ratio or stress amplitude?

3.10　What is the brittle fracture?

3.11　What are the main factors that cause brittle fracture of steel structures?

3.12　What are the similarities and differences between fatigue fracture and brittle fracture?

# Chapter 4 CONNECTIONS OF STEEL STRUCTURES

**GUIDE TO THIS CHAPTER:**

➢ *Knowledge points*

*Types of connections of steel structure; welded defects and quality inspection; welded representation method; method of reducing residual welding stress and residual deformation; details and calculation of welded connection; details and calculation of ordinary bolt, performance and calculation of high-strength bolt.*

➢ *Emphasis*

*Calculation of fillet weld; calculation of bolted connections.*

➢ *Nodus*

*The reason for the welded residual stress; calculation of bolts subject to combined tensile and shear.*

## 4.1 Introduction

A structure may be considered an assemblage of the various column, beam, and beam-column assemblages that must be fastened together to make the finished product. Irrespective of how scientifically or efficiently the basic structural members may have been designed, the result could be a catastrophic collapse if the necessary connections are inadequate. The importance of economical and structurally adequate connections can not be overemphasized. Connection behavior is so complex that numerous simplifying assumptions must be made so that connection design is brought to a practical level. It is generally agreed among designers that the basic members' design is simple compared to the connections between those members. Appropriate design and appropriate construction of the connection are very important for the safe bearing of the structure.

The most common types of structural steel connections currently being used are welded connections and bolted connections. This chapter will explain the performance and design calculations of these two types of connections, shown in Figure 4.1.

Figure 4.1  Two common structural steel connections
(a) welded connection; (b) bolted connection and welded connection

## 4.2  Methods and Characteristics of Welded Connection

### 4.2.1  The method of welding and the form of the weld

Welding is a process in which two metal pieces are fused together by heat to form a connection or joint. In practical structural steel welding, this process is usually accompanied by the addition of filler metal from an electrode. In practical steel structures, structural steel welds are usually made either by the manual shielded metal-arc process or by the submerged arc process.

(1) The manual shielded metal-arc welding

The manual metal-arc welding process, commonly called stick welding, is designed primarily for manual application and is used both in the workshop and on the construction site.

An electric arc is formed between the end of a coated metal electrode and the steel components to be welded. This arc generates an approximate temperature of 6,500°F, which melts a small area of the base metal. The tip of the electrode also melts, and this metal is forcibly propelled across the arc. The small pool of molten metal that is formed is called a crater. As the electrode moves along the joint, the crater follows it, solidifying rapidly as the pool's temperature behind it drops below the melting point. Figure 4.2 illustrates this process. During the welding process, as the electrode coating decomposes, it forms a gas shield to prevent the absorption of impurities from the atmosphere. Besides, the coating contains a material, commonly called flux, which will prevent or dissolve oxides and other undesirable substances in the molten metal or facilitate removing these substances from the molten metal.

The electrode used for manual welding should meet the current national standard *Cov-*

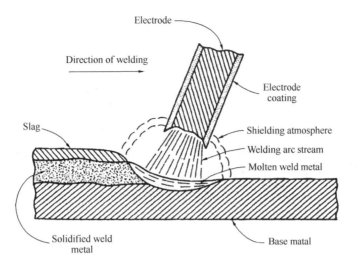

Figure 4.2  Manual shielded metal-arc process

*ered Electrode for Manual Metal Arc Welding of Non-alloy and Fine Grain Steels* GB/T 5117. In addition, the selected electrode should be compatible with the mechanical properties of the structural steel members.

The commonly used electrode models in steel structures are the E43, E50, E55, and E60 series. The letter E represents the electrode, and the following two digits indicate the minimum tensile strength of the molten metal in $N/mm^2$. For example, the E43 type electrode its tensile strength is $430 \ N/mm^2$. When steel members with different strengths are welded, the electrode must be compatible with lower-strength steel.

(2) The submerged arc welding

The submerged arc welding process is primarily a shop-welding process per formed by either an automatic or a semi-automatic method. The principle is similar to manual shielded metal-arc welding, but a bare metal electrode is used instead of a coated electrode. Loose flux is supplied separately in granular form and is placed over the joint to be welded. The electrode is pushed through the flux, and as the arc is formed, part of the flux melts to form a shield that coats the molten metal. This welding process is faster and results in deeper weld penetration. An electrically controlled machine supplies the flux and metal electrode through separate nozzles as it moves along a track in the automatic process. Figure 4.3 illustrates this process.

The welding wire for automatic or semi-automatic welding should comply with the following current national standard: *Steel Wires for Melt Welding* GB/T 14957, *Wire Electrodes and Weld Deposits for Gas Shielded Metal Arc Welding of Non Alloy and Fine Grain Steels* GB/T 8110, *Tubular Cored Electrodes for Non-alloy and Fine Grain Steels* GB/T 10045, and *Tubular Cored Electrodes for Creep-resisting Steels* GB/T 17493.

(3) The form of welds

The weld form can be divided according to different methods. Two types of welds

Chapter 4  CONNECTIONS OF STEEL STRUCTURES

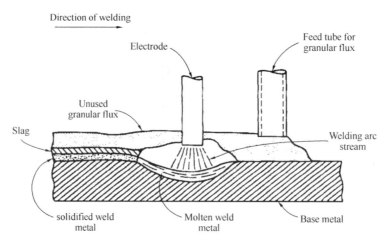

Figure 4.3  Submerged arc welding process

predominate in structural applications with different stress characteristics: fillet weld and groove weld. Other structural welds are the plug weld and slot weld, which are generally used only under the circumstance in which fillet welds lack adequate load carrying capacity. These four weld types are shown in Figure 4.4. The most commonly used weld for structural connection is the fillet weld. Groove welds may require extensive edge preparation as well as precise fabrication and, as a result, are more costly.

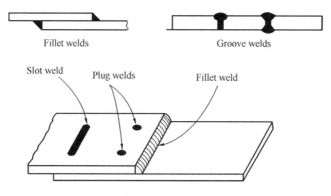

Figure 4.4  Welds types

In any given weld structure, the adjoining components may be placed to each other in the following ways: butt, lap, tee, corner, and edge. They are illustrated in Figure 4.5.

According to the welding position, welds are also classified as flat weld, horizontal weld, vertical weld, and overhead weld, shown in Figure 4.6. The position of the joint when welding is performed is of economic significance. The flat weld is the most economical, and the overhead weld is the most expensive. Moreover, the welding quality of the flat weld is the easiest to control, and the welding quality of the overhead weld is the most difficult to control.

60

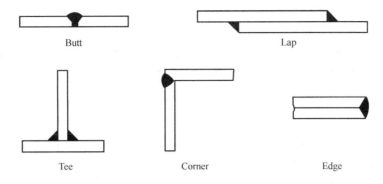

Figure 4.5  Joints types

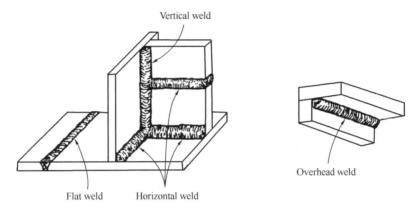

Figure 4.6  Weld types

**4.2.2  Weld defects and inspection**

(1) Weld defects

Weld defects refer to defects on the surface or inside of the weld or nearby heat-affected zone metal during the welding process. Common weld defects include cracks, weld slags, burning through, craters, pores, dirt, undercuts, incomplete fusion, incomplete penetration and weld size that does not meet the requirements, poor weld formation, etc. They are illustrated in Figure 4.7. Cracks are the most dangerous defect in welded connections. There are many reasons for cracks, such as the improper chemical composition of steel, the inappropriate choice of the welding process, uncleaned oil stain on the welded metal surface, etc.

(2) Weld inspection

The existence of weld defects will weaken the force area of the weld and cause stress concentration at the defect to hurt the strength of the connection, impact toughness, and cold bending performance. Therefore, the quality of the weld is very important.

Weld quality inspection generally can be performed by visual inspection and internal non-destructive inspection. The former inspect appearance defects and geometric dimen-

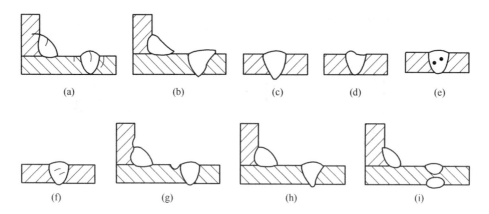

Figure 4.7 Weld defects

(a) cracks; (b) weld slags; (c) burning through; (d) craters; (e) pores; (f) dirt;
(g) undercuts; (h) incomplete fusion; (i) incomplete penetration

sions, and the latter inspects internal defects. Ultrasonic inspection is currently widely used in internal non-destructive inspections. Besides, X-ray and γ-ray transillumination films are also available, but their application is not as extensive as ultrasonic.

The current design standard, *Standard for Acceptance of Construction Quality of Steel Structures* GB 50205, stipulates that welds are divided into first, second, and third levels according to their inspection methods and quality requirements. Third-level welds only require visual inspection of all welds and meet the third-level quality standards. In addition to visual inspection, first-level and second-level welds also require a certain amount of ultrasonic inspection and meet the corresponding quality standards. The first-level weld detection ratio is 100%, and the second-level weld detection ratio is 20%. The third-level welds may not be inspected for flaw detection. Because there are unfused parts between the steel plates at the fillet welds joints, the visual inspection is generally performed according to the third-level welds. In special circumstances, inspections may be required according to the second-level welds.

### 4.2.3 Residual welding stress and residual deformation

(1) Causes and classification of residual welding stress

The welding process is a process of uneven heating and cooling. During welding, an uneven temperature field is generated on the weldment, shown in Figure 4.8. The uneven temperature field produces uneven expansion, and the high-temperature steel expands greatly. Still, it is limited by the lower surrounding temperature, the steel with a smaller expansion amount, resulting in thermal plastic compression. When the weld is cooled, the weld zone plastically compressed tends to shorten but is restricted by the surrounding steel and generates tensile stress. In low-carbon steel and low-alloy steel, this tensile stress often reaches the yield strength of the steel. Residual welding stress is the internal stress under no load. Therefore, it will be self-balanced inside the weldment, which will inevitably produce compressive stress in the section slightly far away from the weld.

Residual welding stress is divided into longitudinal residual stress along the weld's length, transverse residual stress perpendicular to the length of the weld, and residual stress along with the thickness of the steel plate. The longitudinal shrinkage of the weld causes the longitudinal welding residual stress. In general, the longitudinal stress in the weld zone and on both sides of the weld is a tensile stress zone, and the two sides away from the weld are a compressive stress zone (Figure 4.8).

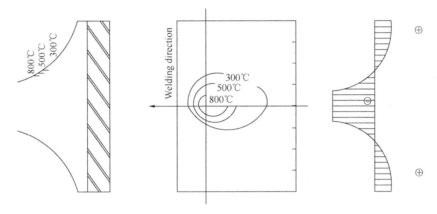

Figure 4.8 Temperature field and residual welding stress in and around the weld during welding

The contraction force of two parts causes the transverse residual welding stress. The first reason is that the weld's longitudinal contraction makes the two steel plates tend to bend in the opposite direction. But in fact, the weld connects the two steel plates as a whole and cannot be separated. Therefore, transverse tensile stress is generated in the middle of the two plates, and compressive stress is generated at both ends (Figures 4.9a, 4.9b). The second reason is that the first weld has already been welded. Solidification prevents the free expansion of the post-weld in the lateral direction and causes the post-weld to undergo lateral plastic compression deformation. When the post-weld is cooled, its shrinkage is limited by the solidified first weld to produce transverse tensile stress, and the

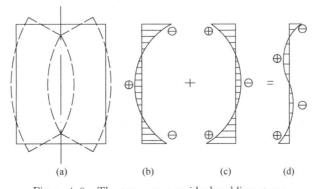

Figure 4.9 The transverse residual welding stress

first part of the weld produces transverse compressive stress. Because the stress is self-balanced, the other end of the weld is subjected to tensile stress (Figure 4.9c). The transverse stress of the weld is the combination of the above two parts of stress (Figure 4.9d).

In the welding of thick steel plates, the weld requires multiple layers of welding. Therefore, in addition to the longitudinal and transverse residual stress $\sigma_x$ and $\sigma_y$, there is also the residual welding stress $\sigma_z$ along with the thickness of the steel plate (Figure 4.10). The stress forms a three-way tensile stress field, which will greatly reduce the plasticity of the connection.

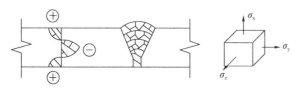

Figure 4.10 Residual welding stress in thick plates

(2) Influence of residual welding stress on structural performance

Under the static load, due to the plasticity of the steel, the residual welding stress is self-balanced on the section. Therefore, when the residual welding stress plus the stress caused by the external force reaches the yield strength, the stress does not increase, and the external force can continue to be borne by the elastic region until the full section reaches yield. Therefore, the residual welding stress does not affect the static strength of the structure.

Residual welding stress will reduce the stiffness of the structure. Since the stiffness of the residual tensile stress region entering the plastic state drops to zero, the external force that continues to increase is only borne by the elastic region. Therefore, the deformation of the component will increase, and the rigidity will decrease.

For the axial compression member, the residual welding stress reduces the flexural rigidity. The inertia moment of the elastic region that resists the increase of external force decreases, resulting in the decrease of the stable bearing capacity of the compression member. In thick plate welding or parts with cross welds, three-way residual welding tensile stress will be generated, hindering the development of plastic deformation in these areas and increasing the brittle fracture tendency of steel at low temperatures. Therefore, reducing or eliminating residual welding stress is an important measure to improve the structure's low-temperature cold brittleness.

The main metal residual welding stress in the weld and its vicinity usually reaches the steel's yield strength, which is the most sensitive area for forming and developing fatigue cracks. Therefore, the residual welding stress has a significant adverse effect on the fatigue strength of the structure.

(3) Form of welding deformation

Due to uneven heating and cooling during the welding process, when the welding zone shrinks in the longitudinal and transverse directions, it can cause welding deformation of the components. Welding deformation includes longitudinal shrinkage deformation, transverse shrinkage deformation, bending deformation, angular deformation, twisting deformation, etc., shown in Figure 4.11.

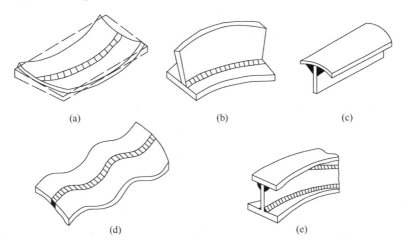

Figure 4.11 Welding deformation
(a) longitudinal and lateral deformation; (b) bending deformation;
(c) angular deformation; (d) wave deformation; (e) twisting deformation

It usually manifests as a combination of several deformations. Therefore, when any welding deformation exceeds the requirements of the *Standard for Acceptance of Construction Quality of Steel Structures* GB 50205, it must be corrected so as not to affect the load-bearing capacity of the component under normal use conditions.

(4) Methods to reduce residual welding stress and welding deformation

Adopting an appropriate welding sequence can reduce welding deformation and residual welding stress. As shown in Figure 4.12, the steel plate butt joint adopts segmented back welding, the thick weld adopts layered welding, the I-section adopts diagonal jump welding, and the steel plate splicing adopts block splicing welding.

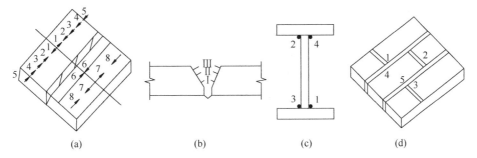

Figure 4.12 Appropriate welding sequence
(a) segmented back welding; (b) layered welding; (c) diagonal jump welding; (d) block splicing welding

As long as the structure allows, the weld's layout should be symmetrical to the component section's centroid axis to reduce welding deformation.

Before welding, give the component a pre-deformation in the opposite direction to the welding deformation to offset the deformation caused by the weldingto reduce the welding deformation. It is preheating before welding or tempering after welding (heating to about 600 ℃ and then slowly cooling) for small weldments, partly eliminating welding stress and welding deformation. A rigid fixing method can also fix the components to limit welding deformation, but it increases welding stress.

## 4.3 Construction Details and Calculation of Groove Welded Connection

### 4.3.1 Construction details of groove welded connection

Groove welds are made in a groove between adjacent ends, edges, or surfaces of two parts joined in a butt, tee, or corner joint. The weld configuration in these joints can be made in various ways. For example, a weld butt joint can be made square, double square, single bevel, double bevel, single V, double V, single J, double J, single U, or double U, as illustrated in Figure 4.13. Except for the square groove weld, some edge preparation is required for either one or both of the members to be connected, as illustrated in Figure 4.14.

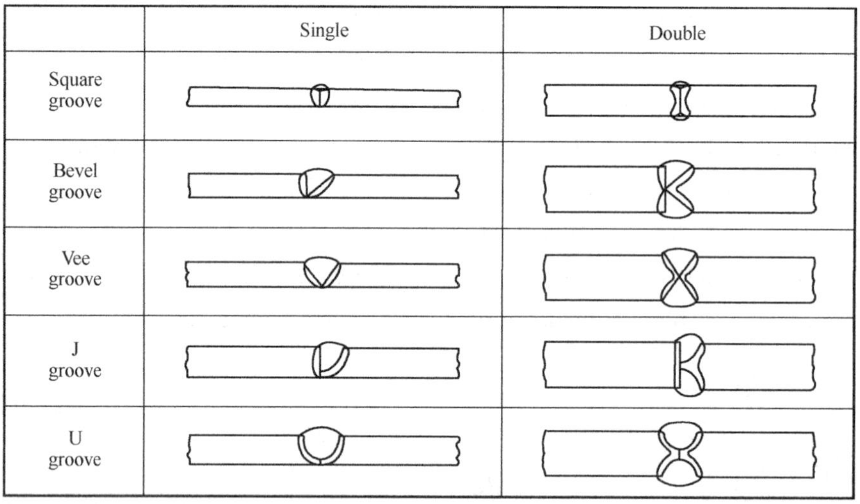

|  | Single | Double |
|---|---|---|
| Square groove | | |
| Bevel groove | | |
| Vee groove | | |
| J groove | | |
| U groove | | |

Figure 4.13 Groove welds

Groove welds are further classified as either complete penetration or partial penetration welds. A complete penetration weld achieves a fusion of weld and base metal throughout the depth of the joint. It is made by welding from both sides of the joint or from one side to a backing bar. The throat dimension of a full penetration groove weld is considered to be the full thickness of the thinner part joined, exclusive of weld reinforcement. Reinforcement is added weld metal over and above the thickness of the weld.

Partial penetration groove weld is used when load requirement does not require full

## 4.3 Construction Details and Calculation of Groove Welded Connection

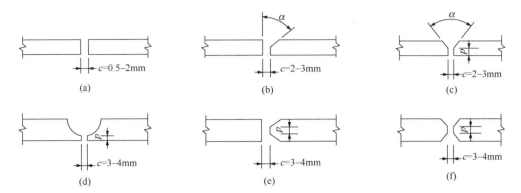

Figure 4.14  Edge preparation details of groove welds
(a) single square groove; (b) single bevel groove; (c) single V groove;
(d) single U groove; (e) double bevel groove; (f) double V groove

penetration or when welding must be done from one side of a joint only without using a backing bar.

At the splicing of groove welds, when the width of the weldment is different, or the thickness differs by more than 4 mm on one side, it is advisable to make an oblique angle with a slope not greater than 1 : 2.5 in the width direction or the thickness direction from one or both sides, respectively, as shown in Figure 4.15, in order to make the section transition smooth and reduce the stress concentration. The transition should be smooth according to this requirement when the groove weld connection of different plate thicknesses is subjected to dynamic load.

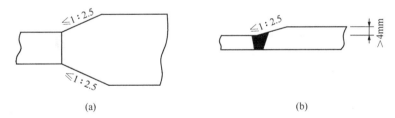

Figure 4.15  Splicing of unequal width and unequal thickness steel plates
(a) changing width; (b) changing thickness

Defects such as arc craters often appear at the arc initiation and extinguishing of groove welds. These defects greatly influence the bearing capacity of the connection, so the arc plates should generally be set during welding, and cut them off after welding. All groove welds that require equal strength should be welded with arc plates to avoid arcing defects at both ends of the weld. Likewise, the structure that bears the static load should be equipped with arc plates. When it is difficult and not allowed to set arc plates, the weld's effective length can be equal to the actual length minus $2t$, where $t$ is the thinner plate's thickness.

## 4.3.2 Calculation of full penetration groove welded connection

Because many allowed defects existin the groove weld under the third level inspection, the weld's tensile strength is 85% of the parent metal's tensile strength. Under the first and second level inspection, the weld's tensile strength is equal to the parent metals.

(1) Calculation of bearing axial force

Since the full penetration groove weld became a part of the weldment, the full penetration groove weld strength's calculation method is the same as that of the weldment under bearing axial force, as shown in Figure 4.16(a).

$$\sigma = \frac{N}{l_w h_e} \leqslant f_t^w, \text{ or } f_c^w \tag{4-1}$$

where  $l_w$ = the calculation length of a groove weld; when it is not to set arc plates, the $l_w$ equals the actual length minus $2t$, where $t$ is the thinner plate's thickness;

$h_e$ = the calculation height of a groove weld, and it equals the thinner plate's thickness $t$;

$f_t^w$ and $f_c^w$ = the tensile and compressive strength design values of groove welds, respectively.

If the straight weld shown in Figure 4.16(a) cannot meet the strength requirements, the oblique groove weld shown in Figure 4.16(b) can be used. The calculation shows that when the angle between the weld and the force is $\theta \leqslant 56°$, and the oblique weld's strength is not lower than the parent metal's strength, the verification calculation can no longer be performed.

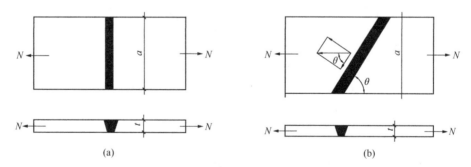

Figure 4.16  Groove welds bearing axial force:
(a) straught groove weld; (b) oblique groove weld

(2) Calculation of bearing bending moment and shear force

Since the weld cross-section shown in Figure 4.17(a) is rectangular, according to the knowledge of *Material Mechanics*, the normal stress and shear stress patterns are triangular and parabolic, respectively, and their maximum values should meet the following strength conditions:

$$\sigma_{\max} = \frac{M}{W_w} = \frac{6M}{l_w^2 t} \leqslant f_t^w \qquad (4\text{-}2)$$

$$\tau_{\max} = \frac{VS_w}{I_w t} = \frac{3}{2} \cdot \frac{V}{l_w t} \leqslant f_t^w \qquad (4\text{-}3)$$

where  $W_w$ = the elastic section modulus of the weld;

$S_w$ = the area moment of the section above the neutral axis of the weld to the neutral axis;

$I_w$ = the weld's moment of inertia about the neutral axis.

A groove weld of I-shaped section under bending and shear, shown in Figure 4.17 (b). Except for checking the maximum normal stress and maximum shear stress of the welds, the stress should be checked and converted by the following formula for locations subject to both greater normal stress and greater shear stress, for example, at the junction of the web and flange.

$$\sigma_{\text{red}} = \sqrt{\sigma_1^2 + 3\tau_1^2} \leqslant 1.1 f_t^w \qquad (4\text{-}4)$$

where  $\sigma_1$ = the normal stress of the weld at the checkpoint;

$\tau_1$ = the shear stress of the weld at the checkpoint;

1.1 = factor is that the maximum reduced stress only appears in a local position, so the strength design value is appropriately increased.

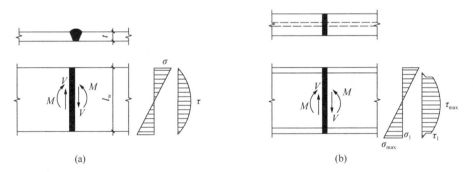

Figure 4.17  Groove welds bearing bending moment and shear force
(a) steel plate weld; (b) I-shape weld

(3) Calculation of bearing bending moment, shear force, and axial force

When the bending moment, the shear force, and the axial force work together, the weld's maximum normal stress equals the sum of the normal stress caused by the axial force and the bending moment and is checked according to Equation (4-5). Then, the maximum shear stress is calculated according to Equation (4-3). Finally, the calculation of the converted stress is still checked according to Equation (4-4).

$$\sigma_{\max} = \frac{M}{W_w} + \frac{N}{A_w} \leqslant f_t^w \qquad (4\text{-}5)$$

69

## 4.4 Construction Details and Calculation of Fillet Welded Connection

### 4.4.1 Construction details of fillet welded connection

Fillet welds are welds of theoretically triangular cross-section, joining two surfaces at right angles, sharp angles, and obtuse angles in lap, tee, and corner joints. The cross-section of a typical fillet weld most commonly used is a right triangle with equal legs. Figure 4.18 illustrates a typical fillet weld together with its pertinent nomenclature. The leg size designates the size of the weld. The root is the vertex of the triangle or the point at which the leg intersect.

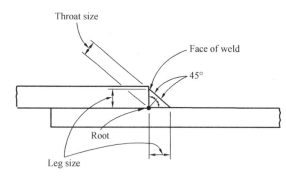

Figure 4.18  Typical fillet weld

The weld face is a theoretical plane since weld faces will be either convex or concave, as shown in Figure 4.19. The convex fillet weld is more desirable since it has less tendency to crack due to shrinking while cooling.

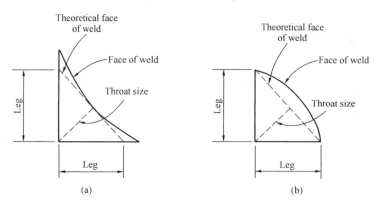

Figure 4.19  Right angle fillet weld:
(a) concave; (b) convex

The distance from the theoretical face of the weld to the root is called the throat size. Variations of this fillet weld are permitted and may be necessary. For example, leg sizes may be unequal. If the pieces to be joined do not intersect at right angles, the welds are considered skewed fillets, as shown in Figure 4.20. For skewed fillets of non-steel pipe structures, the degree of sharp angle should be greater than 60°, and the degree of obtuse angle should be less than 135°.

After welded, the shrinkage stress is caused by cooling. Therefore, the larger the leg size of the fillet weld was, the greater the shrinkage stress. Therefore, in order to avoid the "overburning" of the base metal in the weld zone, reduce the residual welding stress and welding deformation, the leg size of the fillet weld need not be too large.

The fillet weld on the edge of the plate is shown in Figure 4.21. When the plate thickness, $t > 6$ mm, the leg size, $h_f = t - (1\text{-}2)$ mm; when the plate thickness $t \leqslant 6$ mm, the leg size, $h_f = t$.

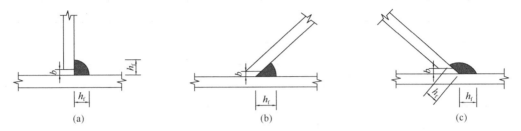

Figure 4.20 Fillet weld
(a) right angle; (b) sharp angle; (c) obtuse angle

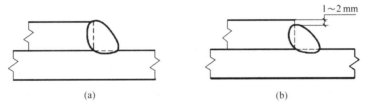

Figure 4.21 The maximum leg size of fillet weld
(a) base metal thickness is less than or equals 6 mm;
(b) base metal thickness is greater than 6 mm

The leg size of the fillet weld should not be too small. The thickness of the weldment will be relatively large due to the input energy being too small, so that the cooling rate will be too fast during welding, resulting in a hardened structure, which will cause the parent material to crack. The minimum leg size of fillet welds is shown in Table 4.1. Among them, the value of the base metal thickness $t$ is related to the welding method. When welding is performed by a non-low hydrogen welding method without preheating, $t$ is equal to the thicker plate thickness. Therefore, it is advisable to use a single weld. When the preheated non-low-hydrogen welding method or low-hydrogen welding method is used for welding, $t$ equals the thinner plate thickness in the welded joint. Besides, the minimum leg size of fillet welds should not be less than 5 mm for the welded connection under dynamic load.

In the lap welded connection, the side fillet weld with the same force acting direction as the length of the weld is not uniformly stressed along the length in the elastic stage, and there is a Shear-lag Effect, and the shear stress is large at both ends and small in the middle. However, if the weld length exceeds a certain limit, the longer the weld, the more obvious the uneven stress, it may be damaged first at both ends of the weld. In order to avoid this, the effective length of the side fillet weld is generally given a certain limit, $l_w \leqslant 60h_f$. When the actual length is greater than the limit mentioned above, the excess part may not be considered in the calculation. However, the fillet weld's effective length between the web and flange of the welded I-section member is not limited.

**The minimum leg size of fillet weld**    Table 4.1

| Base metal thickness(mm) | $h_f$(mm) |
| --- | --- |
| $t \leqslant 6$ | 3 |
| $6 < t \leqslant 12$ | 5 |
| $12 < t \leqslant 20$ | 6 |
| $t > 20$ | 8 |

In order to make the fillet weld bear the load well, the fillet weld's effective length, $l_w \geqslant 8h_f$, and not less than 40 mm. The fillet weld's effective length is consistent with the groove weld's effective length, which is the actual weld length minus the length of the arc craters at both ends.

When only two side fillet welds are connected at the end of the plate, as shown in Figure 4.22, in order not to reduce the connection strength excessively, the length of each side weld should not be less than the distance between the welds on both sides. That is $b/l_w \leqslant 1$. The distance $b$ between the fillet welds on both sides should not be greater than 200 mm. When $b > 200$ mm, transverse fillet welds or middle plug welds should be added.

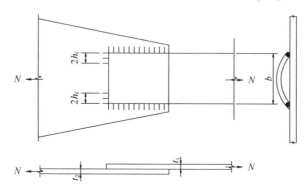

Figure 4.22 The length of the weld and the distance between the welds on both sides

In the lap connection, when only using the front fillet weld whose force is vertical to weld length, as shown in Figure 4.23, the lap length shall not be less than five times the weldment's smaller thickness and shall not be less than 25 mm.

When three-sided welding is used at the ends of the rods, the cross-section at the corners changes suddenly, which will cause stress concentration. In addition, if the arc is extinguished here, defects such as arc craters or undercuts may occur, which will increase

*4.4 Construction Details and Calculation of Fillet Welded Connection*

the impact of stress concentration. Therefore, all corners of the surrounding welding must be welded continuously. For non-surround welding, when the end of the fillet weld is at the corner of the component, it can be continuously welded with a length of two times leg size, as shown in Figure 4.22.

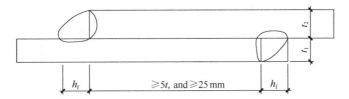

Figure 4.23 Requirements for double fillet welds in lap joints

### 4.4.2 Force performance of fillet weld

The strength of a fillet weld depends on the applied load direction, which may be parallel or perpendicular to the weld axis. According to the force direction, fillet welds can be divided into side fillet welds parallel to the force direction and transverse fillet welds perpendicular to the force direction. In parallel loading, the applied load is transferred parallel to the weld from one leg face to the other, as shown in Figure 4.24(a). The minimum resisting area in the fillet occurs at the throat and is equal to 0.707 times the leg size. Therefore, it assumes equal leg sizes for the weld, which generally is the truth in the right angle

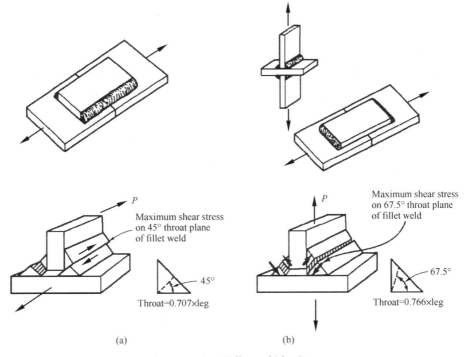

Figure 4.24 Fillet weld loading
(a) parallel loading/side fillet weld; (b) perpendicular loading/transverse fillet weld

fillet weld. However, the side fillet weld with too long length has a Shear-lag Effect, so its length will be limited, see section 4.4.1 for details.

Tests indicate that the strength of a transverse fillet weld is approximately one-third stronger than a side fillet weld. The GB 50017 permits this to be considered when designing welds under static load and dynamic load during non-fatigue design. Furthermore, the fillet weld loaded in the perpendicular direction has greater strength because the failure plane develops an angle other than 45°; hence the resisting area of the fillet is greater than the throat area, which is perpendicular to the theoretical face of the weld. Besides, transverse fillet welds are more uniformly stressed than side fillet welds.

### 4.4.3 The basic formula for calculating the strength of a right-angle fillet weld

(1) The effective section of fillet welds

The cross-section of the right-angle fillet weld is shown in Figure 4.25. The length of the right-angle side is $h_f$, which is the size of the fillet weld. The test shows that the fillet weld's damage often occurs in the weld throat, so the minimum thickness of the right-angle fillet weld in the 45° direction is taken, $h_e = 0.707 h_f$, which is the effective thickness of the fillet weld and called throat. The product of the effective thickness and the effective length of the weld is taken as the product. The effective section or calculated section is shown in Figures 4.24a and 4.26 when the fillet weld is broken. If there is a gap, $g$, between the two weldments in a Tee joint (Figure 4.24), the calculation of $h_e$ should be considered the influence of the gap. When $g \leqslant 1.5$ mm, $h_e = 0.7 h_f$; when $1.5$ mm $< g \leqslant 5$ mm, $h_e = 0.7(h_f - g)$.

(2) The basic calculating formula of a right-angle fillet weld

The stress on the effective section of the weld is shown in Figure 4.26. These stresses include normal stress $\sigma_\perp$ perpendicular to the effective section of the weld, shear stress $\tau_\perp$ perpendicular to the length of the weld, and parallel to the length of the weld, shear stress $\tau_{//}$. Equation (4-6) can be used to illustrate the relationship between the three stresses and weld strength. In Equation (4-6), $f_f^w$ is the design value of weld strength, and the unit is N/mm².

$$\sqrt{\sigma_\perp^2 + 3(\tau_\perp^2 + \tau_{//}^2)} \leqslant \sqrt{3} f_f^w \tag{4-6}$$

As shown in Figure 4.27, a Tee joint bears the oblique axial force of $2N$, borne by the two fillet welds. The resultant force $2N$ is decomposed into two mutually perpendicular components, $2N_x$ and $2N_y$, respectively. $N_y$ is the axial force perpendicular to the weld length, and $N_x$ is the axial force parallel to the weld length. Consider the force of a weld, and it can be seen that the normal stress $\sigma_f$(N/mm) is perpendicular to the right-angle side of a weld produced by $N_y$ on the effective section of the weld and calculated by Equation (4-7).

$$\sigma_f = N_y / h_e l_w \tag{4-7}$$

The stress $\sigma_f$ should be the combined stress of the normal stress $\sigma_\perp$ and the shear stress $\tau_\perp$.

$$\sigma_\perp = \tau_\perp = \sigma_f/\sqrt{2} \tag{4-8}$$

The stress $\tau_f = \tau_{/\!/}$ (N/mm²) generated by $N_x$ on the effective cross-section of the weld, which is parallel to the weld length, is calculated by Equation (4-9).

$$\tau_f = \tau_{/\!/} = N_x/h_e l_w \tag{4-9}$$

Incorporate Equations (4-8) and (4-9) into Equation (4-6). The basic formula for calculating the strength of right-angle fillet welds can be obtained after simplification, as shown in Equation (4-10). In the formula, $\beta_f$ is the strength increase coefficient of the transverse fillet weld, and the value is 1.22. For fillet welds in structures directly subjected to dynamic loads, $\beta_f = 1.0$

$$\sqrt{\left(\frac{\sigma_f}{\beta_f}\right)^2 + \tau_f^2} \leqslant f_f^w \tag{4-10}$$

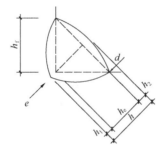

Figure 4.25 Cross-section of a right-angle fillet weld

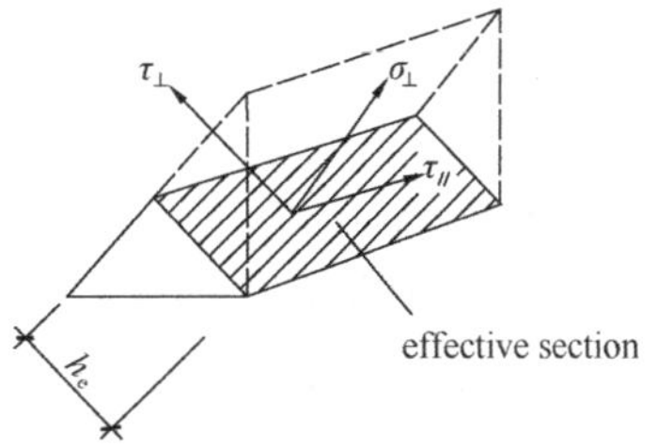

Figure 4.26 Stress on the effective section of a fillet weld

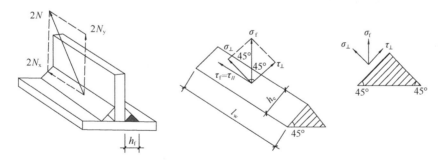

Figure 4.27 Calculation of a right-angle fillet weld

### 4.4.4 Calculation of right-angle fillet welded connection

(1) Subject to axial force

① Steel plate connection

The fillet weld connection of steel plate parts is common in over-lap and Tee joints (Figure 4.24). In the calculation, it is generally assumed that the stress distribution on the weld's effective section is uniform under the axial force. The following will give several common fillet weld calculation formulas for typical connections. They are all the deformed bodies of the basic calculation formula of fillet weld under different circumstances.

Only set the transverse fillet weld—the axial force N is perpendicular to the weld length, and the shear stress on the effective section of the fillet weld is zero. In Equation (4-10), $\tau_f = 0$. Then, Equation (4-10) can be simplified to:

$$\sigma_f = N/(h_e \sum l_w) \leqslant \beta_f f_f^w \qquad (4\text{-}11)$$

where  $l_w$ = the calculation length of a right-angle fillet weld; when it is not to set arc plates, the $l_w$ equals the actual length minus $2h_f$, where $h_f$ is the leg size of a fillet weld;

$h_e$ = the effective thickness or the calculation height of a fillet weld.

Only set side fillet welds—the axial force N is parallel to the weld length, and the normal stress on the effective section of the fillet weld is zero. In Equation (4-10), $\sigma_f = 0$. Then, Equation (4-10) can be simplified to:

$$\tau_f = N/(h_e \sum l_w) \leqslant f_f^w \qquad (4\text{-}12)$$

The axial force N forms an angle $\alpha$ with the fillet weld, and there are both normal stress $\sigma_f$ and shear stress $\tau_f$ on the effective section of the fillet weld. Therefore, they are calculated according to Equations (4-13a) and (4-13b), respectively. However, the calculation of fillet weld is still according to Equation (4-10).

$$\sigma_f = N \cdot \cos\alpha/(h_e \sum l_w) \qquad (4\text{-}13a)$$

$$\tau_f = N \cdot \sin\alpha/(h_e \sum l_w) \qquad (4\text{-}13b)$$

② Angle and gusset plate connection

In steel trusses, the connection fillet welds between single-angle or double-angle webs and gusset plates generally use two-side welding (Figure 4.28a) or three-side welding (Figure 4.28b) circumstances are also allowed. In addition, L-shaped surrounding welding is adopted (Figure 4.28c).

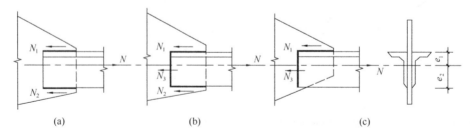

Figure 4.28  Fillet weld between angle and gusset plate under axial force

Although the axial force passes through the cross-section centroid of the angle steel connector, the distance from the cross-section centroid to the angle limb tip and the limb back is not equal; the forces shared by the limb back weld and the limb tip weld are not equal. Therefore, according to the balance conditions of force and bending moment, the force of each weld can be solved separately. In order to simplify the calculation process, the internal force distribution coefficient of the angle steel fillet weld can be directly determined according to the internal force distribution coefficient of the angle steel fillet weld in Table 4.2. In this table, $K_1$ and $K_2$ are the internal force distribution coefficients of the limb back and the limb tip, respectively.

For three-side welding (Figure 4.28b), we can first calculate the force that the transverse fillet weld can bear:

$$N_3 = 0.7h_f \sum l_{w3} \beta_f f_f^w \quad (4\text{-}14a)$$

where $l_{w3}$ is the calculation length of the transverse fillet weld; in three-side welding, $l_{w3}$ is equal to the angle's leg width.

Then, the forces shared by the limb back and the limb tip are:

$$N_1 = K_1 N - N_3/2 \quad (4\text{-}14b)$$
$$N_2 = K_2 N - N_3/2 \quad (4\text{-}14c)$$

where $K_1$ and $K_2$ are the internal force distribution coefficients of the limb back and the limb tip, listed in Table 4.2.

**Internal force distribution coefficient of angle fillet weld**　　　　Table 4.2

| Type of Angles | Connection type | The internal force distribution coefficient | |
|---|---|---|---|
| | | Limb back $K_1$ | Limb tip $K_2$ |
| Angles with equal legs | | 0.7 | 0.3 |
| Unequal-leg angle with short-leg connection | | 0.75 | 0.25 |
| Unequal-leg angle with long-leg connection | | 0.65 | 0.35 |

For two-side welding (Figure 4.28a), the transverse fillet weld force $N_3$ is equal to zero. Then,

$$N_1 = K_1 N \quad (4\text{-}15a)$$
$$N_2 = K_2 N \quad (4\text{-}15b)$$

For L-shaped welding (Figure 4.28c), we can first calculate the force that the transverse fillet weld can bear:

$$N_3 = 0.7h_f \sum l_{w3} \beta_f f_f^w \quad (4\text{-}16a)$$

where $l_{w3}$ is the calculation length of the transverse fillet weld; in three-side welding, $l_{w3}$

is equal to the angle's leg width.

Then, $N_1 = 0$, the force shared by limb back is:
$$N_1 = N - N_3 \tag{4-16b}$$

③ Design of the cover-plate connection

Cover-plate connections are often encountered in actual projects. The two steel plates are butted, and the upper and lower double cover plates and fillet welds are used to connect the connected plates. Before designing, it is necessary to determine whether to use a two-side welding or three-side welding connection according to the actual situation. When calculating the connection, we can first assume the fillet weld's leg size, find the weld length, and then determine the spliced cover plate's size from the weld length. The cover-plate size also needs to be based on the principle that the splicing cover plate's bearing capacity is not less than the bearing capacity of the main steel plate (connected steel plate). When the material properties are the same, ensure that the cover plate's total cross-sectional area on the splicing side is not less than the connected steel plate area.

(2) Subject to bending moment

Under the separate action of the bending moment, $M$, according to the basic knowledge of material mechanics, the stress on the effective section of the fillet weld is triangularly distributed (the central axis passes through the centroid of the effective section of the weld). The calculation formula of the maximum bending normal stress of the fiber at its edges (upper and lower) is:

$$\sigma_f = \frac{M}{W_w} \leqslant \beta_f f_t^w \tag{4-17}$$

(3) Subject to torque

As shown in Figure 4.29, the steel plate corbel is connected to the three sides of the flange of the steel column by welding, and the fillet weld is solely affected by the torque $T$. The calculation is based on the following assumptions: a. Assume that the fillet weld is elastic and the connected part is absolute rigid; b. The connected part rotates around the weld centroid $O$, and the stress direction at any point on the fillet weld is perpendicular to the line of the point to the centroid $O$, and the stress is proportional to the line length $r$. The formula for calculating the stress of a fillet weld under the sole action of torque $T$ is:

$$\tau_A = \frac{T \cdot r_A}{J} \tag{4-18}$$

where $J$ = the polar moment of inertia;

$r_A$ = distance of point A to the centroid $O$.

The stress $\tau_A$ given by the above formula is at an oblique angle to the weld length, and it can be decomposed to the $x$-axis and $y$-axis, as shown in Figure 4.29. The decomposed component stresses are shown in Equations (4-19a) and (4-19b), respectively. Among them, $\tau_{Ax}^T$ is the forced nature of the side fillet weld, and $\sigma_{Ay}^T$ is the forced nature of the

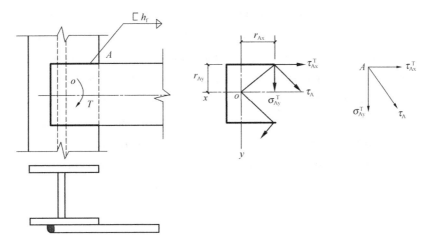

Figure 4.29 Fillet weld under torque

transverse fillet weld.

$$\tau_{Ax}^T = \frac{T \cdot r_{Ay}}{J} \quad (4\text{-}19a)$$

$$\sigma_{Ay}^T = \frac{T \cdot r_{Ax}}{J} \quad (4\text{-}19b)$$

## 4.5 Form of Bolted Connection and Construction Detail Requirements

### 4.5.1 Characteristics and form of bolted connections

Several types of bolts can be used for connecting structural steel members. The two types generally used in structural applications are unfinished/ordinary bolts and high-strength bolts. Proprietary bolts incorporating ribbed shanks, end splines, and slotted ends are also available, but in reality, these are only modifications of the high-strength bolt. The use of these bolts is allowed subject to the approval of the responsible engineer.

(1) Unfinished/ordinary bolts

Unfinished bolts are made of 35 steel and 45 high-quality carbon steel. Unfinished bolts are also known as machine, common, or ordinary bolts. Unfinished bolts are divided into three grades: A, B and C. Among them, A and B are refined bolts, and C is rough bolts. The performance grades of A and B ordinary bolts are 5.6 and 8.8. The performance grades of C grade ordinary bolts are 5.6 and 4.8. Take 4.8 as an example to illustrate the meaning of bolt performance grade: the number 4 before the decimal point indicates that the minimum tensile strength of the bolt is 400 MPa, and the number 0.8 after the decimal point and the decimal point indicates that the ratio of the yield strength of the bolt material to the tensile strength is 0.8.

Grade A and Grade B refined bolts need to be machined, and the connected parts need to be holed with the type of I. The difference between the bolt diameter and the hole diameter is 0.2-0.5 mm. Therefore, the manufacturing and installation costs are high, and it is rarely used in civil constructions at present. Grade C unfinished bolts are pressed from unprocessed round steel. The surface is rough, and the connected parts are the only type of II holes. The bolt diameter and the hole diameter differ by 1.0-1.5 mm. Grade C ordinary bolts have better performance in transmitting tensile force than transmitting shear force. Therefore, they are suitable for tensile connection or shear connection and temporary connection of secondary structures and detachable structures.

(2) High-strength bolts

The high-strength bolt is undoubtedly the most commonly used mechanical fastener for structural steel. Bolting with high-strength bolts has become the primary means of connecting steel members in the field as well as in the workshop.

The two primary types are grades 8.8 and 10.9. The meaning of high-strength bolt performance grade is the same as the unfinished bolt. Grade 8.8 high-strength bolts are made of 35 steel, 45 steel, and 40B steel. Grade 10.9 high-strength bolts are made of 35VB steel and 20MnTiB steel. The two primary types of holes are used for high-strength bolts. One is that its diameter is 1.5-2 mm larger than the bolt diameter used in the friction-type connection; the other is that its diameter is 1.0-1.5 mm larger than the bolt diameter used in the bearing-type connection. The common types of high-strength bolts are *Large hexagon head bolts* and *Tor-shear type high-strength bolts*, shown in Figure 4.30.

(a)  (b)

Figure 4.30  Types of high-strength bolts
(a) large hexagon head bolts; (b) tor-shear type high-strength bolts

### 4.5.2  Forms of bolted connections

Connections serve primarily to transmit the load from or to intersecting members; hence, connections must be based on structural principles. Thus, it involves creating a detail that is both structurally adequate and economical as well as practical.

The simplest form of bolted connection is the ordinary lap connection shown in

## 4.5 Form of Bolted Connection and Construction Detail Requirements

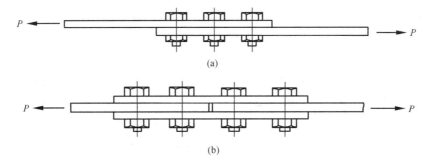

Figure 4.31 Types of bolted connections
(a) Lap connection; (b) Butt connection

Figure 4.31(a). Some connections in structures are of this general type, but they are not commonly used to connect members' tendency to deform.

A more common type of connection, the butt connection, is shown in Figure 4.31 (b). It is a type that may be used for tension member splices, replacing the member at the point where it is cut.

### 4.5.3 Construction detail requirements of bolted connections

(1) Bolt arrangement requirements

The arrangement of bolts on the components should comply with the principles of simple and tidy, uniform specifications, and compact layout. The connection center should be consistent with the center of gravity of the connected component section. Commonly used arrangements are parallel and staggered, as shown in Figure 4.32. The arrangement is simple and tidy, the size of the connecting plate is small, but the member's cross-section is weakened greatly. The staggered arrangement weakens the cross-section less, but the bolt arrangement is not as compact as the parallel arrangement, and the connecting plate size is larger.

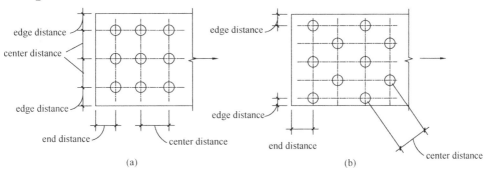

Figure 4.32 Bolt arrangement of steel plate
(a) Side by side; (b) staggered

The distance requirements for bolts arranged on the component shall meet the requirements of Table 4.3. The minimum center distance and edge distance (end distance) of the specified bolts are based on force and installation requirements; the maximum center distance and edge distance (end distance) of the specified bolts ensure that the steel plate is connected tightly enough.

In Table 4.3, $d_0$ is the diameter of the bolt hole, and $t$ is the thickness of the outer thinner plate. That is because the edge of the steel plate is connected with rigid components, such as angle steel and channel steel, and the maximum spacing of high-strength bolts can be adopted according to the value of the middle row. Therefore, when calculating the weakening of the section caused by the bolt hole, the larger value of $d+4$ mm and $d_0$ can be taken.

(2) Construction detail requirements

In addition to meeting the allowable distance of the above arrangement, the bolt connection should meet the following structural requirements according to different situations:

① To make the connection reliable, the number of permanent bolts at one end of each member should not be less than two. For the strips of composite members, one bolt can be used for the end connection, and the web rod of some tower mast structures can also use one bolt.

② Ordinary bolt tension connection for components that directly bearthe dynamic load, double nut, or other effective measures to prevent the nut from loosening should be adopted, such as spring washer or the nut and shank welding.

③ When high-strength bolts connect the section steel components, the flexural rigidity is relatively large due to the component itself. Therefore, in order to ensure close contact of the friction surface of the high-strength bolts, the splicing parts should be made of steel plates with weaker rigidity.

④ The endplates or flange plates in the bolt connection tensioned along the shank should be used. Properly increasing its rigidity, such as adding stiffeners, reduces the prying force's adverse effect on the bolt's tensile bearing capacity.

**The maximum and minimum allowable distance of bolts**  Table 4.3

| Item | Location direction | | | Maximum allowable distance (take the minimum of the two) | Minimum allowable distance |
|---|---|---|---|---|---|
| Center spacing | Outer row (perpendicular to the direction of internal force or parallel to the direction of internal force) | | | $8d_0$ or $12t$ | $3d_0$ |
| | Middle row | Perpendicular to the direction of internal force | | $16d_0$ or $24t$ | |
| | | Parallel to the direction of internal force | Members subject to compression | $12d_0$ or $18t$ | |
| | | | Members subject to tensile | $16d_0$ or $24t$ | |
| | Along the diagonal | | | | |

| Item | Location direction | | | Maximum allowable distance (take the minimum of the two) | Minimum allowable distance |
|---|---|---|---|---|---|
| Distance from center to edge | Parallel to the direction of internal force | | | $4d_0$ or $8t$ | $2d_0$ |
| | Perpendicular to the direction of internal force | Cut edges or hand-cut edges | | | $2d_0$ |
| | | Rolling edge, automatic gas cutting, or sawing edge | High-strength bolt | | |
| | | | Other bolts | | $2d_0$ |

## 4.6 Calculation of Bolted Connection

### 4.6.1 Calculation of unfinished/ordinary bolt connections

The commonly used bolted connections are shown in Figure 4.33. However, the bolts in these bolted connections can be divided into three categories according to the force: bolts only subject to shear (Figures 4.33a, 4.33d); bolt only subject to tension (Figure 4.33b); bolt subject to both shear and tension at the same time (Figures 4.33c, 4.33e).

(1) Mechanical properties of bolts under shear

Figures 4.34 and 4.35 show the working schematic diagram of a single ordinary bolt connection under shearing force. In the initial stage of load application, the shear force is very small. The load is transmitted by the friction force of the contact surface between the plates, the gap between the bolt shank and the hole-wall remains unchanged, and the connection is in the elastic working stage. As the load increases and exceeds the frictional force, relative slippage occurs between the connected parts, the bolt and the hole-wall begin to contact, the bolt shank is sheared, and the hole-wall is compressed at same time, shown in Figure 4.36.

(2) Failure modes of bolted connections under shear

There are five possible failure forms when the ordinary bolt shear connection reaches the ultimate bearing capacity.

① When the bolt diameter is small and the plate is thick, the bolt may be sheared first. This type of failure is called shear failure (Figure 4.37a).

② When the bolt diameter is larger and the plate is thinner, the plate may be squeezed first. This type of damage is called the bearing (or crushing) failure (Figure 4.37b).

③ When the plate's net cross-sectional area is too weak due to the bolt holes, the plate may be broken. The strength calculation of the member can avoid this type of failure (Figure 4.37c).

④ When the bolt arrangement's end distance is too small, the plate within the end

# Chapter 4 CONNECTIONS OF STEEL STRUCTURES

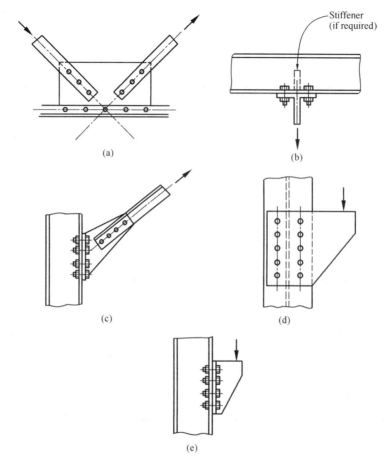

Figure 4.33 Common types of bolted connections
(a) Connection at a truss joint; (b) Hanger connection; (c) Bracing connection(bracket type);
(d) Eccentrically loaded bracket connection; (e) Eccentrically loaded connection

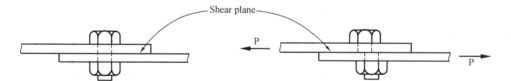

Figure 4.34 Bolt in single shear connection

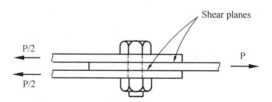

Figure 4.35 Bolt in double shear connection

4.6 Calculation of Bolted Connection

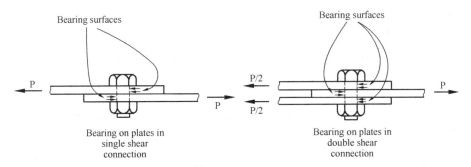

Figure 4.36  Bearing pressures

distance may be damaged by the bolt punching shear (Figure 4.37d). But if it meets the bolt arrangement requirements'end distance specified in the specification, this form of damage will not occur.

⑤ The shank's bending failure may occur when the connecting plate is too thick and the shank is too long (Figure 4.37e).

It can be known from the possible failure modes of ordinary bolt shear connection that the connection calculation only needs to consider the first type of shear failure and the second type of bearing failure.

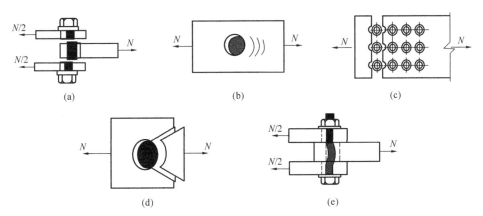

Figure 4.37  Failure modes of bolted connections under shear

(3) The design value of bearing capacity of an unfinished bolt

① The design value of allowable shear for one bolt:

$$N_v^b = n_v \frac{\pi \cdot d^2}{4} f_v^b \qquad (4\text{-}20)$$

where  $n_v =$ the number of shearing planes. $n_v$ equals one when it is a single shear(Figure 4.38a), $n_v$ equals two when it is a double shear(Figure 4.38b);

$f_v^b =$ the design value of bolt allowable shear strength.

② The design value of allowable bearing for one bolt:

$$N_c^b = d \cdot \Sigma t \cdot f_c^b \qquad (4\text{-}21)$$

85

where   $d$ = the nominal bolt diameter;

$\Sigma t$ = the minimum thickness of the plate or connected part; for example, in Figure 4.38(c), $\Sigma t$ is the smaller between $(a+c+e)$ and $(b+d)$;

$f_c^b$ = the design value of bolt allowable bearing strength.

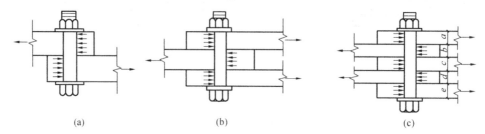

Figure 4.38 The number of shearing planes in bolted connections
(a) single shear connection; (b) double shear connection; (c) four shear connection

In a high-strength bolt shear connection, the design value of the shear capacity of a single bolt should be the smaller value of $N_v^b$ and $N_c^b$.

$$N_{min}^b = \min\{N_v^b, N_c^b\} \qquad (4\text{-}22)$$

The test shows that when the shear connection of the bolt group bears the axial force. The force of each bolt in the length direction of the bolt group is uneven (Figure 4.39), which is manifested by the large force on the two ends and the small force on the middle bolt.

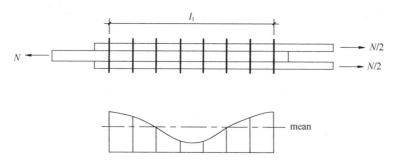

Figure 4.39 Internal force distribution of long joint bolts

When the connection length is $l_1 \leqslant 15d_0$ ($d_0$ is the bolt hole diameter) since the internal force is redistributed after the connection work enters the elastoplastic stage, the force of each bolt in the bolt group is gradually approaching, so it can be considered that the axial force $N$ is determined by each bolt share equally.

When the connection length is $l_1 > 15d_0$, even if the internal force is redistributed after the connection work enters the elastoplastic stage, the force of each bolt in the bolt group is not easy to be uniform. Therefore, the *Standard for Design of Steel Structures* GB 50017 stipulates: *at the joint of the component or one end of the splicing joint, when the bolt (including ordinary bolts and high-strength bolts) in the axial direction of the*

## 4.6 Calculation of Bolted Connection

connection length $l_1 > 15d_0$, the bolts should multiply the design value of the bearing capacity by the long joint reduction factor $\eta$ (Equation 4-23); when $l_1 > 60d_0$, the reduction factor is taken as a fixed value of 0.7.

$$\eta = 1.1 - l_1/(150d_0) \tag{4-23}$$

③ The design value of allowable tensile for one bolt

Tensile bolt connection is under external force action; the component's contact surface tends to disengage. At this time, the bolt is subjected to the pulling force along the shank axis, so the tensile bolt connection's failure mode is shown as the shank is broken.

$$N_t^b = \frac{\pi \cdot d_e^2}{4} f_t^b = A_e f_t^b \tag{4-24}$$

where  $d_e$ = the effective bolt diameter;

$A_e$ = the effective bolt cross-sectional area;

$f_t^b$ = the design value of bolt allowable tensile strength.

④ The design equation of one bolt under shear and tensile

For bolts subjected to both shear and tensile, divide the shear and tensile by the respective bearing capacity when they act individually, so that the dimensionless correlation Equation (4-25) can be used as the calculation formula for bolts under both shear and tensile.

$$\sqrt{\left(\frac{N_v}{N_v^b}\right)^2 + \left(\frac{N_t}{N_t^b}\right)^2} \leqslant 1 \tag{4-25}$$

where  $N_v$ = the design value of a single bolt's shear force;

$N_t$ = the design value of the tensile force of a single bolt;

$N_v^b$ = the design values of the shear of a single bolt;

$N_v^b$ = the design values of the tensile bearing capacity of a single bolt.

(4) Bolt group connection calculation

① Subject to axial force

a. Number of bolts required under axial shear

When the shear force passes through the bolt group's centroid, within the range of the connection length, it is assumed that all bolts are subjected to the same force in the calculation, and the following equation calculates the number of bolts required. $n$ takes an integer.

$$n = \frac{N_v}{\eta \times N_{\min}^b} \tag{4-26}$$

where  $N_v$ = the design value of the axial shear acting on the bolt group;

$\eta$ = the long joint reduction factor.

b. Number of bolts required under axial tensile force

When the tensile force passes through the bolt group's centroid, within the range of the connection length, it is assumed that all bolts are subjected to the same force in the calculation, and the following Equation calculates the number of bolts required. $n$ takes an integer.

$$n = \frac{N}{N_t^b} \quad (4\text{-}27)$$

where $N_b$ = the design value of the axial shear acting on the bolt group.

    c. Calculation of net section strength of the plate or connected part

$$\sigma = \frac{N}{A_n} \leqslant f \quad (4\text{-}28)$$

where $f$ = the design strength of the plate or connected part;

    $A_n$ = the net cross-section area of the plate or connected part.

② Bolt group subjects to torque

As shown in Figure 4.40, the bolt group is subjected to a torque $T$, and each bolt is sheared, but the magnitude or direction of the shear force is different.

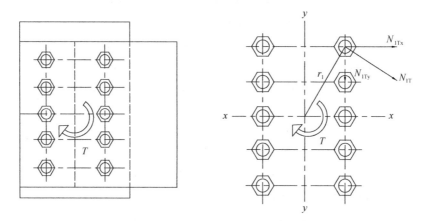

Figure 4.40  Torque action diagram of the bolt group

When analyzing the bolt group subjected to a torque, the following calculation assumptions are used: a. the bolt is an elastic body, and the connected part is absolutely rigid; b. the connecting plate rotates around the bolt group's centroid, and each bolt's shear force is proportional to the distance $r_i$ from the centroid to the bolt, and the direction of the shear force is perpendicular to the line $r_i$.

Bolt 1 is the farthest from the centroid $O$ in the bolt group, and its shear force $N_{1T}$ is the largest. To facilitate calculation, $N_{1T}$ can be decomposed into two components on the x-axis and y-axis:

$$N_{1x}^T = T \cdot y_1 / (\Sigma x_i^2 + \Sigma y_i^2) \quad (4\text{-}29a)$$

$$N_{1y}^T = T \cdot x_1 / (\Sigma x_i^2 + \Sigma y_i^2) \quad (4\text{-}29b)$$

Therefore, the resultant force of bolt 1 with the largest force should be less than the design value $N_{\min}^b$ of the shear capacity of a single bolt:

$$\sqrt{(N_{1x}^T)^2 + (N_{1y}^T)^2} \leqslant N_{\min}^b \quad (4\text{-}30)$$

③ Bolt group subjects to bending moment

The upper bolt of ordinary bolt group under the action of bending moment $M$, is tensioned, shown in Figure 4.41, so there is a tendency to separate the upper part of the connection. As a result, the bolt group's centroid moves downward. Therefore, it is difficult to determine the position of the neutral axis accurately. It is usually assumed that the neutral axis is on the axis of the bottom row of bolts. However, when the direction of the bending moment shown in Figure 4.41 is counter clockwise, the neutral axis is on the axis of the uppermost row of bolts.

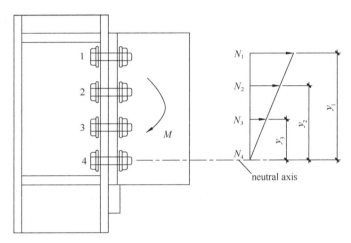

Figure 4.41 Bending moment action diagram of the ordinary bolt group

Therefore, under the action of the bending moment $M$, the maximum tensile force of bolt 1 in the uppermost row is:

$$N_1^M = M \cdot y_1 / \Sigma y_i^2 \leqslant N_t^b \qquad (4\text{-}31)$$

where $\Sigma y_i^2$ = the sum of the squares of the distance between each bolt and the neutral axis.

④ Bolt group subjects to both shear and tensile

For bolt group subjected to both shear and tensile, calculate the bolt with the greatest force according to Equation (4-25), and the bearing failure of the bolt group under the sole action of shearing force shall be checked as follows:

$$N_v \leqslant N_c^b \qquad (4\text{-}32)$$

where $N_v$ = the design value of a single bolt's shear force;

$N_c^b$ = the design value of a single bolt's bearing capacity, calculated by Equation (4-21).

### 4.6.2 Calculation of high-strength bolt friction-type connection

It is easy to visualize the plates slipping in the applied force's direction until they bear against the bolt. This is called a bearing-type connection. It is sometimes referred to as a shear-bearing connection. For this connection to existing, a small movement (slip) must bring the bolts into bearing. A small amount of slip in a connection is not detrimental in many applications and is sometimes even desirable. Both used in a bearing connection need only be tightened to the snug-tight condition. In other connections, even the smallest amount of slip is undesirable. These connections are called slip-critical (friction-type) connections, and the high-strength bolts used in them must be tightened to the full pre-tensioning load.

(1) Pre-tensioning load of the high-strength bolt

The following equation calculates the design pre-tensioning load $P$ of the high-strength bolt:

$$P = \frac{0.9 \times 0.9 \times 0.9}{1.2} f_u \cdot A_e \qquad (4\text{-}33)$$

where  $f_u$ = the minimum tensile strength of the bolt material. For grade 8.8 bolts, $f_u$ = 830 MPa; for grade 10.9 bolts, $f_u$ = 1,040 MPa;

$A_e$ = the effective cross-section area of high-strength bolts (mm²).

Coefficients in Equation (4-33) are explained as follows:

① The first reduction factor of 0.9 is used to consider the variability of bolt material.

② The second reduction factor of 0.9 is considered the loss of prestressing over tension by 5% to 10%.

③ The third reduction factor of 0.9 is used to consider that the strength of steel is based on tensile strength.

④ 1.2 in the denominator is the reduction in the bolt's tensile bearing capacity reduced by the shear caused by the torque when the bolt is tightened.

The pre-tensioning load principle calculation formula is Equation (4-33), and the actual design value is directly taken according to Table 4.4.

Design pre-tensioning load of a high-strength bolt (kN)　　　　Table 4.4

| Bolt bearing performance grade | Nominal diameter of bolt (mm) | | | | | |
|---|---|---|---|---|---|---|
| | M16 | M20 | M22 | M24 | M27 | M30 |
| Grade 8.8 | 80 | 125 | 150 | 175 | 230 | 180 |
| Grade 10.9 | 100 | 155 | 190 | 255 | 290 | 355 |

(2) Anti-slip coefficient of high-strength bolt connection

Domestic and foreign research and engineering practice have shown that the anti-slip coefficient of the friction surface of the high-strength connection is mainly related to the surface treatment process of the steel and the thickness of the coating. Table 4.5 specifies the anti-slip coefficient values corresponding to different contact surface treatment methods.

## 4.6 Calculation of Bolted Connection

**Anti-slip coefficient $\mu$ of friction surface**  Table 4.5

| Treatment method of component contact surface at joint | Steel grades of components | | |
|---|---|---|---|
| | Q235 | Q355 or Q390 | Q420 or Q460 |
| Spray hard quartz sand or cast steel angular sand | 0.45 | 0.45 | 0.45 |
| Shotblasted (sandblasted) | 0.4 | 0.4 | 0.4 |
| Wire brush to remove rust or untreated clean rolled surface | 0.3 | 0.35 | — |

(3) The design value of bearing capacity of high-strength bolt friction-type connection

① The design value of shear capacity of a single high-strength bolt friction-type connection

$$N_v^b = 0.9\kappa n_f \mu P \qquad (4\text{-}34)$$

where 0.9 is reciprocal of *Resistance Partial Coefficient*. $\kappa$ is the hole type coefficient, 1.0 for standard holes; 0.85 for oversized holes; 0.7 when the internal force is perpendicular to the length of the long-slotted hole; 0.6 when the internal force is parallel to the direction of the long-slotted hole. $n_f$ is the number of shear planes. $P$ is the design value of the pre-tensioning load according to Table 4.4.

② The design value of tensile capacity of a single high-strength bolt friction-type connection

$$N_t^b = 0.8P \qquad (4\text{-}35)$$

(4) The calculation of the high-strength bolt group

① Subject to axial shear

a. Number of bolts required under axial shear

Under the action of the axial shear, the calculation method of the number of bolts required for high-strength bolt friction-type connection still adopts the form of Equation (4-26), but it is necessary to replace $N_{min}^b$ in Equation (4-26) with equal $N_v^b$ in Equation (4-34).

b. Calculation of net section strength of the plate or connected part

In the high-strength bolt friction-type connection shown in Figure 4.42, part of the shear will be transmitted by the contact surface before the first row of bolts. Generally,

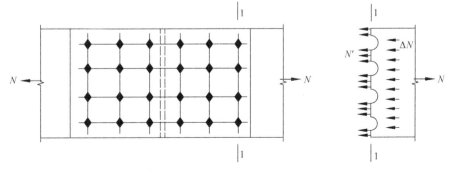

Figure 4.42  Force transmission of high-strength bolt friction-type connection in the front of holes

the transmitted force is accounted for 50% of the bolts' transmission force. In this way, the net section transmission force of section 1-1 is:

$$N' = N - 0.5 \frac{N}{n} \times n_1 = N\left(1 - \frac{0.5n_1}{n}\right) \qquad (4\text{-}36)$$

where $n$ is the total number of bolts on one side of the connection; $n_1$ is the number of bolts on the calculating cross-section 1-1.

Calculation of net section strength of the plate or connected part:

$$\sigma = \frac{N'}{A_n} \leqslant f \qquad (4\text{-}37)$$

② Bolt group subjects to bending moment

High-strength bolt group under the action of bending moment $M$ is shown in Figure 4.43. Due to the pre-tensioning load, it is usually assumed that the neutral axis is on the centroid axis of the bolt group.

Therefore, under the action of the bending moment $M$, the maximum tensile force of bolt 1 in the uppermost row is:

$$N_1^M = M \cdot y_1 / \Sigma y_i^2 \leqslant N_t^b \qquad (4\text{-}38)$$

Where $\Sigma y_i^2$ is the sum of the squares of the distance between each bolt and the neutral axis.

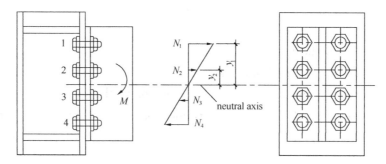

Figure 4.43　Bending moment action diagram of the high-strength bolt group

③ Bolt group subjects to both shear and tensile

For high-strength bolt group subjected to both shear and tensile, calculate the bolt with the greatest force according to Equation (4-39):

$$\frac{N_v}{N_v^b} + \frac{N_t}{N_t^b} \leqslant 1 \qquad (4\text{-}39a)$$

$$N_t \leqslant 0.8P \qquad (4\text{-}39b)$$

where $N_v^b$ and $N_t^b$ are the design values of the shear and tensile bearing capacity of a single high-strength bolt in friction-type connection and calculated by Equations (4-34) and (4-35), respectively.

### 4.6.3　Calculation of high-strength bolt bearing-type connection

The high-strength bolt bearing-type connection takes the shear failure or bearing failure as the bearing capacity limit state, and the damage form is the same as that of ordinary bolts.

In the connection that bears the axial shear force, the calculation method of the design value of the shear capacity is the same as that of the ordinary bolt, except that the bolt allowable shear strength or the bolt allowable bearing strength of the high-strength bolt is adopted. It should be noted that when the shear plane is at the thread, the effective diameter $d_e$ of the bolt needs to be used to replace $d$ in equations.

In the connection that bears the axial tension, the calculation method of the design value of the tensile bearing capacity is the same as that of the ordinary bolt. Therefore, according to Equation (4-24), the design value of high-strength bolt allowable tensile strength is required.

High-strength bolt bearing-type connections bear both shear and tension and should be calculated according to Equation (4-40):

$$\sqrt{\left(\frac{N_v}{N_v^b}\right)^2 + \left(\frac{N_t}{N_t^b}\right)^2} \leqslant 1 \quad (4\text{-}40a)$$

$$N_v \leqslant N_c^b/1.2 \quad (4\text{-}40b)$$

where 1.2 is a reduction factor considers the bolt's axial external pull reducing the design value of bolt allowable bearing strength of the high-strength bolt.

## 4.7 Working Examples

**[Example 4-1]**

In a beam-to-column butt connection, the groove (butt) weld is subjected to the combined action of the moment, shear, and axial force.

As shown in Figure 4.44, an I-section beam and column flanges are connected by penetration groove welds. The steel is Q235B, the type of electrode is E43, manual welding, the third level of weld quality is used when welding uses arc ignition board. The design value of bearing static load is $F=700$ kN, $N=760$ kN. Check the connection strength of the welds.

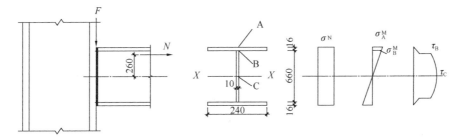

Figure 4.44 A groove weld connection of I-section beam and column

## Chapter 4  CONNECTIONS OF STEEL STRUCTURES

**Solution:**

External force on the connection:
$$N = 760 \text{ kN}$$
$$M = N \cdot e = 760 \times 260 \times 10^{-3} = 197 \text{kN} \cdot \text{m}$$
$$V = 700 \text{ kN}$$

Geometric characteristics of weld cross-section:
$$A = 2 \times 16 \times 240 + 10 \times 660 = 14,280 \text{ mm}^2$$
$$I_x = \frac{1}{12}(240 \times 692^3 - 230 \times 660^3) = 1,117,137,760 \text{ mm}^4$$
$$W_x = \frac{I_x}{h} = \frac{1,117,137,760}{346} = 3,228,722 \text{ mm}^3$$
$$S_C = (240 \times 16) \times 338 + (330 \times 10) \times 165 = 1,842,420 \text{ mm}^3$$
$$S_B = (240 \times 16) \times 338 = 127,920 \text{ mm}^3$$

Strength check of each critical point:

① Maximum tensile stress (point A)
$$\sigma_{A,\max} = \sigma_A^N + \sigma_A^M = \frac{N}{A_n} + \frac{M}{W_x} = \frac{760 \times 10^3}{14,280} + \frac{197.6 \times 10^3}{3,228,722} = 114.2 \text{ N/mm}^2 < f_t^w$$
$$= 185 \text{ N/mm}^2 \quad \textbf{O. K.}$$

② Maximum shear stress (point C)
$$\tau_C = \frac{VS_C}{I_x t_w} = \frac{700 \times 10^3 \times 1,842,420}{1,117,137,760 \times 10} = 115.45 \text{ N/mm}^2 < f_v^w = 125 \text{ N/mm}^2 \quad \textbf{O. K.}$$

③ Converted stress (point B)
$$\tau_B = \frac{VS_B}{I_x t_w} = \frac{700 \times 10^3 \times 127,920}{1,117,137,760 \times 10} = 81.33 \text{ N/mm}^2$$
$$\sigma_B = \frac{N}{A_n} + \frac{M}{W_x} \cdot \frac{h}{h_0} = \frac{760 \times 10^3}{14,280} + \frac{197.6 \times 10^3}{3,228,722} \times \frac{330}{346} = 111.6 \text{ N/mm}^2$$
$$\sqrt{\sigma_B^2 + 3\tau_B^2} = \sqrt{111.6^2 + 3 \times 81.33^2} = 179.7 \text{ N/mm}^2 < 1.1 f_t^w = 1.1 \times 185 = 203.5 \text{ N/mm}^2 \quad \textbf{O. K.}$$

**[Example 4-2]**

In a double-angle web and gusset plate connection with three-side welding, the three-side fillet weld is subjected to the axial force.

As shown in Figure 4.45, a connection of double-angle webs and gusset plate with three-side welding in the steel truss and bears the static axial force. Try to determine the bearing capacity of the connection and the length of the limb tip weld. It is known that the angles are two ∟ 125 × 10, which is connected to the gusset plate with a thickness of 8 mm, the lap length (the weld length of the limb back) is 300 mm, the weld leg size is $h_f = 8$ mm, the steel is grade Q355B, manual welding, and the type of electrode is E50.

**Solution:**

The internal force $N_3$ that the transverse fillet weld can bear is:

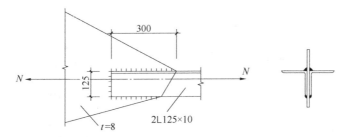

Figure 4.45 A three-side fillet weld connection of double-angle webs and gusset plate

$$N_3 = 2 \times 0.7 h_f b \beta_f f_f^w = 2 \times 0.7 \times 8 \times 125 \times 1.22 \times 200 = 341.6 \text{ kN}$$

The internal force $N_1$ of the limb back fillet weld is:

$$N_1 = 2 \times 0.7 h_f b \beta_f f_f^w = 2 \times 0.7 \times 8 \times (300-8) \times 200 = 651.1 \text{ kN}$$

From the equation:

$$N_1 = \alpha_1 N - \frac{N_3}{2} = 0.67N - \frac{341.6}{2} = 654.1 \text{ kN}. \text{ It can be obtained that}$$

$$N = 1,231.2 \text{ kN}.$$

The internal force $N_2$ of the weld of the limb tip is calculated as:

$$N_2 = \alpha_2 N - \frac{N_3}{2} = 0.33 \times 1,231.2 - \frac{341.6}{2} = 235.5 \text{ kN}$$

From this, the actual length of the weld of the limb tip can be calculated as:

$$l'_{w2} = \frac{N_2}{2 \times 0.7 \times 8 \times 200} + 8 = 113 \text{ mm}$$

It can be set to 115 mm.

[**Example 4-3**]

In a corbel and steel column joint, the connection fillet weld is subjected to shear and moment.

As shown in Figure 4.46 for the connection joint between the corbel and the steel column, the static load design value $N=365$ kN, the eccentricity $e=350$ mm, and the leg size $h_{f1}=8$ mm and $h_{f2}=6$ mm. Check the strength of the connection fillet weld. The steel is grade Q355B, the type of electrode is E50, manual welding. Figure 4.46(b) is a schematic diagram of the effective cross-section of the weld.

**Solution:**

The vertical force $N$ causes a shear force and a moment at the centroid of the fillet weld:

$$V = N = 365 \text{ kN}$$
$$M = V \cdot e = 365 \times 0.35 = 127.8 \text{ kN} \cdot \text{m}$$

**Method 1: Considering that the web welds participate in the transfer of the moment**

The moment of inertia of the effective cross-section of all welds on the neutral axis is:

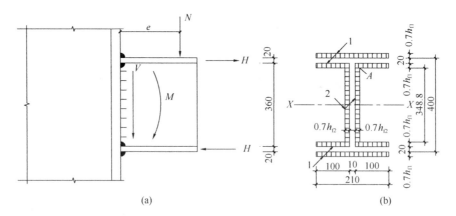

Figure 4.46 A connection joint of a corbel and steel column

$$I_x = 2 \times \frac{4.2 \times 348.8^3}{12} + 2 \times 210 \times 5.6 \times 202.8^2 + 4 \times 100 \times 5.6 \times 177.2^2$$
$$= 196.8 \times 10^6 \, \text{mm}^4$$

It can be seen that the maximum stress of the flange weld is:

$$\sigma_{f1} = \frac{M}{I_x} \cdot \frac{h}{2} = \frac{127.8 \times 10^6}{196.8 \times 10^6} \times 205.6 = 133.5 \, \text{N/mm}^2 < \beta_f f_f^w = 1.22 \times 200 = 244 \, \text{N/mm}^2$$

It can be seen that the flange weld meets the strength requirements.

From the proportional relationship, the maximum stress of the web weld caused by the moment $M$ (point $A$ in Figure 4.46b) is:

$$\sigma_{f2} = 133.5 \times \frac{174.4}{205.6} = 113.2 \, \text{N/mm}^2$$

It can be seen that the average shear stress generated by the shear force $V$ in the web weld is:

$$\tau_f = \frac{V}{\Sigma(h_{e2} l_{w2})} = \frac{365 \times 10^3}{2 \times 0.7 \times 6 \times 348.8} = 124.6 \, \text{N/mm}^2$$

Substituting the obtained $\sigma_{f2}$ and $\tau_f$ into Equation (4-10), the strength of the web weld (point $A$ is the design control point) is:

$$\sqrt{\left(\frac{\sigma_{f2}}{\tau_f}\right)^2 + \tau_f^2} = \sqrt{\left(\frac{113.2}{1.22}\right)^2 + 124.6^2} = 155.4 \, \text{N/mm}^2 < 200 \, \text{N/mm}^2$$

It can be seen that the web welds also meet the strength requirements.

So, the strength of the connection fillet weld is safety.

**Method 2: The web welds are not considered to participate in the transfer of the moment**

The horizontal force afforded by the flange welds is:

$$H = \frac{M}{h} = \frac{127.8 \times 10^6}{380} = 336.3 \, \text{kN}$$

From Equation (4-7), it can be seen that the normal stress generated by horizontal force $H$ in the flange welds is:

$$\sigma_f = \frac{H}{h_{e2} l_{w2}} = \frac{336.3 \times 10^3}{0.7 \times 8 \times (210 + 2 \times 100)} = 146.5 \, \text{N/mm}^2 < \beta_f f_f^w = 244 \, \text{N/mm}^2$$

It meets the requirements.

The web welds only bear the shear force, which has been calculated in Method 1 and meets the requirements.

**[Example 4-4]**

The fillet weld group connecting the column and the corbel bears the combined action of torque and shear.

Check the strength of the fillet weld connection between the corbel and the steel column, as shown in Figure 4.47. Known to use three-side fillet weld, $h_f=8$ mm, steel is Q235B, the type of electrode is E43, manual welding. The design load value on the component is $F=217$ kN, and the eccentricity is $e=300$ mm (distance to the edge of the column), the overlap size is $l_1=400$ mm, $l_2=300$ mm.

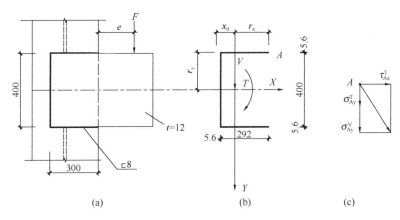

Figure 4.47  A connection joint of a corbel plate and steel column

**Solution:**

Determine the core position of the effective cross-section of the fillet weld.

$$x_0 = \frac{2\times(300-8)\times 5.6\times(146+5.6)+(400+2\times 5.6)\times 5.6\times 2.8}{2\times 292\times 5.6+411.2\times 5.6} = 9\text{cm}$$

$$I_x = \frac{1}{12}\times 0.7\times 0.8\times 40^3 + 2\times 0.7\times 0.8\times 29.76\times 20.28^2 = 16,695 \text{ cm}^4$$

$$I_x = \frac{2}{12}\times 0.7\times 0.8\times 29.76^3 + 2\times 0.7\times 0.8\times 29.76\times (14.88-9)^2$$
$$+0.7\times 0.8\times 40\times 6.2^2$$
$$= 2,460+1,152+861 = 4,473 \text{ cm}^4$$

The polar moment of inertia of the effective section of the fillet weld is:

$J = I_x + I_y = 21,168 \text{ cm}^4$

The distance from the weld point A to the x-axis and y-axis is:

$r_x = 20.76 \text{ cm}, r_y = 20.28 \text{ cm}$

Simplifying the external force $F$ toward the weld centroid point $O$, it gets:

Shear $V = F = 217$ kN

Toque $T = F \times \dfrac{(30+30-9)}{100} = 110.67 \text{ kN} \cdot \text{m}$

For the calculation of weld strength, point A of the weld is the design control point, and the stresses are:

$\tau_{Ax}^T = \dfrac{T \cdot r_y}{J} = \dfrac{110.67 \times 202.8 \times 10^6}{21,168 \times 10^4} = 106 \text{ N/mm}^2$

$\tau_{Ay}^T = \dfrac{T \cdot r_x}{J} = \dfrac{110.67 \times 207.8 \times 10^6}{21,168 \times 10^4} = 108 \text{ N/mm}^2$

$\sigma_{Ay}^T = \dfrac{V}{\Sigma h_e l_w} = \dfrac{217 \times 10^3}{0.7 \times 0.8 \times (2 \times 297.6 + 400)} = 38.9 \text{ N/mm}^2$

$\sqrt{\left(\dfrac{\sigma_{Ay}^T + \sigma_{Ay}^V}{\beta_f}\right)^2 + (\tau_{Ax}^T)^2} = \sqrt{\left(\dfrac{104.4 + 38.9}{1.22}\right)^2 + 102^2} = 160.4 \text{ N/mm}^2 \approx f_f^w$

$= 160 \text{ N/mm}^2$

Therefore, the leg size of welds is 8 mm to meet the connection safety requirements.

**[Example 4-5]**

The unfinished/ordinary bolt group bears shear force in the splicing connection of double cover plates.

Two steel plates with a cross-section of $-14 \text{ mm} \times 400 \text{ mm}$ are connected by double cover plates and C-grade ordinary bolts, as shown in Figure 4.48. The steel is grade Q235B, the bolt is grade 4.6, M20, and the bearing axial force's design value is $N = 935$ kN (static load). Try to design this connection.

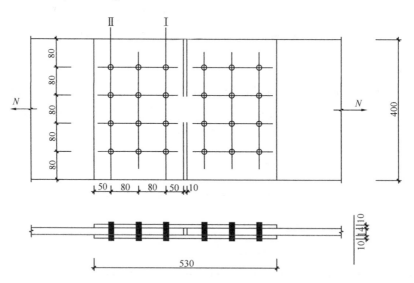

Figure 4.48  The splicing connection of double cover plates and unfinished bolts

**Solution:**

Double cover plates are used for splicing, and the cross-sectional size is 10 mm×400 mm. Therefore, the sum of the cross-sectional area of the cover plate is greater than the cross-sectional area of the steel plate to be connected.

The design value of single bolt shear capacity:

$$N_v^b = n_v \frac{\pi \cdot 20^2}{4} f_v^b = 2 \times \frac{\pi \cdot 20^2}{4} \times 140 \times 10^{-3} = 87.92 \text{ kN}$$

The design value of bearing capacity of single bolt:

$$N_c^b = d \cdot \Sigma t \cdot f_c^b = 20 \times 14 \times 305 \times 10^{-3} = 85.4 \text{ kN}$$

$$N_{\min}^b = \min\{N_v^b, N_c^b\} = 85.4 \text{ kN}$$

Then the number of bolts required on one side of the connection is:

$$n \geq \frac{N}{N_{\min}^b} = \frac{935}{85.4} = 11 \text{, take } 12.$$

The side-by-side arrangement shown in Figure 4.47 is adopted, and the size of the connecting cover plate is $2-10 \times 400 \times 530$, and the middle distance, side distance, and end distance of the bolts meet the structural requirements.

The connecting steel plate receives the greatest force at section I-I, and the cover plate receives the greatest force at section II-II. Still, because the two steel plates are the same, and the sum of the cross-sectional area of the cover plate is greater than the section area of the connected steel plate, only the net section strength of the connected steel plate is checked. Set the bolt diameter $d_0 = 21.5$ mm.

$$A_n = (b - n_1 d_0)t = (100 + 4 \times 21.5) \times 14 = 4,396 \text{ mm}^2$$

$$\sigma = \frac{N}{A_n} = \frac{935 \times 10^3}{4,396} = 212.7 \text{ N/mm}^2 < f = 215 \text{ N/mm}^2$$

The strength of the member meets the design standard.

**[Example 4-6]**

The unfinished/ordinary bolt group in the connection between the column and the corbel bears eccentric shear force.

Check the ordinary bolt connection shown in Figure 4.49. The thickness of the column flange plate is 10 mm, the thickness of the connecting plate is 8 mm, the steel is grade Q235B, the load design value is $F = 150$ kN, the eccentricity $e = 250$ mm, and the bolts are M22.

**Solution:**

Simplifying the external force $F$ toward the bolt group centroid point $O$, it gets:

Shear $V = F = 150$ kN

Toque $T = F \cdot e = 150 \times 0.25 = 37.5$ kN·m

Calculation of the design bearing capacity of a single bolt:

$$N_v^b = n_v \frac{\pi \cdot 22^2}{4} f_v^b = 1 \times \frac{3.14 \cdot 22^2}{4} \times 140 \times 10^{-3} = 53.2 \text{ kN}$$

$$N_c^b = d \cdot \Sigma t \cdot f_c^b = 22 \times 8 \times 305 \times 10^{-3} = 53.7 \text{ kN}$$

$$N_{\min}^b = \min\{N_v^b, N_c^b\} = 53.2 \text{ kN}$$

# Chapter 4  CONNECTIONS OF STEEL STRUCTURES

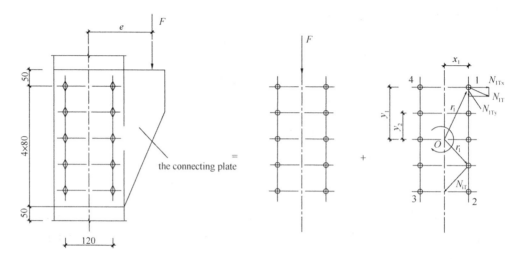

Figure 4.49  A connection joint of a corbel plate and steel column

Bolt strength check:

$$\Sigma x_i^2 + \Sigma y_i^2 = 10 \times 60^2 + 4 \times 60^2 + 4 \times 80^2 = 164,000 \text{ mm}^2$$

$$N_{1x}^T = \frac{T \cdot y_1}{\Sigma x_i^2 + \Sigma y_i^2} = \frac{37.5 \times 10^6 \times 160}{0.164 \times 10^6} = 36.6 \text{ kN}$$

$$N_{1y}^T = \frac{T \cdot x_1}{\Sigma x_i^2 + \Sigma y_i^2} = \frac{37.5 \times 10^6 \times 60}{0.164 \times 10^6} = 13.7 \text{ kN}$$

$$N_{1F} = \frac{V}{n} = \frac{150}{10} = 15 \text{ kN}$$

$$N_1 = \sqrt{(N_{1x}^T)^2 + (N_{1y}^T + N_{1F})^2} = \sqrt{36.6^2 + (13.7 + 15)^2}$$
$$= 46.5 \text{ kN} < N_{min}^b = 53.2 \text{ kN}$$

The strength of the connection meets the design standard.

**[Example 4-7]**

The unfinished/ordinary bolt group in the connection between the column and the corbel with support bears eccentric tension.

As shown in Figure 4.50, the corbel is connected to the steel column with M22 and 4.6 C-grade ordinary bolts. The beam and column are all made of grade Q235B steel to withstand eccentric tension. The design value $F = 150$ kN, $e = 150$ mm, check whether this connection is safe.

**Solution:**

Simplifying the external force $F$ toward the bolt group centroid point $O$, it gets:

Tensile $N = F = 150$ kN

Moment $M = F \cdot e = 150 \times 0.15 = 22.5$ kN · m

Calculation of the design bearing capacity of a single bolt:

$$N_t^b = \frac{\pi \cdot d_e^2}{4} f_t^b = 303 \times 170 \times 10^{-3} = 51.51 \text{ kN}$$

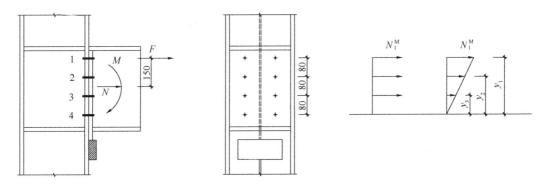

Figure 4.50  A connection joint of a corbel and steel column with support

Under the action of $N$:

$$N_1^N = \frac{N}{n} = \frac{150}{8} = 18.75 \text{ kN}$$

Under the action of $M$, the uppermost row of bolts receives the greatest force.

$$N_{1,\max}^M = \frac{My_1}{\Sigma y_i^2} = \frac{22.5 \times 10^6 \times 240}{2 \times (240^2 + 160^2 + 80^2)} = 30.13 \text{ kN}$$

The resultant force of the maximum tensile force on bolt 1:

$$N_{1,\max}^{N,M} = N_1^N + N_{1,\max}^M = 18.75 + 30.13 = 48.88 \text{ kN} < N_t^b = 51.51 \text{ kN}$$

Therefore, the bolt connection is safe.

[**Example 4-8**]

The unfinished/ordinary bolt group in the connection between the column and the corbel without support bears eccentric tension.

Figure 4.51 shows the connection between the short beam and the column flange. The design value of the shear force is $V=250$ kN, $e=120$ mm. The bolts are grade C and M20, and the steel is grade Q355B. Check this connection.

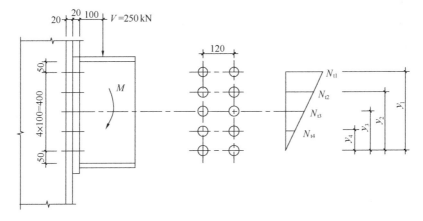

Figure 4.51  A connection joint of a corbel and steel column without support

**Solution:**

The bolt group bears the action of shear force $V=250$ kN and bending moment $M=$

## Chapter 4  CONNECTIONS OF STEEL STRUCTURES

$V \cdot e = 250 \times 0.12 = 30$ kN·m at the same time. Therefore, the design value of the bearing capacity of a single bolt is:

$$N_v^b = n_v \frac{\pi \cdot d^2}{4} f_v^b = 1 \times \frac{3.14 \cdot 20^2}{4} \times 140 = 43,960 \text{ N} \approx 44.0 \text{ kN}$$

$$N_c^b = d \cdot \Sigma t \cdot f_c^b = 20 \times 20 \times 385 = 154,000 \text{ N} = 154 \text{ kN}$$

$$N_t^b = 41.7 \text{ kN}$$

The tensile of the bolt with the greatest force is:

$$N_1 = \frac{My_1}{\Sigma y_i^2} = \frac{30 \times 10^3 \times 400}{2 \times (100^2 + 200^2 + 300^2 + 400^2)} = 20 \text{ kN}$$

Shear of a single bolt:

$$N_v = \frac{V}{n} = \frac{250}{10} = 25 \text{ kN} < 44 \text{ kN}$$

Under the combined action of shear and tension:

$$\sqrt{\left(\frac{N_v}{N_v^b}\right)^2 + \left(\frac{N_t}{N_t^b}\right)^2} = \sqrt{\left(\frac{25}{44.0}\right)^2 + \left(\frac{20}{41.7}\right)^2} = 0.744 < 1$$

Therefore, the bolt connection is safe.

**[Example 4-9]**

The high-strength bolt group bears shear force in the splicing connection of double cover plates.

Try to design a steel plate connection with a pair of cover plates spliced. The steel is grade Q355B, the high-strength bolt is M20 of grade 8.8, and the bolt hole is a standard hole. The contact surface of the connection member is sandblasted. The design value of the axial tension acting on the centroid of the bolt group connection is $N=800$ kN.

**Solution:**

(1) Design friction-type connection

It is found in Table 4.4 that the pre-tension $P=125$ kN of the 8.8-grade M20 high-strength bolts, and in Table 4.5, $\mu=0.40$ when the Q355 steel contact surface is sandblasted.

The design value of the shear capacity of a single bolt is:

$$N_v^b = 0.9 k n_f \mu P = 0.9 \times 1 \times 2 \times 0.40 \times 125 = 90 \text{ kN}$$

The number of bolts required:

$$n = \frac{N}{N_v^b} = \frac{800}{90.0} = 8.9, \text{ take } 9.$$

The bolt arrangement is shown in Figure 4.52(a).

(2) Design bearing-type connection

The design value of the shear capacity of a single bolt is:

$$N_v^b = n_v \frac{\pi \cdot d^2}{4} f_v^b = 2 \times \frac{3.14 \cdot 20^2}{4} \times 250 = 157,000 \text{ N} = 157 \text{ kN}$$

$$N_c^b = d \cdot \Sigma t \cdot f_c^b = 20 \times 20 \times 590 = 236,000 \text{ N} = 236 \text{ kN}$$

The number of bolts required:

$$n = \frac{N}{N_{\min}^b} = \frac{800}{157.0} = 5.1, \text{ take } 6.$$

The bolt arrangement is shown in Figure 4.52(b).

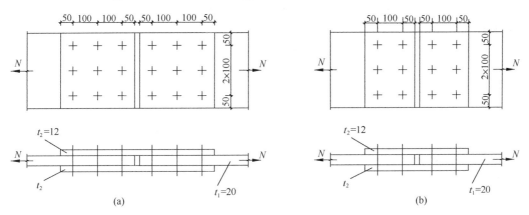

Figure 4.52 The splicing connection of double cover plates and high-strength bolts

[**Example 4-10**]

The high-strength bolt group bears combine of axial force, shear, and moment.

The high-strength bolt friction-type connection is shown in Figure 4.53. The internal forces in the figure are all design values. The steel of the connected component is grade Q355B, the bolt is 10.9 grade M20, the bolt hole is a standard hole, and the contact surface is sandblasted. Check the bearing capacity of this connection.

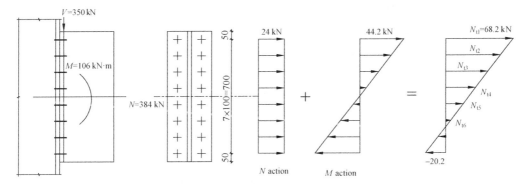

Figure 4.53 A connection joint of a corbel and steel column

**Solution:**

It is found in Table 4.4 that the pre-tension $P = 155$ kN of the 10.9 grade M20 high-strength bolts, and in Table 4.5, $\mu = 0.40$ when the Q355 steel contact surface is sandblasted.

The maximum external tension of a single bolt is:

$$N_{t1} = \frac{N}{n} + \frac{My_1}{\Sigma y_i^2} = \frac{384}{16} + \frac{106 \times 10^3 \times 350}{2 \times 2 \times (350^2 + 250^2 + 150^2 + 50^2)}$$
$$= 24.0 + 44.2 = 68.2 \text{ kN} < 0.8P = 124 \text{ kN}$$

Check the single bolt with the most unfavorable force, and it can be seen that the most unfavorable bolt is the outermost row of bolts on the upper part of the bolt group. The corresponding design value of internal force and the bearing capacity of a single bolt are:

$$N_{v1} = 350/16 = 21.875 \text{ kN}$$
$$N_{t1} = 0.8P = 0.8 \times 155 = 124 \text{ kN}$$
$$N_v^b = 0.9kn_f \mu P = 0.9 \times 1 \times 1 \times 0.40 \times 155 = 55.8 \text{ kN}$$

Substituting the above results into Equation (4.39a) to calculate, it can get:

$$\frac{N_{v1}}{N_v^b} + \frac{N_{t1}}{N_t^b} = \frac{21.875}{55.8} + \frac{68.2}{124.0} = 0.94 < 1$$

Therefore, the high-strength bolt connection is safe.

## 4.8 Chapter Summary

This chapter has introduced the working principle, construction details, and calculation of groove/butt weld, right-angle fillet weld, unfinished/ordinary bolt connection, and high-strength bolt connection. The demonstration of the working examples shows that the calculation of the connection of the steel structure needs a lot of related knowledge of the mechanics, such as the material mechanics and structural mechanics. Therefore, the student should combine the knowledge of material mechanics and structural mechanics and should be proficient in the force analysis methods of various commonly used connections under the action of external forces in order to fully understand the strength conditions that the connection should meet and the structural measures and design to ensure the safety of the connection.

(1) Welded and bolted Connections are the two main connections commonly used in steel structures. The student must recognize their respective advantages and restrictions to choose connection methods reasonably in the design.

(2) Defects in the weld connection will cause brittle damage to the joint. In order to reduce welding defects and ensure the reliability of the weld connection, in addition to appropriate welding processes and measures, the quality level of the weld must be strictly controlled. The weld quality level should be based on appropriate selection and control of the importance of the structure, load characteristics, weld form, working environment, and stress state. The uneven temperature field and welding rigidity during the welding process, and the local plastic deformation caused by it are the root causes of residual stress and residual deformation. Residual stress and deformation will reduce the structure's stiff-

ness and stability and seriously affect the structure's fatigue strength and cause brittle fracture of the structure or component. Therefore, the residual stress and residual deformation of welding must be strictly controlled during the construction process.

(3) Understand the concept that fillet welds are divided into side fillet welds and transverse fillet welds based on the different directions of the external force, and correctly understand that the strength of transverse fillet welds is higher than that of side fillet welds and plastic deterioration. Then, understand the meaning of each symbol in the basic calculation formula of fillet weld correctly.

(4) The force of a single bolt in a bolted connection can only be under tension, under shear, and under both tension and shear at the same time. Therefore, for all kinds of connections under the action of external forces, find which bolt controls the design. And is it a tension calculation or a shear calculation? Or is it the combined calculation of tension and shear? Generally, the calculation under the action of axial shear or torque is a matter of shear calculation. Under the action of axial tension or bending moment, it is generally a matter of tension calculation. Under the combined action of several other forces or moments, it is a joint calculation problem for tension and shear.

(5) When the moment acts alone, the neutral axis of the ordinary bolt connection and the neutral axis of the high-strength bolt connection is different. The neutral axis of the ordinary bolt connection is the first row of bolt-shaped mandrels whose moment points to one side, while the high-strength bolt connection is the centroid of the bolt group.

(6) According to the connection's ultimate state, the high-strength bolted connection is divided into a bearing-type connection and friction-type (slip-critical) connection.

(7) The structural design and calculation of connections and joints are important parts of the entire steel structure design. In actual projects, such as the connections or joints of multi-story moment-resistance frames, the welds and high-strength bolts are often used in combination. Therefore, because of the characteristics of different connections or joints, it is necessary to analyze the force state of the individual welds or bolts accurately and apply the knowledge of this chapter to solve the design problem.

# Questions

4.1   What are the main connection methods for steel structures? What are the characteristics?

4.2   According to the relative position of welding, what are the welding methods?

4.3   What are the effects of residual welding stress and residual deformation on structural performance? How to reduce the influence of residual stress and residual deformation?

4.4   Why should the groove weld be cut edge?

4.5　What are the construction details for fillet welds?

4.6　How should the weld's effective length be taken when there is an arc ignition plate or not?

4.7　What are transverse fillet welds and side fillet welds? What are their characteristics?

4.8　What is the effective section height of fillet welds?

4.9　What are the structural requirements for bolted connections?

4.10　What are the bolt failure forms in ordinary bolt shear connections? How to avoid them?

4.11　What is the difference between ordinary bolts and high-strength bolts?

4.12　Why do high-strength bolts need to be pre-tensioned?

4.13　What are the factors related to the anti-slip coefficient?

4.14　What are the similarities and differences between a high-strength bolt friction connection and a high-strength bolt bearing-type connection?

4.15　What are the similarities and differences between ordinary bolt connections and high-strength bolt bearing-type connections?

4.16　Under the moment action, what is the difference between an ordinary bolt group and a high-strength bolt group?

## Exercises

4.1　As shown in Figure 4.54, the T-shaped corbel is connected with the column by the groove weld. The design value of the load is $N=200$ kN, the steel is grade Q235 steel, manual welding, the type of electrode is E43, and the welding quality grade is level three, add arc ignition board. Check whether the strength of this connection is satisfied.

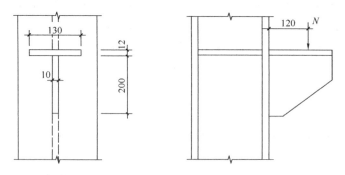

Figure 4.54　Calculation exercise 4.1

4.2　The I-shaped beam is provided with a splicing butt weld on the web (Figure 4.55). The design value of the moment is $M=1,122$ kN·m, and the design value of shear is $V=374$ kN. The steel is grade Q355B, the type of electrode is E50, semi-automatic welding, and level-three inspection standard. Check the strength of the weld.

4.3 Try to design the fillet weld connection between the double angles and the gusset (Figure 4.56) with a three-side weld and a two-side weld, respectively. The steel is grade Q235, manual welding, the type of electrode is E43, and the welding quality grade is level three. The design value of the axial force is $N=800$ kN.

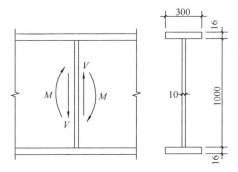

Figure 4.55 Calculation exercise 4.2

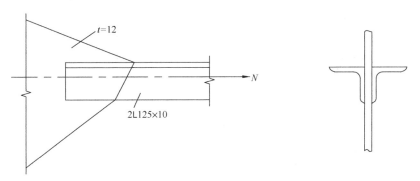

Figure 4.56 Calculation exercise 4.3

4.4 Try to calculate the maximum design load that the connection shown in Figure 4.57 can withstand. The steel is grade Q355B, and the electrode is the type of E50, manual welding, fillet weld leg size $h_f=8$ mm, $e_1=300$ mm.

4.5 As shown in Figure 4.58, the steel plate and the column flange are connected by right-angle fillet welds, and the steel is grade Q235, manual welding, the type of electrode is E43. The oblique force $F=390$ kN (static load), $h_f=8$ mm. Check the construction details of this weld and verify whether the weld is safe.

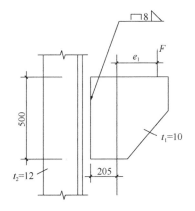

Figure 4.57 Calculation exercise 4.4

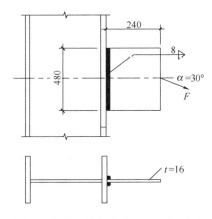

Figure 4.58 Calculation exercise 4.5

*107*

4.6 C-grade ordinary bolt connection is shown in Figure 4.59. The steel is grade Q235B, the bolt diameter is $d=20$ mm, the bolt-hole diameter is $d_0=21.5$ mm, and the static load's design value is $V=240$ kN. Try to check whether this connection is safe according to the following conditions:

(1) It is assumed that the support bears shear force;

(2) It is assumed that the support does not bear shear force.

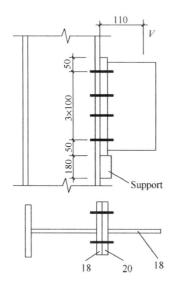

Figure 4.59  Calculation exercise 4.6

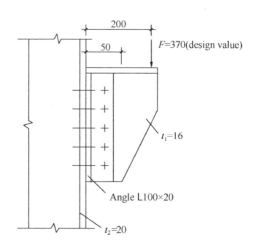

Figure 4.60  Calculation exercise 4.7

4.7 The corbel shown in Figure 4.60 is connected to the column with double angle ∟ 100×20 and 10.9 grade M22 high-strength bolts. It is required to be designed as a friction-type connection. The component steel is Q355B, and the contact surface is sandblasted. Try to determine the number of bolts connecting the two limbs of the angle.

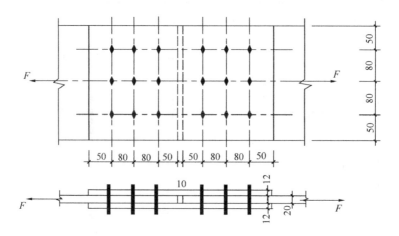

Figure 4.61  Calculation exercise 4.8

4.8 As shown in Figure 4.61, the steel plates are connected by double cover plates. The steel is grade Q345. The bolts are 10.9 grade high-strength bolts, and the connection is a friction-type connection. The contact surface is sandblasted, bolt diameter $d=20$ mm, aperture $d_0=22$ mm. Try to calculate the design value of the maximum axial force $F$ that this connection can withstand.

# Chapter 5  AXIALLY LOADED MEMBERS

**GUIDE TO THIS CHAPTER:**

➢*Knowledge points*

*The calculation method of the strength of axially loaded members; the concept of stiffness and slenderness ratio; the three global buckling modes of ideal axially loaded compression members, the influence of initial defects on the overall stable bearing capacity of axially loaded compression members; the overall stable bearing capacity of axially loaded compression members and the calculation method and design simplification; the critical force and local stability of the axially loaded compression rectangular plate; the allowable width-to-thickness ratio of the plate and the application of the post-buckling strength of the web; the design method of solid web and laced (built-up section with lacing) axially loaded compression columns.*

➢*Emphasis*

*The concept of overall stability and local stability of basic structural steel members.*

➢*Nodus*

*The concept of the overall stability of the axially loaded compression member, initial defects influence the overall stability of the column and the local stability of the plate.*

## 5.1  Introduction

Axially loaded members are widely used in structures of bar systems, referring to spatial truss and grid structures, towers, and bracings(Figure 5.1). The connection of joints of the se structures is assumed as the hinged connection. Under no internode load, members bear axial tension and axial compression basically, called axially loaded tension members and axially loaded compression members.

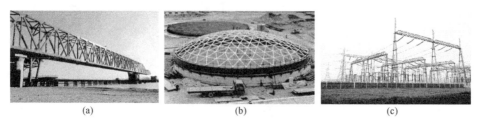

Figure 5.1  Application of axially loaded members in engineering
(a) spatial truss; (b) spatial reticulated shell; (c) steel transmission towers

The section forms of axially loaded members are divided into solid web and laced (built-up section with lacing). Solid web members have many cross-section forms due to simple manufacturing and the convenience of connecting with other members. The single hot-rolled shape is often used as a cross-section of solid web members, including I-sections, H-shapes, round steel, pipes, angles, channels, and T-shapes (Figure 5.2a). The common built-up sections composed of hot-rolled shapes or steel plates are also employed to cross-section solid web members (Figure 5.2b). Except for T-shapes, the built-up sections of single-angles or double-angles are often applied to chords and web members in steel trusses (Figure 5.2c). In light structures, cold-formed thin-walled steel sections can be used (Figure 5.2d).

In the sections mentioned above, compact cross-section (such as the round steel and the composite plate with a small width-to-thickness ratio) or the cross-section with stiffness difference between the two principal axes (such as single-channel, I-section) are generally only used for axial tension members. In addition, axially loaded compression members usually use open sections with wide and thin-walled components to improve section stiffness (moment of inertia).

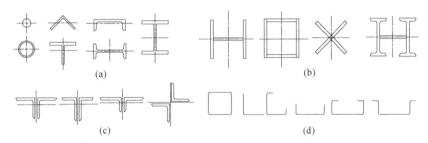

Figure 5.2 Sectional forms of solid web axially loaded members

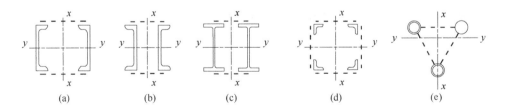

Figure 5.3 Sectional forms of laced axially loaded members

Laced (built-up section with lacing) members have many advantages, such as large stiffness, good torsion resistance, and low steel consumption. The sections of laced members consist of two or more steel limbs (Figure 5.3). The limbs are connected with lacing bars (Figure 5.4a) or batten plates (Figure 5.4b) as a whole; lacing bars or batten plates are collectively referred to as lacings.

The calculation of axial bearing members needs to meet the ultimate bearing capacity and normal service limit state. For the ultimate state of bearing capacity, axial tension members need to meet strength, and axially loaded compression members need to meet the requirements of stability and strength. In addition, axially loaded members ensure the stiffness of the member, that is, restricts the slenderness ratio to meet the normal service limit state. Therefore, according to different bearing abilities, the design of axial tension members is required to verify the strength and stiffness of members. However, except for strength and stiffness, axially loaded compression members need to verify stability.

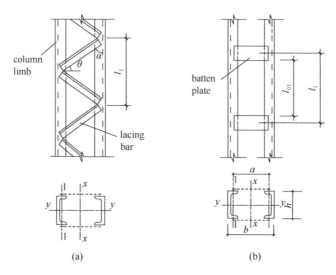

Figure 5.4  Lacings of laced axially loaded members
(a) lacing bar; (b) batten plate

## 5.2  Strength and Stiffness of Axially Loaded Members

### 5.2.1  Strength calculation of axially loaded members

The limit strength of axially loaded members is defined based on section stress reaching the yield strength of the steel. Namely, when holes do not weaken the section, the yield strength of the gross cross-section can be calculated by the following equation:

$$\sigma = \frac{N}{A} \leqslant f \qquad (5\text{-}1)$$

where  $N$ = the design value of axial tensile force or compressive force;
 $A$ = the gross cross-sectional area of members;
 $f$ = the design value of tensile or compressive strength of the steel used in members.

Under tensile force, when the section of the member is weakened locally, the stress distribution of the section is no longer uniform. There is a stress concentration phenomenon near the hole, as shown in Figure 5.5a. In the elastic state, the maximum stress of the hole edge maybe has three times the average stress on the gross section of the member. For ideal elastoplastic material, when the maximum stress of hole edge reaches the yield strength of the material, with the continuous increase of tensile force, the stress will not continue to increase. However, plastic deformation continues to increase. And the stress of the section will produce the redistribution of plastic internal force. Finally, when the net section of members appears to fracture, the stress reaches the uniform distribution basically (Figure 5.5b). So the average stress reaching the yield strength limit is generally taken as the control value during calculation. Now, high-strength steels have become more and more popular in building steel structures. But the stress-strain curve of high-strength steel has no obvious yield platform. Therefore, when the section of axially loaded members breaks due to weakened holes, the ultimate bearing capacity of the net section has reached the minimum value of tensile strength $f_u$. Suppose this is used as the control value in the calculation, considering that the consequences of the net section fracture are more serious than the section yielding. In that case, the partial coefficient of resistance needs to be larger. According to the current national design standard GB 50017, multiply the partial coefficient of resistance by 1.3 based on the average value of 1.1 to obtain the strength limit when the net section is broken. Namely, the limit strength of the fractured net section is equal to $1/(1.1 \times 1.3) f_u = 0.7 f_u$. The strength calculation of the net section of axis tensile members can be obtained:

$$\sigma = \frac{N}{A_n} \leqslant 0.7 f_u \tag{5-2}$$

where  $A_n$ = the net cross-sectional area of the member. When the section of the members has multiple holes, the most unfavorable section should be taken;

$f_u$ = the minimum tensile strength of the steel used in members.

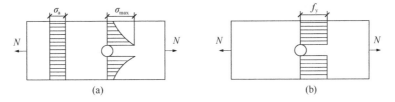

Figure 5.5 Stress distribution of a tensile member with a hole
(a) elastic state stress; (b) ultimate state stress

The cross-section strength can be calculated for the axially loaded compression members according to Equation (5-1) because bolts transmit force directly.

When axially loaded compression members use ordinary bolts arranged in parallel, the net cross-sectional area is calculated based on the most dangerous orthogonal section. However, when axially loaded compression members use ordinary bolts and arrange staggered, the members have two failure modes, including the failure of the orthogonal section and dentate section. Therefore, the net cross-sectional area of members can be calculated

# Chapter 5  AXIALLY LOADED MEMBERS

with a smaller area.

When using high-strength bolts and calculating the net section strength of the axial tension member, the internal force bore by the net section should subtract the transmitted force because the contact surface has transmitted part of the force in front of the first row of bolts. The details have been given in Section 4.6.2.

## 5.2.2  Calculation of stiffness

In order to meet the normal service limit state of structures, axially loaded members need enough stiffness to ensure no excessive deformation.

The slenderness ratio $\lambda$ is used to measure the stiffness of tension members and compression members. When the slenderness ratio is too large, it will have an adverse effect.

Because the ultimate bearing capacity state of the compression members generally controlled by the overall stability, the too-large slenderness ratio also makes the ultimate bearing capacity of members reduce significantly. Simultaneously, the deformation caused by the initial bending and weight will not be conducive to the overall stability of the member.

To ensure that the axially loaded member has enough stiffness, the maximum slenderness ratio $\lambda$ of the member should be required.

$$\lambda = \frac{l_0}{i} \leqslant [\lambda] \tag{5-3}$$

where  $l_0 = $ the effective length of members;

$i = $ the sectional radius of gyration;

$[\lambda] = $ the allowable slenderness ratio of the member, which is generally given based on summarizing the experience of long-term use of the steel structure.

Tables 5.1 and 5.2 show the requirements of the design standard GB 50017 for allowable slenderness ratio.

Allowable slenderness ratio of axially loaded tension members    Table 5.1

| Member details | Structures bearing static load or indirectly bearing dynamic load | | | Structures directly subjected to dynamic loads |
|---|---|---|---|---|
| | General building structure | A chord that provides out-of-plane support to the web | Workshops with heavy-duty cranes | |
| Truss members | 350 | 250 | 250 | 250 |
| Bracings between columns below the crane beam or crane truss | 300 | — | 200 | — |
| Other tension rods, supports, thin rods, etc. except for the tightened round steel | 400 | — | 350 | — |

**Allowable slenderness ratio of axially loaded compression members**  Table 5.2

| Member details | Allowable slenderness ratio |
| --- | --- |
| Axial compression columns, trusses, and compression rods in skylight frames | 150 |
| Bracings between columns below the crane beam or crane truss, lacing bar of columns | 150 |
| Bracings | 200 |
| Rods used to reduce the slenderness ratio of compression members | 200 |

Under direct or indirect dynamic loads, when verifying the allowable slenderness ratio, the minimum radius of gyration is used to calculate the slenderness ratio of a single-angle steel tensile member. And the radius of gyration parallel to the angle steel limb edge is employed to calculate the slenderness ratio outside the plane of cross members. A series of requirements about allowable slenderness ratio proposed by the design standard GB 50017 are as follows: (1) Except for the chord that provides the external support for the web, the tension member under static load only needs to calculate the slenderness ratio in the vertical plane; (2) The slenderness ratio of the lower chord of the intermediate or heavy-duty crane truss should not exceed 200; (3) The slenderness ratio of support should not exceed 300 in the workshop with clamps or rigid rake and other hard hook cranes; (4) Under permanent and wind load, the slenderness ratio of the tensile members under compression should not exceed 250; (5) For trusses with a span equal to or greater than 60 m, the slenderness ratio of tension chord and web should not exceed 300 under static load or indirect dynamic load. Under direct dynamic load, the slenderness ratio of tension chord and web should not exceed 250; (6) The slenderness ratio of tensile members should not exceed allowable values provided by Table 5.1.

Lack of stiffness has a worse effect on compression members than tension members. The minimum radius of gyration is used to calculate the slenderness ratio of a single-angle member. And the radius of gyration parallel to the angle steel limb edge is employed to calculate the slenderness ratio outside the plane of cross members. The torsion effect is not considered when calculating the allowable slenderness ratio. The allowable slenderness ratio of axially loaded compression members needs to meet the following: (1) For trusses with a span equal to or greater than 60 m, the slenderness ratio of compression chords, end compression members, and compression web members bearing direct dynamic load should not exceed 120; (2) The slenderness ratio of compression members should not exceed allowable values provided by Table 5.2. But the slenderness ratio of members can be 200 when the design value of internal force is less than 50% of bearing capacity.

## 5.3 Stability of Axially Loaded Compression Member

When the slenderness ratio of the axially loaded compression member is relatively large, and holes do not weaken the section, it will generally not lose the bearing capacity even the average stress of the section reaches the design value of compressive strength. Therefore, the strength calculation is not necessary. However, in recent decades, due to the continuous development of structures and high-strength steel, members have become lighter and thinner, contributing to the appearance of instability. Furthermore, the failure of members caused by instability often happens in engineering accidents of steel structures. Therefore, it is critical to choose an appropriate cross-section for the overall stability of members.

### 5.3.1 Calculation of overall stability

(1) Critical stress of overall stability

Various factors affect the overall stability critical stress of axially loaded compression members, which causes the complex calculating of the bearing capacity of compression members. Four methods will determine overall stability critical stress:

① Buckling criterion

The buckling criterion is developed following the assumption of ideal axially loaded compression members. The so-called ideal axially loaded compression member assumes that the member is completely straight, the load is along its centroid axis, without initial stress before the loading, without initial bending and initial eccentricity defects; the section along the member is uniform. The overall instability mode of the ideal member is described as global buckling. There are three kinds of buckling modes:

a. Flexural buckling

The axially loaded compression members only appear bending deformation. The member section rotates about only one principal axis, and the longitudinal axis of the member changes from a straight line to a curve. It is the most common buckling mode of biaxial symmetric sections.

Figure 5.6a shows the flexural buckling of the I-section compression member with hinged support at both ends (that is, the supporting end can freely rotate around the main axis of the section but cannot move sideways and twist) around the weak axis ($y$-axis).

b. Torsional buckling

The torsional buckling of the crosssection member with a small length is shown in Figure 5.6b. When instability, each section of the member except the supporting end is twisted around the longitudinal axis, which is the possible buckling mode of some biaxial symmetric crosssection compression members.

c. Flexural-torsional buckling

The flexural-torsional buckling of the T-shape is as shown in Figure 5.6c. When the uniaxial symmetrical section buckles around the axis of symmetry, the member will inevitably be twisted while bending and deforming.

The most basic and simplest of the three global buckling modes mentioned above is

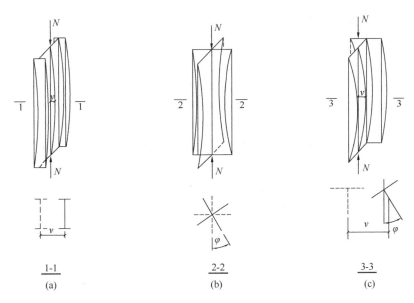

Figure 5.6 Buckling modes of the axially loaded compression members
(a) flexural buckling; (b) torsional buckling; (c) flexural-torsional buckling

flexural buckling. When the slim ideal straight compression member appears flexural buckling at the elastic stage, the Euler critical force $N_{cr}$ and the Euler critical stress $\sigma_{cr}$ of the slim ideal straight compression member can be obtained by the Euler Equation.

$$N_{cr} = \frac{\pi^2 EI}{l^2} \tag{5-4}$$

$$\sigma_{cr} = \frac{\pi^2 E}{\lambda^2} \tag{5-5}$$

where   $\pi$ = mathematical constant (3.1416);
$E$ = material modulus of elasticity;
$I$ = the least moment of inertia of the cross section;
$l$ = the length of the column from pin end to pin end;
$\lambda$ = the slenderness ratio of the column.

During the derivation of the Euler Equation, the member material is assumed to be an ideal elastomer. When the slenderness ratio $\lambda$ of members is less than $\lambda_p$ ($\lambda_p = \pi\sqrt{E/f_p}$), the critical stress of members exceeds the proportional limit of the material, which causes members to enter the elastic-plastic stage. At the same time, the stress-strain relationship of the material becomes nonlinear. In 1889, German scientist Engesser put forward tangent modulus theory; the calculation formula of critical stress proposed by this theory is as follows:

$$\sigma_{cr} = \frac{\pi^2 E_t}{\lambda^2} \tag{5-6}$$

where   $E_t$ = the tangent modulus of the inelastic region.

The formula of tangent modulus examined by tests is considered to be in good agreement

## Chapter 5 AXIALLY LOADED MEMBERS

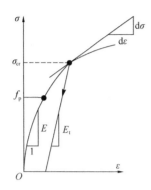

Figure 5.7 Stress-strain relationship curve

with the actual critical stress of the compression member. Still, it is only applicable when the material has a definite stress-strain curve (Figure 5.7).

A stability calculation method is based on the buckling criterion. The elastic stage is based on Euler's critical force, and the elastic-plastic stage is based on the tangent modulus critical force. The initial eccentricity, initial bending, and other adverse effects are considered by improving the safety factor.

② Edge yield criterion

The mechanical performance of actual axially loaded compression members differs from ideal members due to the initial imperfection. The edge yield criterion takes the compression member with initial eccentricity and initial bending as the calculation model. When the edge stress of the section reaches the yield strength, it is regarded as the ultimate bearing capacity of the actual axially loaded compression members.

The axially loaded compression member with pin supports at both ends is shown in Figure 5.8. Considering the influences of initial bending, initial eccentricity, and residual stress, the maximum equivalent initial bending deflection in the middle span is $v_0$. Once the compression member is loaded, the deflection increases to $v$. As the actual compression member is not infinite elastomer, as long as the deflection increases to a certain extent, the mid-span section of the member begins to yield under the action of axial force $N$ and bending moment $Nv$. Subsequently, the plastic zone of the cross-section continues to increase, and the compression member enters the elastic-plastic stage, which results in the loss of bearing capacity before the compressive force reaches the critical force $N_{cr}$. As shown in Figure 5.9, the dotted line is the force-deflection curve at the elastic-plastic stage. The highest point of the dotted line is the limit pressure point at the elastic-plastic stage of the compression member.

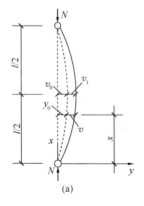 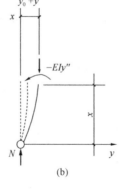 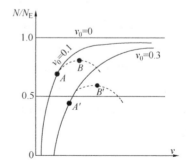

Figure 5.8 An axially loaded compression member with initial bending

Figure 5.9 Load-deflection relationship curve of a column with initial bending

According to elastic theory, the yielding condition of axially loaded compression members with initial bending and without residual stress can be defined as:

$$\frac{N}{A} + \frac{Nv}{W} = \frac{N}{A} + \frac{Nv_0}{W} \cdot \frac{N_E}{N_E - N} = f_y \tag{5-7}$$

or

$$\frac{N}{A}\left(1 + v_0 \frac{A}{W} \cdot \frac{\sigma_E}{\sigma_E - \sigma}\right) = f_y$$

$$\sigma\left(1 + \varepsilon_0 \cdot \frac{\sigma_E}{\sigma_E - \sigma}\right) = f_y \tag{5-8}$$

where $\varepsilon_0$ = the ratio of initial bending, which is defined as $\varepsilon_0 = v_0 \frac{A}{W}$;

$A$ = the cross-sectional area;

$W$ = the cross-section modulus;

$\sigma_E$ = the Euler critical stress.

According to the edge yield criterion, the critical stress can be expressed as:

$$\sigma_{cr} = \frac{f_y + (1+\varepsilon_0)\sigma_E}{2} - \sqrt{\left[\frac{f_y + (1+\varepsilon_0)\sigma_E}{2}\right]^2 - f_y \sigma_E} \tag{5-9}$$

The above Perry Equation, a strength equation considering the second-order effect of compressive forces, is derived from the edge yield criterion.

③ Maximum strength criterion

Perry Equation derived from the edge yield criterion is essentially a strength formula rather than a stability formula. It does not express the ultimate bearing capacity of the axially loaded compression member. The compressive forces can continue to increase because the edge fibers yield after the plastic can go deep into the section. However, when the compressive load exceeds the maximum bearing capacity $N_A$ at the edge yield (Figure 5.10), the member enters the elastoplastic stage. As the plastic zone of the cross-section expands, the $v$ value increases faster. After reaching point B, the resistance of the compression member begins to be less than that of the external force, and it cannot maintain stability. The compressive load $N_B$ at the highest point B of the curve is the axially loaded compression member's real ultimate stability bearing capacity with initial defects. That is the criterion to calculate the stability of the column with initial defects, which is called the maximum strength criterion.

The maximum strength criterion which takes maximum axially compressive load as ultimate stability bearing capacity of columns and considers plasticity entering the section, is developed based on columns with initial defects. Taking residual stress and initial bending into ac-

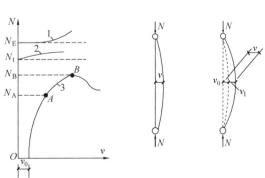

Figure 5.10 Load-deflection relationship curves of columns

count, when the maximum strength criterion is used to calculate the overall stability of columns, the stress-strain relationship of each point along the cross-section and each section along the longitudinal axis of columns is variable. Therefore, it is difficult to formulate analytical formulas for critical stress based on the above conditions. Therefore, the critical stress can only be obtained by the numerical calculation method with the help of calculation software. At present, many commercial general-purpose finite element analysis software can perform this analysis, such as ABAQUS, ANSYS, etc.

④ Empirical formula

Due to little research on the stability theory of the elastoplastic stage of the columns early, the critical stress can be obtained from the regression of experimental data, which can be used as the design standard of stability bearing capacity of columns.

(2) The column strength curves of axially loaded compression members

The relationship curve between critical stress and slenderness ratio can be defined as a column curve. The maximum strength criterion defines the curve of axial compression columns applied to the design standard GB 50017. The calculation results of curves have good agreement with test results, which shows that the calculation theory is appropriate. The single-column curves applied to early design code for design of steel structure considered that the ultimate bearing capacity of column was only related to slenderness ratio. Even if the slenderness ratio of members is the same, the cross-section shape, bending direction, level and distribution of residual stress have a different effect on the ultimate bearing capacity of members. As shown in Figure 5.11, the column strength curves of axial compression columns distribute with a banded pattern in the area surrounded by the dotted line. There is a big difference between the upper and lower limits of the above range, especially in the common medium slenderness ratio. Therefore, the application of a single curve is not appropriate. Combine the above calculation data and engineering practices, the current *Standard for Design of steel Structures* GB 50017 merges four kinds of curves, as shown in Figure 5.11. Column curve a is 4%-15% higher than curve b. However, column

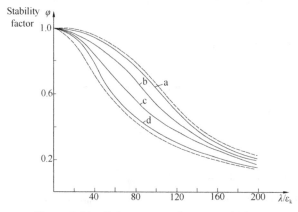

Figure 5.11 Column strength curves of China

curve c is 7%-13% lower than curve b. The curve d is even lower and is mainly used for thick plate sections. Table 5.3 shows the section classification of axially loaded compression members (columns) whose plate thickness is less than 40 mm. Table 5.4 shows the section classification of sections whose plate thickness is greater than 40 mm. The general section situation almost belongs to class b.

The residual stress has little effect on the hot-rolled circular pipe and hot-rolled ordinary I-sections and H-shapes when it loses stability around the $x$-axis (sometimes called the strong axis). So their sections belong to class a.

For the stability calculation of laced members around the virtual axis, the $\varphi$ determined by edge yield criterion is close to curve b by referring to the *Technical Code for Cold-formed Steel Structures* GB 50018, curve b is used.

When the channel is used for the limb of the laced column, curve b can be used to calculate the stability of the limb around its symmetric axis because the lacing elements restrain the torsional deformation of the limb. For the welded I-shaped section with rolled or sheared flange edges, when instability occurs around its weak axis, the edge is the residual compressive stress, which reduces the load-bearing capacity, so it is included in curve c.

The stability research of axially loaded compression members of high-strength steel at home and abroad shows that the peak residual stress of hot-rolled section steel is not related to steel strength. On the contrary, its adverse effect decreases with the increase of steel strength. Therefore, for the H-shapes and angles with equal legs, when yield strength reaches or exceeds 355 N/mm², and the ratio of width to height exceeds 0.8, the coefficient $\varphi$ should be higher than that of Q235 steel.

When the plate thickness of I-section members and welding solid web members is greater than 40 mm, the residual stress changes along the plate's width direction and significantly changes along the thickness direction. Besides, the poor quality of plates harms stability. Therefore, it should be classified according to Table 5.4.

(3) The calculation of the overall stability of the axially loaded compression member

The stress of the axially loaded compression member should not be greater than the critical stress of overall stability. After considering partial factor for resistance, the cross-sectional stress of the member is

$$\sigma = \frac{N}{A} \leqslant \frac{\sigma_{cr}}{f_y} \cdot \frac{f_y}{\gamma_R} = \varphi f \qquad (5\text{-}10)$$

According to the design standard GB 50017, the overall stability of axially loaded compression members is calculated by the following equation.

$$\frac{N}{\varphi A f} \leqslant 1.0 \qquad (5\text{-}11)$$

where  $N = $ the compressive load;

## Chapter 5  AXIALLY LOADED MEMBERS

$\varphi$ = the overall stability factor of axially loaded compression members, is defined as $\sigma_{cr}/f_y$ ;

$A$ = the gross cross-sectional area;

$f$ = the design strength of steel.

The overall stability factors of axially loaded compression members are related to section classification and slenderness ratio. They can be found from Table D. 1 to Table D. 4 of Appendix D.

**Section classification of axially loaded compression members ($t<$40 mm)**  Table 5.3

| Type of section | | $x$ | $y$ |
|---|---|---|---|
| Rolled (circular) | | a | a |
| Rolled (H-section) | $b/h \leqslant 0.8$ | a* | b* |
| | $b/h > 0.8$ | | |
| Rolled equal leg angle | | a* | a* |
| Rolling or welded with flame cutting edges | Welded (circular) | b | b |
| (cross sections) | Rolled | | |
| Welded or rolled (Width-to-thickness ratio >20) | Welded or rolled | b | b |
| Welded | Rolled or welded with flame cutting edges | | |
| (box sections) | Rolled or welded with flame cutting edges | | |

122

## 5.3 Stability of Axially Loaded Compression Member

continued

| Type of section | | x | y |
|---|---|---|---|
| (welded H/T sections) Welded, flanges are rolled or sheared edges | | b | c |
| Welded, the edges are rolled or sheared | Welded or rolled (Width-to-thickness ratio ≤ 20) | c | c |

Notes: 1. Class a* means that Q235 steel takes class b; Q355, Q390, Q420, and Q460 steels take class a. Class b* means that Q235 steel takes class c; Q355, Q390, Q420, and Q460 steels take class b;

2. For the section without symmetry axis and the shear center and centroid do not coincide, and the section classification can be determined according to the similar section with symmetry axis. For example, angles with equal legs are used for angles with unequal legs. When there is no similar cross-section, it can be class c.

The overall stability factor $\varphi$ can be expressed in the form of Perry Equation, that is:

$$\varphi = \frac{\sigma_{cr}}{f_y} = \frac{1}{2}\left\{\left[1+(1+\varepsilon_0)\frac{\sigma_E}{f_y}\right] - \sqrt{\left[1+(1+\varepsilon_0)\frac{\sigma_E}{f_y}\right]^2 - 4\frac{\sigma_E}{f_y}}\right\} \quad (5\text{-}12)$$

At this time, the calculation of the overall stability factor $\varphi$ is no longer based on the edge yield of the section, but the ultimate bearing capacity of the column is determined first according to the maximum strength theory and then $\varepsilon_0$ is calculated. Therefore, $\varepsilon_0$ in the equation is essentially the equivalent initial bending rate considering the comprehensive influence of initial bending and residual stress.

**Section classification of axially loaded compression members ($t \geqslant 40$ mm)**      Table 5.4

| Type of section | | | x | y |
|---|---|---|---|---|
| Rolled | | $t < 80$ mm | b | c |
| Rolled | | $t \geqslant 80$ mm | c | d |
| Welded | | Flanges are flame cut | b | b |
| Welded | | Flanges are rolled or sheared edges | c | d |
| Welded | | The width-to-thickness ratio of the plate is greater than 20 | b | b |
| Welded | | The width-to-thickness ratio of the plate is less than or equal to 20 | c | c |

## Chapter 5　AXIALLY LOADED MEMBERS

According to four *Column Curves* used in the design standard GB 50017, the equivalent initial bending rate ($\varepsilon_0$) can be expressed as:

Class a cross-section: $\varepsilon_0 = 0.152\bar{\lambda} - 0.014$

Class b cross-section: $\varepsilon_0 = 0.300\bar{\lambda} - 0.035$

Class c cross-section: $\varepsilon_0 = 0.595\bar{\lambda} - 0.094$ ($\bar{\lambda} \leqslant 1.05$);

$$\varepsilon_0 = 0.302\bar{\lambda} - 0.216 \ (\bar{\lambda} > 1.05)$$

Class d cross-section: $\varepsilon_0 = 0.915\bar{\lambda} - 0.132$ ($\bar{\lambda} \leqslant 1.05$);

$$\varepsilon_0 = 0.432\bar{\lambda} - 0.375 \ (\bar{\lambda} > 1.05)$$

where $\bar{\lambda} = \dfrac{\lambda}{\pi}\sqrt{\dfrac{f_y}{E}}$ is defined as dimensionless slenderness ratio.

The value mentioned above $\varepsilon_0$ is only suitable for the condition that $\bar{\lambda}$ is not less than 0.215 (equivalent to $\lambda > 20\sqrt{f_y/235} = 20\varepsilon_k$). Therefore, the value mentioned above $\varepsilon_0$ is substituted into Equation (5-12), which is the equation of value $\varphi$ in Tables D.1 to D.4 of Appendix D.

When $\bar{\lambda}$ is less than 0.215, the Perry Equation is suitable for calculation no longer. The value $\varphi$ cannot be obtained by finding out from Appendix D. The design standard GB 50017 adopts an approximate curve that connects $\bar{\lambda} = 0.215$ with $\bar{\lambda} = 0$ ($\varphi = 1.0$).

$$\varphi = 1 - \alpha_1 \bar{\lambda}^2 \tag{5-13}$$

The coefficients $\alpha_1$ are 0.41 (section of class a), 0.65 (section of class b), 0.73 (section of class c), and 1.35 (section of class d), respectively.

When calculating the overall stability bearing capacity of a xially loaded compression members, the slenderness ratio of the members shall be determined according to the buckling modeand the following provisions:

① A cross-section centroid of a member coincides with its shear center, such as a member whose cross-section is a biaxial symmetric section or polar symmetry section.

a. When members occur flexural buckling around two principal axes, the calculation equations can be expressed as follows:

$$\lambda_x = \frac{l_{0x}}{i_x} \tag{5-14}$$

$$\lambda_y = \frac{l_{0y}}{i_y} \tag{5-15}$$

where　$l_{0x}$ = calculation lengths of members for $x$-axis;

　　　$l_{0y}$ = calculation lengths of members for $y$-axis;

　　　$i_x$ = radius of gyration about $x$-axis;

　　　$i_y$ = radius of gyration about $y$-axis.

b. The following equation describes the torsional buckling of members.

$$\lambda_z = \sqrt{\dfrac{I_0}{\dfrac{I_t}{25.7} + \dfrac{I_\omega}{l_\omega^2}}} \qquad (5\text{-}16)$$

where  $I_0$ = the polar moment of inertia of a member's gross cross-section to the shear center;

$I_t$ = the uniform torsional constant, also called the torsional moment of inertia of the gross cross-section to the shear center;

$I_\omega$ = the warping constant and is also known as the sector moment of inertia of the gross cross-section to the shear center, and $I_\omega$ is approximately equal to 0;

$l_\omega$ = the effective length of torsional buckling. If both ends are pin-end, and the end section can warp freely, $l_\omega$ takes the geometric length $l$; if both ends are fixed, and the warpage of the end section is completely restrained, $l_\omega$ take the half of geometric length as $0.5l$.

When the width-to-thickness ratio of the biaxial symmetrical crosssection plate does not exceed $15\varepsilon_k$, the critical load of torsional buckling is larger than that of flexural buckling. Therefore, the torsional buckling is not considered in the process of calculation.

② The members with a single axisymmetric section

a. When the flexural buckling of members around the asymmetric principal axis is calculated, the slenderness ratio can be calculated by Equation (5-14).

b. Due to the eccentricity between the centroid and shear center of the cross-section, the members occur flexural-torsional buckling around the symmetric axis. Under the same condition, the critical stress of flexural-torsional buckling is lower than that of flexural buckling. Therefore, after a flexural-torsional buckling analysis of a single axisymmetric section, such as a double-plate T-shape section and channel section, the equivalent slenderness ratio calculated by the following equation is used to replace $\lambda_y$.

$$\lambda_{yz} = \dfrac{1}{\sqrt{2}}\left[(\lambda_y^2 + \lambda_z^2) + \sqrt{(\lambda_y^2 + \lambda_z^2)^2 - 4\left(1 - \dfrac{y_s^2}{i_0^2}\right)\lambda_y^2\lambda_z^2}\right]^{\frac{1}{2}} \qquad (5\text{-}17)$$

$$i_0^2 = y_s^2 + i_x^2 + i_y^2$$

where  $y_s$ = the distance from section centroid to shear center;

$i_0$ = the polar radius of gyration of the section relative to the shear center;

$\lambda_z$ = the conversion slenderness ratio of torsional buckling and calculated by Equation (5-16).

c. When the effective lengths of axially loaded equal leg single-angle are equal, the calculation analysis and experimental research show that the bearing capacity of flexural-torsional buckling around the strong axis is always higher than that of bending buckling around the weak axis. Therefore, flexural-torsional buckling is not calculated for this type of member.

d. The converted slenderness ratio $\lambda_{yz}$ of built-up T-shape with double-angle around

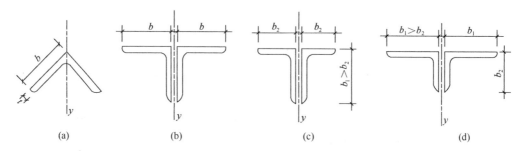

Figure 5.12 Single-angle and double-angle combined T-shaped section
$b$—leg length of equal leg angle; $b_1$—long leg length of unequal leg angle;
$b_2$—short leg length of unequal leg angle

the axis of symmetry can be determined by the following simplified method.

Equilateral double-angle section (Figure 5.12b)

When $\lambda_y \geqslant \lambda_z$

$$\lambda_{yz} = \lambda_y \left[1 + 0.16 \left(\frac{\lambda_z}{\lambda_y}\right)^2\right] \tag{5-18}$$

When $\lambda_y < \lambda_z$

$$\begin{cases} \lambda_{yz} = \lambda_y \left[1 + 0.16 \left(\frac{\lambda_y}{\lambda_z}\right)^2\right] \\ \lambda_z = 3.9 \frac{b}{t} \end{cases} \tag{5-19}$$

where $t$ = the leg thickness of the angle.

Double-angle section with long legs connected (Figure 5.12c)

When $\lambda_y \geqslant \lambda_z$

$$\lambda_{yz} = \lambda_y \left[1 + 0.25 \left(\frac{\lambda_z}{\lambda_y}\right)^2\right] \tag{5-20}$$

When $\lambda_y < \lambda_z$

$$\begin{cases} \lambda_{yz} = \lambda_y \left[1 + 0.25 \left(\frac{\lambda_y}{\lambda_z}\right)^2\right] \\ \lambda_z = 5.1 \frac{b_2}{t} \end{cases} \tag{5-21}$$

Double-angle section with short legs connected (Figure 5.12d)

When $\lambda_y \geqslant \lambda_z$

$$\lambda_{yz} = \lambda_y \left[1 + 0.06 \left(\frac{\lambda_z}{\lambda_y}\right)^2\right] \tag{5-22}$$

When $\lambda_y < \lambda_z$

$$\begin{cases} \lambda_{yz} = \lambda_y \left[1 + 0.06 \left(\frac{\lambda_y}{\lambda_z}\right)^2\right] \\ \lambda_z = 3.7 \frac{b_1}{t} \end{cases} \tag{5-23}$$

(4) According to the following simplified equations, the converted slenderness ratio

of unequal leg single-angle axially loaded compression members can be determined.

When  $\lambda_x \geqslant \lambda_z$

$$\lambda_{xyz} = \lambda_x \left[1 + 0.25 \left(\frac{\lambda_z}{\lambda_x}\right)^2\right] \quad (5\text{-}24)$$

When  $\lambda_x < \lambda_z$

$$\begin{cases} \lambda_{xyz} = \lambda_x \left[1 + 0.25 \left(\frac{\lambda_x}{\lambda_z}\right)^2\right] \\ \lambda_z = 4.21 \frac{b_1}{t} \end{cases} \quad (5\text{-}25)$$

where $b_1$ is the long leg length of the unequal leg angle; subscript $x$ is the symbol of the $x$-axis.

The cross-section does not have any symmetry axis, and the shear center and the centroid does not overlap, except for the single-angle with unequal-leg connected on one side. It is not suitable for the axially loaded compression member.

For one side connection of single-angle axially loaded compression members, the strength calculation and stability calculation consider the reduction coefficient (see GB 50017). The flexural-torsional buckling effect may not be considered.

When the channel is used for the split limbs of the laced column and calculating the stability of the limb around the axis of symmetry (y-axis), the torsion effect does not need to be considered, and $\varphi_y$ is directly detected by $\lambda_y$.

### 5.3.2 The calculation of local stability

(1) The local stability of panels

Due to the axially loaded compression members made of steel plates with a large width-to-thickness ratio, the local stability of members should be considered in the design. Figure 5.13 shows the deformation of I-section axially loaded compression members in case of local instability. The instability of webs and flanges is shown in Figures 5.13(a) and 5.13(b).

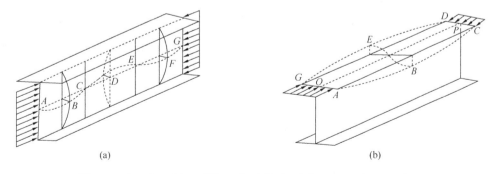

Figure 5.13  Local instability of axially loaded compression members

After the member loses its local stability, it may continue to maintain the overall equilibrium state. However, due to the withdrawal of some plates after buckling, the effective cross-section is reduced, accelerates the member's overall instability, and loses the load-

## Chapter 5  AXIALLY LOADED MEMBERS

bearing capacity.

(2) Limit of the width-to-thickness ratio of plates

Based on elastic stability theory, the shape, size, support, and stress of plates affect critical stress, that is, the maximum stress that the plate can withstand in a stable state. Therefore, the critical stress of plates can be expressed as:

$$\sigma_{cr} = \frac{\sqrt{\eta}\,\chi\,\beta\pi^2 E}{12(1-\nu^2)}\left(\frac{t}{b}\right)^2 \tag{5-26}$$

where  $\chi=$ the elastic constraint coefficient of the plate edge;
$\beta=$ the buckling coefficient;
$\nu=$ the Poisson's ratio of steel;
$E=$ the modulus of elasticity of the steel;
$\eta=$ the reduction coefficient of elastic modulus, defined as $E_t/E$. According to tests of local stability, the $\eta$ is described as:

$$\eta = 0.1013\lambda^2(1-0.0248\lambda^2 f_y/E)f_y/E \tag{5-27}$$

where  $E_t=$ the tangent modulus of steel.

The calculation of local stability considers equal stability conditions, ensuring that the critical stress of local buckling of plates is larger than or equal to that of the global buckling of columns. That is:

$$\frac{\sqrt{\eta}\,\chi\,\beta\pi^2 E}{12(1-v^2)}\left(\frac{t}{b}\right)^2 \geqslant \varphi f_y \tag{5-28}$$

Perry Equation [Equation (5-12)] expresses the overall stability factor $\varphi$. The overall stability factor $\varphi$ is related to the slenderness ratio of members. The limit of the width-to-thickness ratio of plates can be defined by Equation (5-28). Take an I-section member as an example.

① Flange

The web thickness of the I-section member is less than its flange thickness, which causes that the webs have no fixing effect on the flange. Therefore, the flange can be regarded as a uniformly compressed plate with three sides simply supported and one side free. The buckling coefficient is defined as $\beta = 0.425$. The elastic constraint coefficient is defined as $\chi = 1.0$.

From Equation (5-28), the relationship curve between the width-to-thickness ratio $b/t$ and the slenderness ratio $\lambda$ of the outstanding part of the flange can be obtained. However, the relationship equation of this curve is more complicated. Therefore, in order to facilitate the application, the following simple linear expression is adopted.

$$\frac{b}{t_f} \leqslant (10+0.1\lambda)\varepsilon_k \tag{5-29}$$

where  $b=$ the outstanding width of the flange;
$t_f=$ the thickness of the flange;

$\lambda$ = the larger value of the slenderness ratio in two principle axes directions of the members. When $\lambda$ is less than 30, the $\lambda$ is taken as 30. When $\lambda$ is not less than 100, the $\lambda$ is taken as 100.

② Web

The web is regarded as a four-side simply supported plate. The buckling coefficient is taken as 4.0. When the web occurred buckling, the flange as the longitudinal edge of the web support has an elastic fixation effect on the web, contributing to improving the critical stress of the web. The elastic constraint coefficient is taken as 1.3 based on tests. The depth-to-thickness ratio of webs can be expressed as:

$$\frac{h_0}{t_w} \leqslant (25 + 0.5\lambda)\varepsilon_k \qquad (5\text{-}30)$$

The limit of the width-to-thickness ratio of other members is shown in Table 5.5. For box-shaped members (including the outer wall-plate with double-layer flange), the limit of the width-to-thickness ratio is approximately borrowed from those of the box-shaped beam flanges (see more details in Chapter 6). For the cross-section of the round pipe, the limit of the width-to-thickness ratio is derived on the premise that the material is an ideal elasto-plastic body, and the axial compressive force reaches the yield strength.

Equations (5-29) and (5-30) are proposed when the overall stability bearing capacity of members arrives at ultimate bearing capacity. Obviously, when the stress of axially loaded compression members is less than stability bearing capacity ($\varphi A f$), the limit of the width-to-thickness ratio of members obtained according to Equation (5-29) can be extended appropriately. The amplification factor $\alpha = \sqrt{\varphi A f / N}$ can be multiplied by the limit of the width-to-thickness ratio in Table 5.5. The amplification factor can be obtained using the normal stress ($N/A$) borne by the member to substitute the right end of the Equation (5-29).

(3) Use of post-buckling strength of plates

When the width-to-thickness ratio of axially loaded compression members does not meet the requirement of Table 5.5, there are two methods adopted. One is to enhance the thickness of plates, and however, for wall-plates of box-shaped sections and webs of H-shaped or I-section, the other more effective method is to install longitudinal stiffeners in the middle of their web.

**Limit width-to-thickness ratio of the axially loaded compression member**　　　Table 5.5

| Section and plate dimensions | the limit ratio of width to thickness |
|---|---|
|  | Flange: $\dfrac{b}{t_f} \leqslant (10 + 0.1\lambda)\varepsilon_k$ <br><br> Web: $\dfrac{h_0}{t_w} \leqslant (25 + 0.5\lambda)\varepsilon_k$ |

continued

| Section and plate dimensions | the limit ratio of width to thickness |
|---|---|
| (T-shape sections) | Flange: $\dfrac{b}{t_f} \leqslant (10+0.1\lambda)\varepsilon_k$<br>Web:<br>Hot-rolled split T-shapes: $\dfrac{h_0}{t_w} \leqslant (15+0.2\lambda)\varepsilon_k$<br>Welded T-shapes: $\dfrac{h_w}{t_w} \leqslant (13+0.17\lambda)\varepsilon_k$ |
| (box sections) | $\dfrac{h_0}{t_w}\left(\text{or }\dfrac{b_0}{t_f}\right) \leqslant 40\varepsilon_k$ |
| (angle section) | $\lambda \leqslant 80\varepsilon_k, \dfrac{w}{t} \leqslant 15\varepsilon_k$<br>$\lambda > 80\varepsilon_k, \dfrac{w}{t} \leqslant 5\varepsilon_k + 0.125\lambda$ |
| (circular tube) | $\dfrac{d}{t} \leqslant 100\varepsilon_k^2$ |

Note: For the web, $h_0/t_w$ is the depth-to-thickness ratio.

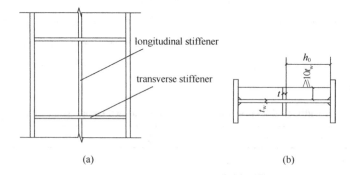

Figure 5.14  Web stiffeners for solid web column

As shown in Figure 5.14, due to longitudinal stiffener and flange forming the support of the longitudinal sides of the web, the effective height of strengthened webs is defined as the distance between flange and longitudinal stiffener. Longitudinal stiffeners should be paired on both sides of the web and should have enough stiffness, so the outstanding width of one side should not be less than $10t_w$, and the thickness should not be less than $0.75t_w$.

The purpose of limiting the plate's width-to-thickness ratio and setting longitudinal stiffeners is to ensure that no local buckling occurs before the overall instability of the member. Therefore, the ideal flat plates with four-side supports have a larger bearing capacity after buckling, called post-buckling strength. The post-buckling strength of plates is derived from the transverse tension in the middle of plates, so plates can continue to bear load after buckling. In this case, the longitudinal compressive force in the plate is not uniform. The web stress distribution of the I-section after buckling is shown in Figure 5.15a.

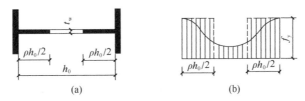

Figure 5.15 The effective section of the web after buckling of the I-section member

Suppose the distribution of the web's longitudinal compressive stress of I-section after buckling is approximated by the stress graph shown by the dotted line in Figure 5.15(b). In that case, the concepts of equivalent width and effective section are introduced. Consider parts of the web out of work, and an equivalent plate can replace the actual web with stress $f_y$ and width $\rho h_0$. The section of the equivalent plate is the effective section. Consider the use of post-buckling strength of plates, and the effective section should be calculated first. Then the strength and overall stability of the members can be calculated according to the following formula.

Strength calculation:

$$\frac{N}{A_{ne}} \leqslant f \tag{5-31}$$

Overall stability calculation:

$$\begin{cases} \dfrac{N}{\varphi A_e f} \leqslant 1.0 \\ A_{ne} = \sum \rho_i A_{ni} \\ A_e = \sum \rho_i A_i \end{cases} \tag{5-32}$$

where  $A_{ne}$ = the effective net cross-sectional area;
$A_e$ = the effective gross cross-sectional area;
$A_{ni}$ = the effective net cross-sectional area of each plate;
$A_i$ = the gross cross-sectional area of each plate;
$\varphi$ = the overall stability factor which can be calculated according to the gross

cross-section;

$\rho_i$ = the effective section coefficient of each plate which can be calculated according to the following methods.

(1) Wall-plate of box-shapes, web of H-shape or I-section

When $\quad h_0/t_w \leqslant 42\varepsilon_k$

$$\rho = 1.0 \tag{5-33}$$

When $\quad h_0/t_w > 42\varepsilon_k$

$$\rho = \frac{1}{\lambda_{n,p}}\left(1 - \frac{0.19}{\lambda_{n,p}}\right)$$
$$\lambda_{n,p} = \frac{h/t_w}{56.2\varepsilon_k} \tag{5-34}$$

When $\quad \lambda > 52\varepsilon_k$

$$\rho \geqslant (29\varepsilon_k + 0.25\lambda)t_w/h_0 \tag{5-35}$$

where $h_0$ and $t_w$ are the net width and thickness of plates or webs.

(2) A single angle

When $\quad \dfrac{w}{l} > 15\varepsilon_k$

$$\rho = \frac{1}{\lambda_{n,p}}\left(1 - \frac{0.1}{\lambda_{n,p}}\right)$$
$$\lambda_{n,p} = \frac{w/t}{16.8\varepsilon_k} \tag{5-36}$$

When $\quad \lambda > 80\varepsilon_k$

$$\rho \geqslant (5\varepsilon_k + 0.13\lambda)t/w \tag{5-37}$$

where $w$ and $t$ are the flat width and thickness of the angle leg, when calculating briefly, the $w$ is equal to $b-2t$, and $b$ is the leg length of the angle.

## 5.4 Design of Axially Loaded Columns

From the geometry of a given structural arrangement, $L_x$ and $L_y$ of a column to be designed are generally known from the length $L$ and unsupported lengths $L_x$ and $L_y$ of a column. The required axial compressive strength (including buckling strength) of that column is obtained from the analysis of the structure under factored loads using the load combinations (Section 1.5.3). Using this information, the designer must select a rolled shape or built-up section capable of bearing the required axial load.

### 5.4.1 Design of solid web column

(1) Section shape selection

We had better select the members with biaxial symmetrical sections as solid web columns to avoid flexural-torsional buckling. The recommended shapes of cross-sections are shown in Figure 5.16.

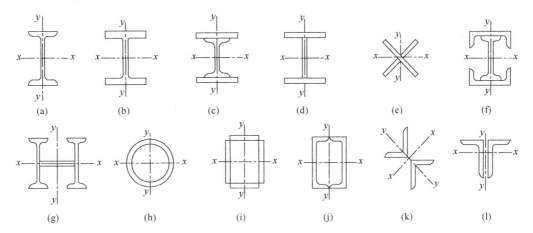

Figure 5.16 Common cross-sectional shapes of solid web columns

When selecting a cross-section of solid web columns, the following rules should be considered.

① The layout of the plates is as far away as possible from the principal axis of the cross-section, and the section follows the thin-walled and expanded principle to increase the moment of inertia and radius of gyration and improve overall stability and column stiffness.

② The overall stability factors of the two principal axis directions need to be approximately equal ($\varphi_x \approx \varphi_y$) to achieve the maximum utilization of steel.

③ The selected section is easy to connect with other members.

④ Choose the section with simple construction details, labor-saving manufacturing, and convenient materials as far as possible.

The selection of sections should generally be based on the axial compressive force, the effective length of the two principal axis directions, and the manufacturing capacity, etc.

For columns with identical end restraint conditions in both principal planes, I-shapes (Figure 5.16a) make uneconomical columns. For such members, the radius of gyration about the weak-axis ($y$-axis) is significantly less than the radius of gyration about the strong-axis ($x$-axis). The added stiffness associated with the strong axis is not utilized since buckling is controlled by the slenderness associated with the weak axis. The HN series of hot-rolled H-shapes has the same reason, and it is not suitable for use as a column.

Given the same cross-sectional area and end restrains, an H-shape having less web depth but greater flange width, such as the HW series of hot-rolled H-shapes (Figure 5.16b), makes for a more efficient column. Sometimes, the HM series of hot-rolled H-shapes is also a good cross-section choice.

The welded I-shape (Figure 5.16d) and welded crosssection (Figure 5.16e) are combination flexible, easy to make the distribution of the cross-sections appropriate, and the manufacturing is not complicated. The built-up sections (Figures 5.16c, 5.16f, and 5.16g) are suitable for columns under high loading.

For a column with identical support conditions about two principal axes at both ends and has no intermediate bracing in either plane, the most efficient cross-sectional shape is a hollow circular pipe (Figure 5.16h). Next is the square hollow structural section (HSS) or welded square tube, close built-up sections (Figures 5.16i and 5.26j). These sectional shapes have a similar radius of gyration in two directions. Note that the torsional rigidities of all close sections are much greater than those of open sections having similar area and width-to-thickness ratios. Therefore, torsional buckling is not a problem for close sections. A practical objection to close sections is the difficulty of connecting the beam to such sections. The ends of the close section should also be sealed against moisture and air penetration to prevent corrosion from inside. In contrast, open sections are very easy to connect with others.

(2) Cross-sectional design

When we design a column, first is to select the appropriate section shape according to the rules mentioned above; secondly, the preliminary determination of section size is selected. Finally, the strength, overall stability, local stability, and stiffness are checked. However, it is not easy to do for a beginner. The slenderness parameter (slenderness ratio $\lambda$) of the column determines the compressive design stress cannot be computed until the radius of gyration ($r$) is unknown until the cross-section has been selected. Consequently, column design is generally an iterative procedure. A recommended procedure is as follows:

① Assume an appropriate value for the slenderness ratio $\lambda$ of columns. Generally, assume $\lambda = 50-100$. When the load is large, and the effective length is small, the smaller value is used; otherwise, the larger value is used. According to $\lambda$, section classification and steel grade, the overall stability factor $\varphi$ can be found, and the approximate required cross-sectional area can be calculated as:

$$A = \frac{N}{\varphi f} \tag{5-38}$$

② The following equation calculates the required radius of gyration for the two principal axes.

$$i_x = \frac{l_{0x}}{\lambda}, i_y = \frac{l_{0y}}{\lambda} \tag{5-39}$$

③ The known cross-sectional area $A$, the radius of gyration $r_x$ and $r_y$ of the two principal axes, rolled shapes are preferred, such as I-shapes and H-shapes. When the existing rolled shape specifications do not meet the requirement, the built-up section can be applied to a solid web column. The radius of gyration can define the width and depth of a trial section.

$$h \approx \frac{i_x}{\alpha_1}, b \approx \frac{i_y}{\alpha_2} \qquad (5\text{-}40)$$

where $\alpha_1$ and $\alpha_2$ are coefficients, the approximate numerical relationship between $h$, $b$, and the radius of gyration $i_x$, $i_y$. Common cross-sections can be obtained from Table 5.6. For example, the coefficients including $\alpha_1$ and $\alpha_2$ of an I-section composed of three plates are 0.43 and 0.24.

**Approximate value of the radius of gyration**      Table 5.6

| Section | I | ⊏⊐ | ⊐⊏ | □ (b=h) | T | T | T |
|---|---|---|---|---|---|---|---|
| $i_x = \alpha_1/h$ | 0.43h | 0.38h | 0.38h | 0.40h | 0.30h | 0.28h | 0.32h |
| $i_y = \alpha_2/b$ | 0.24b | 0.44b | 0.60b | 0.40b | 0.215b | 0.24b | 0.20b |

④ The primary size of the trial cross-section is determined by considering the required $A$, $h$, $b$, etc., construction detail requirements, local stability, rolled shapes specifications, etc.

⑤ Check the strength, stability, and stiffness of columns.

a. When the cross-section is weakened, the strength must be checked by the following equation. If the cross-section does not be weakened, go directly to the overall stability check of the second step.

$$\sigma = \frac{N}{A_n} \leqslant f_u \qquad (5\text{-}41)$$

where $A_n$ is the net cross-sectional area of columns.

b. The check of overall stability.

$$\frac{N}{\varphi A f} \leqslant 1.0 \qquad (5\text{-}42)$$

c. The check of local stability

As mentioned above, the local stability of the axially loaded compression member is guaranteed by limiting the width-to-thickness ratio of its constituent plates. Due to the small width-to-thickness ratio, it can generally meet the local stability requirements for hot-rolled shapes. For the built-up section, the width-to-thickness ratio should be calculated according to the provisions of Table 5.5.

d. The calculation of stiffness

The slenderness ratio of the solid web column needs to meet the allowable slenderness ratio proposed in design specifications. Furthermore, in checking overall stability, the slenderness ratio of columns has been calculated to define the overall stability factor. Therefore, the check of stiffness and overall stability can be carried out simultaneously.

(3) Construction detail requirements

When the depth-to-thickness ratio of the web is larger than $80\varepsilon_k$, the transverse stiffeners should be set in the web of solid web columns to avoid the deformation of webs during construction and transportation and improve the torsional rigidity of columns. The spacing of transverse stiffeners is smaller than $3h_0$. The outstanding width of the bilateral stiffeners is larger than $h_0/30+40$ mm. Therefore, the thickness $t_s$ is larger than 1/15 outstanding width.

Due to the longitudinal weld of the solid web column bearing small force, the connection weld between flange and web is unnecessary to calculate. The weld leg size can be determined according to the construction detail requirements.

## 5.4.2 Laced column design

(1) The section form of a laced column

Biaxial symmetrical cross-sections are used for axially loaded compression laced columns. For example, channels or H-shapes are selected as limbs, and the two limbs are connected by lacing bars (see Figure 5.4a) or batten plates (see Figure 5.4b). It is convenient for laced columns to adjust the distance between two limbs, which is easy to realize equal stability of two principal axes. For example, flanges of channel limbs can be inward (see Figure 5.3a) or outward (see Figure 5.3b), and the former has a flat appearance.

In the built-up cross-section of a laced column, the axis passing limb web is called the real axis (the $y$-axis in Figure 5.4), and the axis passing lacing is called the virtual axis (the $x$-axis in Figure 5.4).

The four-limb columns of four angles are suitable for the column with a large length and small loading (see Figure 5.3d). Lacing connects the four sides, and the two principal axes are all virtual axes. The round pipe is usually selected as a limb of the three-limb column (see Figure 5.3e). This type of laced column with a geometrically invariant triangular section has better mechanical performance. The lacings of the four-limb column and three-limb column are the lacing bars commonly rather than batten plates.

The lacing bars are made of single-angles, while batten plates are made of flat bars.

(2) Equivalent slenderness ratio of a laced column about the virtual axis

The stability calculation of the laced column about the real axis is the same as that of the solid web columns. Still, the critical force of the overall stability about the virtual axis is lower than that of the same solid web columns when their slenderness ratio is the same.

After the column bending, there will be a bending moment and shear on each section along the longitudinal axis of the column. For solid web columns, the additional deformation caused by shear has little effect on critical stress. Therefore, the definition of critical stress of overall stability of solid web columns only considers the deformation caused by bending moment, which ignores the deformation caused by shear.

For the laced column, when it loses overall stability about the virtual axis, the situation is different because the limbs are not continuous plates but are connected by lacings at certain distances. As a result, the shear deformation of the column is large, and the additional deflection effect caused by the shear cannot be ignored. Therefore, in the design of laced columns, the influence of shear deformation is usually considered by increasing the slenderness ratio, which is called the equivalent slenderness ratio.

The design standard GB 50017 adopts different equivalent slenderness ratios to calculate columns with lacing bars or batten plates.

① Double-limb columns with lacing bars

According to the elastic stability theory, the critical load considering the effect of shear can be expressed by the following equation.

$$N_{cr} = \frac{\pi^2 EA}{\lambda_x^2} \cdot \frac{1}{1 + \frac{\pi^2 EA}{\lambda_x^2} \cdot \gamma} = \frac{\pi^2 EA}{\lambda_{0x}^2} \tag{5-43}$$

where $\lambda_{0x}$ is the equivalent slenderness ratio from the critical load of a laced column about the virtual axis to the critical load of the solid web column.

$$\lambda_{0x} = \sqrt{\lambda_x^2 + \pi^2 EA\gamma} \tag{5-44}$$

where $\gamma$ is the axis rotation angle under unit shear.

The unit shear angle is calculated by analyzing a section of Figure 5.17(a). As shown in Figure 5.17(b), the lacings on one side bear shear force under unit shear. Suppose that the sum area of diagonal lacing bars in the internode is $A_1$, the internal force $N_d$ is equal to $1/\sin \alpha$. Consequently, the length $l_d$ of diagonal lacing bars is equal to $l_1/\cos \alpha$. Therefore, the axial deformation of diagonal lacing bars can be calculated by the following equation.

$$\Delta_d = \frac{N_d l_d}{EA_1} = \frac{l_1}{EA_1 \sin\alpha \cos\alpha} \tag{5-45}$$

Supposing that deformation and shear angles are tiny, the horizontal displacement caused by $\Delta_d$ is described as:

$$\Delta = \frac{\Delta_d}{\sin\alpha} = \frac{l_1}{EA_1 \sin^2\alpha \cos\alpha} \tag{5-46}$$

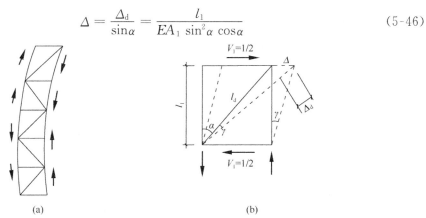

Figure 5.17 Shear deformation of a laced column with lacing bars

The shear angle $\gamma$ can be expressed as:

$$\gamma = \frac{\Delta}{l_1} = \frac{1}{EA_1 \sin^2\alpha \cos\alpha} \quad (5\text{-}47)$$

where $\alpha$ is the angle between diagonal lacing bars and column axis, which the following equation can obtain:

$$\lambda_{0x} = \sqrt{\lambda_x^2 + \frac{\pi^2}{\sin^2\alpha\cos\alpha} \cdot \frac{A}{A_1}} \quad (5\text{-}48)$$

The angle between diagonal lacing bars and column axis is between 40° and 70°. The $\pi^2/(\sin^2\alpha \cos\alpha)$ is simplified as 27 (Figure 5.18). Thus, the equivalent slenderness ratio of double-limb columns can be expressed as:

$$\lambda_{0x} = \sqrt{\lambda_x^2 + 27\frac{A}{A_1}} \quad (5\text{-}49)$$

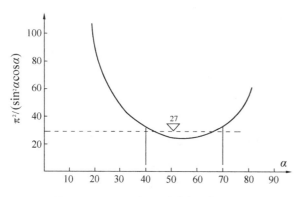

Figure 5.18 Values of $\pi^2/(\sin^2\alpha\cos\alpha)$

where $\lambda_x$ is the slenderness ratio of the laced column to the virtual axis, $A$ is the gross cross-sectional area of the laced column.

$\pi^2/(\sin^2\alpha\cos\alpha)$ will be much larger than 27 when the angle $\alpha$ beyond the range of 40°-70°, especially less than 40°. Equation (5-49) is slightly unsafe, so the equivalent slenderness ratio should be calculated according to Equation (5-48).

② Double-limb columns with batten plates

The connection between the batten plate and the limb of the double-limb columns can be regarded as a rigid connection, so the limb and the batten plate form a multi-story frame. It is assumed that the reverse bending point is at the midpoint of each intersection during deformation (Figure 5.19a). If only considering the bending deformation of the split limbs and batten plates under the action of transverse shear, take the isolated body as shown in Figure 5.19(b). It can be obtained that the displacement $\Delta_1$ of the limbs caused by the bending deformation of the batten plates under the action of unit shear is:

$$\Delta_1 = \frac{l_1}{2}\theta_1 = \frac{l_1}{2} \cdot \frac{al_1}{12EI_b} = \frac{al_1^2}{24EI_b} \quad (5\text{-}50)$$

The displacement $\Delta_2$ caused by the bending deformation of the limb itself is:

$$\Delta_2 = \frac{l_1^3}{48EI_1} \quad (5\text{-}51)$$

The shear angle $\gamma$ can be expressed as:

$$\gamma = \frac{\Delta_1 + \Delta_2}{0.5l_1} = \frac{al_1}{12EI_b} + \frac{l_1^2}{24EI_1} = \frac{l_1^2}{24EI_1}\left(1 + 2\frac{I_1/l_1}{I_b/a}\right) \quad (5\text{-}52)$$

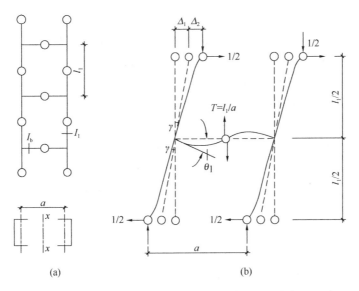

Figure 5.19 Shear deformation of a laced column with batten plates

Take Equation(5-52) into Equation (5-44), the equivalent slenderness ratio is defined by the following formula:

$$\lambda_{0x} = \sqrt{\lambda_x^2 + \frac{\pi^2 A l_1^2}{24 I_1}\left(1 + 2\frac{K_1}{K_b}\right)} \qquad (5\text{-}53)$$

where $K_1$ is equal to $I_1/l_1$, $K_b$ is equal to $I_b/a$.

Assuming that the cross-sectional area $A_1$ of the limb is equal to $0.5A$ and $A_1 l_1^2 / I_1 = \lambda_1^2$, the following equation can be obtained:

$$\lambda_{0x} = \sqrt{\lambda_x^2 + \frac{\pi^2}{12}\left(1 + 2\frac{K_1}{K_b}\right)\lambda_1^2} \qquad (5\text{-}54)$$

where  $\lambda_1 = l_{01}/i_1$, it is the slenderness ratio of the limb, $i_1$ is the radius of gyration of the limb about its weak axis, $l_{01}$ is the net distance between the batten plates;

$K_1 = I_1/l_1$, it is the linear stiffness of a limb, $l_1$ is the center distance of batten plates, $I_1$ is the moment of inertia of the limb about its weak axis;

$K_b = I_b/a$, it is the sum of the linear stiffness of batten plates on both sides, $I_b$ is the moment of inertia of batten plates on both sides, $a$ is distance between the axes of the limbs.

According to the design standard GB 50017, it must guarantee $K_b/K_1 \geqslant 6$. The sum of the linear stiffness of batten plates on both sides should be greater than six times the linear stiffness of a limb.

Assume $K_b/K_1 = 6$, then can get $\frac{\pi^2}{12}\left(1 + 2\frac{K_1}{K_b}\right) \approx 1$. Therefore, the equivalent slenderness ratio of the double-limb laced column with batten plates can be obtained:

139

## Chapter 5  AXIALLY LOADED MEMBERS

$$\lambda_{0x} = \sqrt{\lambda_x^2 + \lambda_1^2} \tag{5-55}$$

If some special circumstances cannot meet the requirements $K_b/K_1 \geqslant 6$, the equivalent slenderness ratio should be calculated according to Equation (5-54).

For the equivalent slenderness ratio of the four-limb laced column and three-limb laced column, please refer to clause 7.2.3 of the design standard GB 50017.

(3) Design of lacings

① The horizontal shear force of the axially loaded compression laced column

When the laced column is buckling about the virtual axis, the lacings should bear the transverse shear. Therefore, it is necessary to calculate the transverse shear first and then design lacings.

As shown in Figure 5.20(a), when a pin-ended column bends about the virtual axis, supposing that the final deflection line is a *Sine Curve* and the maximum mid-span deflection is $v_0$, then the deflection at any point along the longitudinal axis of the column is described as:

$$y = v_0 \sin \frac{\pi z}{l} \tag{5-56}$$

The following equation can express the bending moment at any point.

$$M = Ny = Nv_0 \sin \frac{\pi z}{l} \tag{5-57}$$

The shear at any point can be expressed by the following equation.

$$V = \frac{dM}{dz} = N \frac{\pi v_0}{l} \cos \frac{\pi z}{l} \tag{5-58}$$

The shear is distributed according to the *Cosine Curve* (Figure 5.20b). Thus, the maximum shear at both ends of the member is:

$$V_{max} = N \frac{\pi}{l} \cdot v_0 \tag{5-59}$$

The deflection $v_0$ of the midpoint of the span can be derived from the edge yield criterion. When the maximum stress at the edge of the cross-section reaches the yield strength, which can be described as:

$$\frac{N}{A} + \frac{Nv_0}{I_x} \cdot \frac{b}{2} = f_y$$
$$\frac{N}{Af_y}\left(1 + \frac{v_0}{i_x^2} \cdot \frac{b}{2}\right) = 1 \tag{5-60}$$

The $N/Af_y$ is defined as $\varphi$, and take $b = i_x/0.44$. The following equation can be obtained:

$$v_0 = 0.88 i_x (1 - \varphi) \frac{1}{\varphi} \tag{5-61}$$

Submit Equation (5-61) into Equation (5-59), the following equation can be obtained:

$$V_{max} = \frac{0.88\pi(1-\varphi)}{\lambda_x} \cdot \frac{N}{\varphi} = \frac{1}{k} \cdot \frac{N}{\varphi} \tag{5-62}$$

where $k = \dfrac{\lambda_x}{0.88\pi(1-\varphi)}$.

## 5.4 Design of Axially Loaded Columns

Through calculation and analysis of double-limb laced column, within the scope of the common slenderness ratio, the slenderness ratio $\lambda_x$ has little relationship with $k$, which can be defined as constant. For Q235 steel columns, $k = 85$; for other grade steel columns, $k = 85\varepsilon_k$.

Therefore, the shear that is parallel to the surface of lacings is defined as:

$$V_{max} = \frac{N}{85\varphi\varepsilon_k} \qquad (5\text{-}63)$$

$\varphi$ is the overall stability factor determined by the equivalent slenderness ratio with respect to the virtual axis.

Let $N = \varphi A f$, and the calculation formula of the maximum shear stipulated in the design standard GB 50017 is obtained:

$$V = \frac{Af}{85\varepsilon_k} \qquad (5\text{-}64)$$

In the design, the shear $V$ is set as a constant value along the longitudinal axis of columns, equivalent to the distribution graph simplified to Figure 5.20(c).

② Design of the lacing bars

The layout of lacing bars generally adopts a single lacing bar, as shown in Figure 5.21(a), or a cross lacing bar, as shown in Figure 5.21(b). The lacing bars can be regarded as the web members of parallel chord trusses with column limbs as chords. The calculation method of internal forces is the same as that of truss web members. Under horizontal shear, the axial force of diagonal lacing bars can be described by the following equation.

$$N_1 = \frac{V_1}{n\cos\theta} \qquad (5\text{-}65)$$

where $V_1$ is the shear assigned to the surface of a lacing; $n$ is the number of diagonal lacing bars bearing shear force $V_1$. for a single lacing bar, $n = 1$; for a cross lacing bar, $n = 2$. $\theta$ is the inclination angle of the diagonal lacing bars (Figure 5.21).

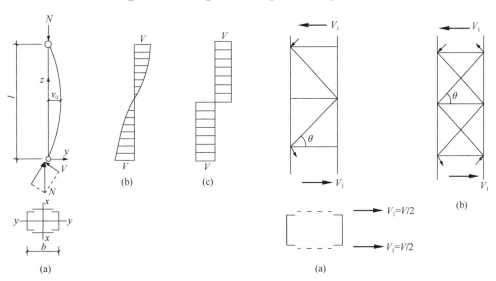

Figure 5.20   Shear calculation diagram        Figure 5.21   The internal force of lacing bars

Due to various directions of shear, the diagonal lacing bars may bear tensile force or compressive force. Therefore, the section should be selected according to the axial compression member.

The lacing bars generally use a single-angle, which is connected with the column on one side. However, considering the eccentricity and flexural-torsion during compression, when designing the axial force member without considering the torsion effect, it should be based on the design value of the steel strength multiply the reduction factor $\eta$, which shall be valued according to the following regulations.

a. According to the axial force, when calculating the strength of members and connections, $\eta$ is taken as 0.85.

b. When calculating the overall stability of a single-angle according to the axially loaded compression member:

For equal leg angles:
$$\eta = 0.6 + 0.0015\lambda \leqslant 1.0$$
For unequal leg angles connected by short legs:
$$\eta = 0.6 + 0.0025\lambda \leqslant 1.0$$
For unequal leg angles connected by long legs:
$$\eta = 0.7$$

$\lambda$ is the slenderness ratio of the lacing bar. For a single-angle with no connection in the middle, calculate according to the minimum radius of gyration. When $\lambda$ is less than 20, take $\lambda$ equal to 20. The horizontal bar of the cross-system (Figure 5.21b) is calculated according to the compressive force $N = V_1$. In order to reduce the effective length of the split limbs, single-system lacing bars (Figure 5.21a) can also be added with horizontal lacing bars, whose size is generally the same as that of diagonal lacing bars and can also be determined by the allowable slenderness ratio.

③ Design of batten plates

The batten plate column can be regarded as a multi-story frame (Figure 5.19a), limb parts are regarded as frame columns, and batten plates are regarded as beams). Supposing that the midpoint of limbs and batten plates of each story is the reverse bending points (Figure 5.19a), when the batten plate column bends, the internal force of the batten plate can be calculated according to Figure 5.22(a).

$$T = \frac{V_1 l_1}{a} \qquad (5\text{-}66)$$

$$M = T \cdot \frac{a}{2} = \frac{V_1 l_1}{2} \qquad (5\text{-}67)$$

where  $l_1 =$ the distance between the center lines of batten plates;
$\quad\quad a =$ the distance between the axes of limbs.

The fillet welds are used to connect batten plates with limbs, which bear the combined action of the shear and bending moment. Since the design strength of fillet weld is less than that of steel, only the above $M$ and $T$ are used to check the connection weld between the batten plates and limbs.

The batten plate should have a certain stiffness. According to relevant specifications, the sum of the linear stiffness of batten plates on both sides should be greater than six times the linear stiffness of a limb. Generally, the width $d$ is not less than $2a/3$ (Figure 5.22b). The thickness $t$ is not less than $a/40$ and 6 mm. The end batten plate should be widened appropriately and letting $d = a$.

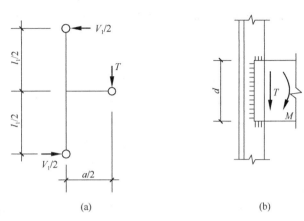

Figure 5.22 The calculation diagram of lacing bars

(4) Design procedures of laced column

Firstly, the design of a laced column should choose the section of column limbs and the form of lacing. The small-sized and medium-sized columns can be used for batten plate columns or lacing bar columns. And large-sized column should be a lacing bar column. The following steps can design the laced column:

① The limb sections of columns are selected according to the overall stability of the real axis ($y$-$y$ axis), and the calculation method is the same as that of the solid web column.

② The distance between the two limbs is determined according to the overall stability of the virtual axis ($x$-$x$ axis).

In order to obtain equal stability, the slenderness ratios with respect to the two principal axes need to be equal, that is $\lambda_{0x} = \lambda_y$.

Lacing bar column (double-limb):

$$\lambda_{0x} = \sqrt{\lambda_x^2 + 27 \frac{A}{A_1}} = \lambda_y$$

$$\lambda_x = \sqrt{\lambda_y^2 - 27 \frac{A}{A_1}}$$

(5-68)

Batten plate column (double-limb):

$$\lambda_{0x} = \sqrt{\lambda_x^2 + \lambda_1^2} = \lambda_y$$

$$\lambda_x = \sqrt{\lambda_y^2 - \lambda_1^2}$$

(5-69)

The cross-sectional area $A_1$ of the diagonal lacing bar should be pre-determined for the batten bar column, and the slenderness ratio $\lambda_1$ of the split limb should be assumed for the batten bar column.

143

According to Equation (5-68) or Equation (5-69), the radius of gyration about the virtual axis can be obtained:

$$i_x = l_{ox}/\lambda_x \qquad (5-70)$$

According to Table 5.6, the column width along lacing bars can be obtained $b \approx i_x/\alpha_1$, and the column width can also be directly calculated from the geometry of the known section.

③ Check the overall stability of the virtual axis. If it is not suitable, the column width $b$ should be modified to recheck the calculation.

④ Design lacing bars or batten plates (including their connection to the split limb).

The calculation mentioned above should pay attention to the following requirements:

a. The slenderness ratio of the column to the real axis and the equivalent slenderness ratio to the virtual axis shall not exceed the allowable slenderness ratio.

b. The slenderness ratio $\lambda_1$ of the split limbs of the column shall not exceed 0.7 time the larger value of the slenderness ratio to two principal axes of the column (which is the equivalent slenderness ratio to the virtual axis). Otherwise, the split limbs may be buckling before the overall instability of the laced column.

c. In order to ensure that the split limb does not lose its bearing capacity before the column's failure, the slenderness ratio of the split limb should not be greater than 40 and should not be greater than 0.5 time the maximum slenderness ratio of the column.

## 5.4.3 Diaphragm of columns

The laced column's cross section is a hollow rectangle in the middle, and its torsional rigidity is poor. In order to improve the torsional rigidity of the laced column and ensure the shape of the column section in the process of transportation and installation, the horizontal diaphragms should be set at intervals. In addition, large solid web columns (I-shaped or box-shaped) should also be provided with diaphragms. The spacing of the diaphragms shall not be greater than nine times larger column width or 8 m, and both ends of each transport unit shall be provided with diaphragms.

When a certain part of the column is subjected to a large horizontal concentrated force, transverse diaphragms should also be set to avoid local bending of the column limbs. Diaphragms may be made of steel plates (Figures 5.23a, 5.23c, and 5.23d) or cross angles (Figure 5.23b).

The diaphragm of the solid web column with the I-shaped cross-section can only be made of steel plate (Figure 5.23c). The difference between it and the transverse stiffener lies in the same width as the flange, while the transverse stiffener is usually narrower. Therefore, if one or both sides of the solid web column of the box-shaped section cannot be welded in advance, the two or three sides can be welded first, and then the other sides can be welded by ESR drilling in the column wall after assembly (Figure 5.23d).

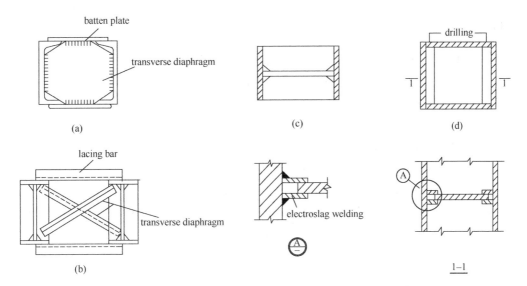

Figure 5.23  Diaphragm of columns

## 5.5  Working Examples

**[Example 5-1]**

Figure 5.24 shows a double-angle tension member of a roof truss, the cross-section is 2L100 × 10, the thickness of the filling plate is 10 mm, staggered ordinary bolt holes on the angles, and the hole diameter is $d_0 = 20$ mm. Try to calculate the maximum tensile force that this member can bear and the maximum effective length that can be reached. The steel is grade Q355B.

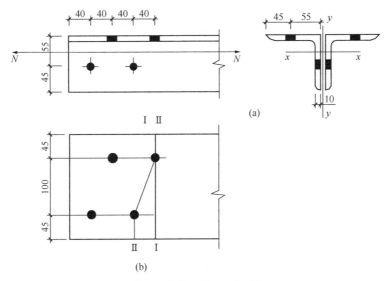

Figure 5.24  Example 5-1

145

## Chapter 5  AXIALLY LOADED MEMBERS

**Solution:**

Check the attached Attached Table F. 4 to know: Q355B- 2L100 × 10
$A = 38.52$ cm$^2$; $i_x = 3.05$ cm; $i_y = 3.05$ cm;
$f = 305$ N/mm$^2$; $f_y = 470$ N/mm$^2$.

(1) Calculation of bearing capacity

① Calculation of the yield-bearing capacity of gross cross-section

$N = Af = 38.52 \times 10^2 \times 305 = 1,174,860$ N $= 1,175$ kN

② Calculation of net section fracture bearing capacity

The thickness of the angle steel is 10 mm. Before determining the dangerous section, it should be unfolded in the middle, as shown in Figure 5.24b.

The area of the orthogonal net section ( I - I ) is:

$$A_{nI} = 38.52 \times 10^2 - 2 \times 10 \times 20 = 3,452 \text{ mm}^2$$

The area of the tooth-shaped net section ( II - II ) is:

$$A_{nII} = 2 \times (45 + \sqrt{100^2 + 40^2} + 45 - 2 \times 20) \times 10 = 3,160 \text{ mm}^2 < A_{nI}$$

The dangerous section is the II - II section. According to Equation (5-2), the net section fracture bearing capacity can be obtained as:

$$N = 0.7 A_{nII} f_n = 0.7 \times 3,160 \times 470 = 1,039,640 \text{ N} \approx 1,040 \text{ kN}$$

Based on the above, the bearing capacity of this tie rod is controlled by the net section fracture bearing capacity, and the maximum tensile force that can be withstood is 1,040 kN.

(2) Maximum calculation length calculation

Table 5.1 shows that the allowable slenderness ratio of the tension member is 350. Therefore, according to Equations (5-14) and (5-15), we can get:

$$\left. \begin{array}{l} l_{0x} = i_x \cdot [\lambda] = 3.05 \times 350 = 1,067.5 \text{ cm} \\ l_{0y} = i_y \cdot [\lambda] = 4.52 \times 350 = 1,582 \text{ cm} \end{array} \right\} \quad l_{0x} < l_{0y}$$

Therefore, the maximum allowable effective length is 1,067.5 cm.

**[Example 5-2]**

Figure 5.25(a) shows a pipe support system, the design value of design compressive force of the pillar is $N=1,600$ kN, the two ends of the column are pin-ends, the steel is grade Q355B, and the cross-section is not weakened. Try to design the cross-section of this pillar: ① using hot-rolled H-shapes; ② using welded I-section, the flange plate is flame-cut.

**Solution:**

Check the Attached Table B. 1 to know: Q355B
$f_1 = 305$ N/mm$^2$ ($t \leqslant 16$ mm);

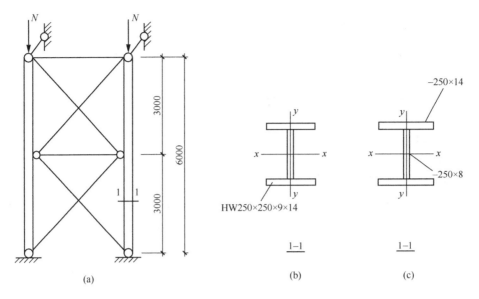

Figure 5.25  Example 5-2

$f_2 = 295 \text{ N/mm}^2 (16 \text{ mm} < t \leqslant 40 \text{ mm})$;

$f_y = 470 \text{ N/mm}^2$;

$\varepsilon_k = \sqrt{235/355} = 0.814$.

From Figure 5.25(a), the effective lengths of the pillars in the two directions are not equal, so the cross-sectional orientation shown in Figure 5.25(b) is taken, and the strong axis ($x$-axis) is perpendicular to the page, and the weak axis ($y$-axis) is parallel to the page. In this way, the column may have a better economy. We get:

$l_{0x} = 600$ cm;

$l_{0y} = 300$ cm.

(1) Hot rolled H-shape pillar

① Trial cross-section

Since the hot-rolled H-shapes can be selected in a wide flange (HW), the cross-sectional width is larger. Therefore, the assumed value of the slenderness ratio can be appropriately reduced and assume $\lambda = 60$. For HW-shapes, because $b/h > 0.8$, it belongs to the type a section on the $x$-axis and belongs to the type b section on the $y$-axis.

From $\lambda/\varepsilon_k = 60 \div 0.814 = 73.7$ and Attached Table D.1 and D.2, get:

$\left. \begin{array}{l} \varphi_x = 0.820 \\ \varphi_y = 0.724 \end{array} \right\} \Rightarrow \varphi_{\min} = 0.724$

The required cross-sectional geometric characteristics are:

$A = \dfrac{N}{\varphi_{\min} f} = \dfrac{1,600 \times 10^3}{0.728 \times 305 \times 10^2} = 72.1 \text{ cm}^2$;

$i_x = \dfrac{l_{0x}}{\lambda} = \dfrac{600}{60} = 10 \text{ cm}$;

# Chapter 5  AXIALLY LOADED MEMBERS

$$i_y = \frac{l_{oy}}{\lambda} = \frac{300}{60} = 5 \text{ cm}.$$

The cross-section of the column selected in the Attached Table F. 2 is HW $250 \times 250 \times 9 \times 14$ (Figure 5.25b).

The cross-sectional geometric characteristics are:

$A = 91.43 \text{ cm}^2$;

$i_x = 10.81 \text{ cm}$;

$i_y = 6.32 \text{ cm}$.

② Cross-section check

Since the cross section has no perforation weakening, the strength may not be checked. Because the hot-rolled section steel does not need to be checked for local stability, only the overall stability and stiffness are checked.

$$\lambda_x = \frac{l_{0x}}{i_x} = \frac{600}{10.81} = 55.5 < [\lambda] = 150; \qquad \lambda_x/\varepsilon_k = 55.5/0.814 = 68.2 \Rightarrow \varphi_x = 0.848;$$

$$\lambda_y = \frac{l_{0y}}{i_y} = \frac{300}{6.32} = 47.5 < [\lambda] = 150. \qquad \lambda_y/\varepsilon_k = 47.5/0.814 = 58.6 \Rightarrow \varphi_y = 0.816.$$

$$\frac{N}{\varphi_{min} A f} = \frac{1,600 \times 10^3}{0.816 \times 91.43 \times 10^2 \times 305} = 0.703 < 1.0$$

Based on the above calculation results, the overall stability and stiffness meet the requirements.

(2) Welded I-section pillar

① Trial section

Refer to the H-section steel section and select the section as shown in Figure 5.25(c). Flange: $2 - 250 \times 14$, web: $1 - 250 \times 8$, and the geometric characteristics of the section are:

$A = 2 \times 250 \times 14 + 250 \times 8 = 9,000 \text{ mm}^2$;

$I_x = \frac{1}{12} \times (250 \times 278^3 - 242 \times 250^3) = 13,250 \times 10^4 \text{ mm}^4$;

$I_y = \frac{1}{12} \times (2 \times 14 \times 250^3 - 250 \times 8^3) = 3,646.9 \times 10^4 \text{ mm}^4$;

$i_x = \sqrt{\frac{I_x}{A}} = \sqrt{\frac{13,250 \times 10^4}{9,000}} = 121.3 \text{ mm} = 12.13 \text{ cm}$;

$i_y = \sqrt{\frac{I_y}{A}} = \sqrt{\frac{3,646.9 \times 10^4}{9,000}} = 63.6 \text{ mm} = 6.36 \text{ cm}$.

② Checking calculation of overall stability bearing capacity and stiffness

$$\lambda_x = \frac{l_{0x}}{i_x} = \frac{6,000}{121.3} = 49.5 < [\lambda] = 150;$$

$$\lambda_y = \frac{l_{0y}}{i_y} = \frac{3,000}{63.6} = 47.2 < [\lambda] = 150.$$

Check Table 5.3, welded H-section steel with flame-trimmed flanges, and when it loses stability around the $x$-axis and $y$-axis, it belongs to the type b section.

Because $\lambda_x > \lambda_y$, we got $\varphi_x = 0.803$ from $\lambda_x/\varepsilon_k = 49.5 \div 0.814 = 60.8$.

$$\frac{N}{\varphi_x A f} = \frac{1,600 \times 10^3}{0.803 \times 9,000 \times 305} = 0.726 < 1.0$$

The overall stability and stiffness meet the requirements.

There is no perforation weakening in the section, so it is unnecessary to check the strength.

③ Local stability check

Width-to-thickness of the outstanding flange:

$$\frac{b}{t} = \frac{(250-8)/2}{14} = 8.6 < (10+0.1\lambda)\varepsilon_k = (10+0.1\times 49.5)\times 0.814 = 12.2$$

Depth-to-thickness of the web:

$$\frac{h_0}{t_w} = \frac{250}{8} = 31.3 < (25+0.5\lambda)\varepsilon_k = (25+0.5\times 49.5)\times 0.814 = 40.5$$

④ Construction requirements

Because the web depth-to-thickness ratio is less than 80 $\varepsilon_k$, it is unnecessary to provide transverse stiffeners. The minimum welding leg size of the connecting weld between the flange and the web is 6 mm, so we use 6 mm.

Comparing the hot-rolled H-shape and the welded I-section, the slenderness ratio in the two directions is very close, equal stability is achieved, and the materials used are more economical. However, the I-section's welding process is increased, and the hot-rolled H-shape should be preferred when the member is designed to be compression column.

[**Example 5-3**]

One axially loaded compression column, the column height is 6 m, the two ends are hinged, and the axially compressive load is 1,000 kN (design value), the steel is grade Q355. Holes do not weaken the cross-section. Try to design a lacing bar column and a batten plate column, respectively.

**Solution:**

Know from the question:

$l_{0x} = l_{0y} = 6,000$ mm.

Check the Attached Table B.1 to know: Q355B

$f_1 = 305$ N/mm² ($t \leqslant 16$ mm);

$f_2 = 295$ N/mm² (16 mm $< t \leqslant 40$ mm);

$f_y = 470$ N/mm²;

$\varepsilon_k = \sqrt{235/355} = 0.814$.

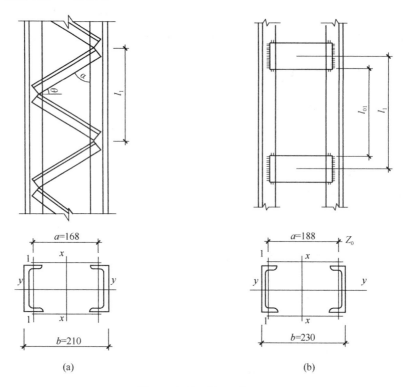

Figure 5.26 Example 5-3

(1) Design lacing bar column

① Select the cross-section of the column according to the overall stability of the real axis (y-axis)

Assume $\lambda = 70$, the column belongs to the type b section on the y-axis. Then, from $\lambda/\varepsilon_k = 70 \div 0.814 = 86.0$ and Attached Table D.2, get $\varphi_x = 0.648$.

The required cross-sectional area is:

$$A = \frac{N}{\varphi_y f} = \frac{1,000 \times 10^3}{0.648 \times 305} = 5,060 \text{ mm}^2$$

We choose Channel 2 [22a, as shown in Figure 5.26(a).

The cross-sectional geometric characteristics are: $A = 63.68 \text{ cm}^2$; $i_y = 8.67 \text{ cm}$.

Check the overall stability:

$$\lambda_y = \frac{l_{0y}}{i_y} = \frac{6,000}{8.67 \times 10} = 69.2 < [\lambda] = 150$$

From $\lambda/\varepsilon_k = 69.2 \div 0.814 = 85.0$ and Attached Table D.2, get $\varphi_x = 0.654$.

$$\frac{N}{\varphi_y A f} = \frac{1,000 \times 10^3}{0.654 \times 63.38 \times 10^2 \times 305} = 0.787 < 1.0$$

The result meets the requirements of the design standard.

② Determine the column width $b$

Initially, select the cross-section of the lacing bar ∟ 45 × 4.
$A'_1 = 3.49 \text{ cm}^2$ ; $i_1 = 0.89 \text{ cm}$.

In order to obtain equal stability, the slenderness ratio of the column around the virtual axis ($x$-axis) should satisfy:

$A_1 = 2A'_1 = 6.98 \text{ cm}^2$ ;

$\lambda_x = \sqrt{\lambda_y^2 - 27 \dfrac{A}{A_1}} = \sqrt{69.2^2 - 27 \times \dfrac{63.6}{6.98}} = 67.4$ ;

$i_x = \dfrac{l_{ox}}{\lambda_x} = \dfrac{600}{67.4} = 8.9 \text{ cm}$.

Using the section form shown in Figure 5.26(a), as can be seen from Table 5.6, the radius of gyration of the section around the imaginary axis is approximate:

$i_x \approx 0.44b \rightarrow b \approx \dfrac{i_x}{0.44} = \dfrac{8.9}{0.44} = 20.23 \text{ cm}$, take $b = 21 \text{ cm}$.

The cross-sectional geometric characteristics of a single Channel [ 22a are:

$A = 31.8 \text{ cm}^2$ ;

$Z_0 = 2.1 \text{ cm}$ ;

$I_1 = 157.8 \text{ cm}^4$ ;

$i_1 = 2.23 \text{ cm}$.

Data of the whole cross-section about the virtual axis ($x$-axis):

$I_x = 2 \times \left[ 157.8 + 31.8 \times \left( \dfrac{21.0 - 2.1 \times 2}{2} \right)^2 \right] = 4,803.2 \text{ cm}^4$ ;

$i_x = \sqrt{\dfrac{48,032}{63.6}} = 8.69 \text{ cm}$ ;

$\lambda_x = \dfrac{600}{8.69} = 69.0$ ;

$\lambda_{0x} = \sqrt{\lambda_x^2 + 27 \dfrac{A}{A_1}} = \sqrt{69.0^2 + 27 \times \dfrac{63.6}{6.98}} = 70.8 < [\lambda] = 150$.

From $\lambda_{0x}/\varepsilon_k = 70.8 \div 0.814 = 87.0$ and Attached Table D.2, get $\varphi_x = 0.641$.

$\dfrac{N}{\varphi_x A f} = \dfrac{1,000 \times 10^3}{0.641 \times 63.38 \times 10^2 \times 305} = 0.803 < 1.0$.

The overall stability of the virtual axis meets the requirements.

③ Check lacing bars

As shown in Figure 5.26a, take $\theta = 45°$.

The shear experienced by the lacing bar is:

$V = \dfrac{Af}{85\varepsilon_k} = \dfrac{63.68 \times 10^2 \times 305}{85 \times 0.814} = 28,071 \text{ N}$

The axial force of an oblique lacing bar is:

$N_1 = \dfrac{V/2}{\cos\theta} = \dfrac{28,071/2}{\cos 45°} = 19,849 \text{ N}$ ;

$a = b - 2Z_0 = 210 - 2 \times 21 = 168$ mm.

The length of the lacing bar:

$$l_0 = \frac{a}{\cos 45°} = \frac{168}{\sqrt{2}/2} = 238 \text{ mm}$$

Slenderness ratio:

$$\lambda = \frac{0.9 l_0}{i_1} = \frac{0.9 \times 238}{0.89 \times 10} = 24 < [\lambda] = 150$$

From $\lambda/\varepsilon_k = 24 \div 0.814 = 29.5$ and Attached Table D.2, get $\varphi_x = 0.938$.

If the single equal angle is connected to the column on one side, the strength should be multiplied by the reduction factor:

$\eta = 0.6 + 0.00152\lambda = 0.64$ ;

$$\frac{N_1}{\eta \varphi_x A_1' f} = \frac{19,849}{0.64 \times 0.9375 \times 3.49 \times 10^2 \times 305} = 0.313 < 1.0.$$

Because ∟ 45×4 is the minimum angle recommended for structural design, the selection of lacing bars meets the requirements.

The fillet weld between the lacing bar and the column limb adopts the low-hydrogen welding method, taking $h_f = 4$ mm.

Limb back:

$$l_{w1} \geqslant \frac{\frac{2}{3} N_1}{0.7 h_f \eta f_f^w} = \frac{\frac{2}{3} \times 19,849}{0.7 \times 4 \times 0.85 \times 200} = 28 \text{ mm}$$

Limb tip:

$$l_{w2} \geqslant \frac{\frac{1}{3} N_1}{0.7 h_f \eta f_f^w} = \frac{\frac{1}{3} \times 19,849}{0.7 \times 4 \times 0.85 \times 200} = 14 \text{ mm}$$

Considering the structural requirements, the actual weld length of the limb back and limb tip are both 50 mm.

④ Stability of the single limb

Length of internodes:

$$l_1 = 2a \times \tan\alpha = 2 \times (210 - 2 \times 21) \times \tan 45° = 336 \text{ mm}$$

The slenderness ratio of a single limb of the column in the plane (around 1-axis):

$i_1 = 2.23$ cm

$$\lambda_1 = \frac{l_1}{i_1} = \frac{336}{22.3} = 15 < 0.7\{\lambda_{0x}, \lambda_{0y}\}_{\max} = 0.7 \times 70.9 = 49.6$$

The stability of a single limb can be guaranteed.

(2) Design batten plate column

① Select the cross-section of the column according to the overall stability of the real axis (y-axis)

Still choose 2 [22a.

② Determine the column width $b$

Assume $\lambda_1 = 35$ (approximately equal to $0.5\lambda_y$).

$\lambda_x = \sqrt{\lambda_y^2 - \lambda_1^2} = \sqrt{69.2^2 - 35^2} = 59.7$;

$i_x = \dfrac{l_{0x}}{\lambda_x} = \dfrac{6,000}{59.7} = 100.5$ mm.

Using the cross-sectional form of Figure 5.26(b), check Table 5.6 to get:

$b \approx \dfrac{i_x}{0.44} = \dfrac{100.5}{0.44} = 228$ mm, taking $b = 230$ mm.

The cross-sectional geometric characteristics of a single Channel [22a are:

$A = 31.8$ cm$^2$;

$Z_0 = 2.1$ cm;

$I_1 = 157.8$ cm$^4$;

$i_1 = 2.23$ cm.

Data of the whole cross-section on the virtual axis ($x$-axis):

$I_x = 2 \times \left[ 157.8 + 3.84 \times \left( \dfrac{18.8}{2} \right)^2 \right] = 5,935$ cm$^4$;

$i_x = \sqrt{\dfrac{5,935 \times 10^4}{2 \times 31.84 \times 10^2}} = 96.5$ mm;

$\lambda_x = \dfrac{6,000}{96.5} = 62.2$;

$\lambda_{0x} = \sqrt{\lambda_x^2 + \lambda_1^2} = \sqrt{62.2^2 + 35^2} = 71.4 < [\lambda] = 150$.

From $\lambda_{0x}/\varepsilon_k = 71.4 \div 0.814 = 87.7$ and Attached Table D.2, get $\varphi_x = 0.636$.

$\dfrac{N}{\varphi_x A f} = \dfrac{1,000 \times 10^3}{0.636 \times 63.38 \times 10^2 \times 305} = 0.810 < 1.0$.

The overall stability of the virtual axis meets the requirements.

① Check batten plates

$l_{01} = \lambda_1 i_1 = 35 \times 22.3 = 781$ mm

Select batten plate: $-180 \times 8$.

$l_1 = 781 + 180 = 961$ mm, taking $l_1 = 960$ mm.

Linear stiffness of the split limb:

$K_1 = \dfrac{I_1}{l_1} = \dfrac{157.8 \times 10^4}{960} = 1.64 \times 10^3$ mm$^3$

The sum of the linear stiffness of the batten plates on both sides:

$K_b = \dfrac{\sum I_b}{a} = \dfrac{2 \times \dfrac{1}{12} \times (8 \times 180^3)}{188} = 41.36 \times 10^3 > 6K_1 = 9.84 \times 10^3$ mm$^3$

The transverse shear experienced by the batten plate is:

$V = \dfrac{Af}{85\varepsilon_k} = \dfrac{63.68 \times 10^2 \times 305}{85 \times 0.814} = 28,071$ N

$$V_1 = \frac{V}{2} = 14,035.5 \text{ N}$$

The internal force at the connection between the batten plate and the split limb is:

$$T = \frac{V_1 l_1}{a} = \frac{14,035.5 \times 960}{188} = 71,670.6 \text{ N}$$

$$M = T \cdot \frac{a}{2} = \frac{V_1 l_1}{2} = \frac{14,035.5 \times 960}{2} = 6.74 \times 10^6 \text{ N} \cdot \text{mm}$$

The batten plate is connected with the column limbs by fillet welds, and the fillet welds bear the combined action of shear and bending moments. Since the design strength of the fillet weld is less than that of the base material, only the above $M$ and $T$ are required to check the connection weld between the batten plate and the limb.

Take fillet weld leg, $h_f = 8$ mm, without considering the length of the corner part (Figure 5.22b), and use $l_w = 180$ mm, the shear stress generated by the shear $T$ (parallel to the length of the weld):

$$\tau_f = \frac{71,670.6}{0.7 \times 8 \times 180} = 71.1 \text{ N/mm}^2$$

Stress generated by bending moment $M$ (perpendicular to the length of the weld):

$$\sigma_f = \frac{6 \times 6.74 \times 10^6}{0.7 \times 8 \times 180^2} = 222.9 \text{ N/mm}^2$$

$$\sqrt{\left(\frac{\sigma_f}{1.22}\right)^2 + \tau_f^2} = \sqrt{\left(\frac{222.9}{1.22}\right)^2 + 71.1^2} = 196.1 \text{ N/mm}^2 < f_f^w = 200 \text{ N/mm}^2$$

It is safe.

② Stability of the single limb

$\lambda_1 = l_{01}/i_1 = 780/22.3 = 35.0 < 40$ and $\lambda_1 < 0.5 \{\lambda_{0x}, \lambda_y\}_{\max} = 0.5 \times 71.4 = 35.7$

Then, the stability of a single limb can be guaranteed.

(3) Diaphragms

The diaphragm is made of steel plates (Figure 5.23a), and the spacing should be less than nine times the column width ($9 \times 23 = 207$ cm). The schematic diagram of the batten plate column is shown in Figure 5.27.

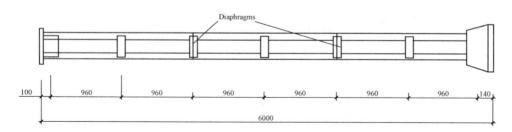

Figure 5.27 Schematic diagram of batten plate column

## 5.6 Chapter Summary

The axially loaded member is one of the basic members of steel structures. Therefore, the axially loaded member's force performance and failure characteristics should be mastered, and the focus should be on the stability design of the axially loaded compression member.

(1) The stiffness performance of axially loaded members is generally measured by the slenderness ratio. A too-large slenderness ratio may affect normal use for axially loaded compression members and reduce their overall stability and bearing capacity. Therefore, it is necessary to limit the slenderness ratio. When choosing the cross-section of the axially loaded compression members, priority should be given to the thinner and wider development section of the plate to ensure that a higher radius of gyration. A larger radius of gyration can be obtained with a small slenderness ratio under the same cross-sectional area.

(2) The strength of steel is relatively high, and the cross-section selected according to the strength conditions is generally small. Therefore, the axially loaded members are generally slender. For the tension member, the slender has little effect on the force, but for the compression member, it may be due to its low bearing capacity of the overall stability, and global buckling occurs when the section stress has not reached its strength design value.

(3) The ideal axial compression column does not exist in actual engineering. The actual axial compression column must have the initial bending, the initial eccentricity of the load, and the residual stress, which inevitably reduces the overall stability bearing capacity of the column. Therefore, the column strength curve of the axially loaded compression member in the actual project is obtained by using the equivalent initial bending rate considering the comprehensive influence of various initial defects. the design standard GB 50017 uses four column strength curves to calculate axially loaded compression members' overall stability factor according to different cross-sectional forms.

(4) In order to improve the overall stability and bearing capacity, the axial compression member is generally designed with a thinner plate and a relatively wide-open section. When the width-to-thickness ratio or depth-to-thickness ratio of these cross-section plates is particularly large, local buckling will occur under pressure stress. The critical load of flanges and webs of axially loaded compression members can be solved by elastic stability theory. In order to facilitate engineering applications, the design standard GB 50017 simplifies the local stability check of compression plates to a method that limits their width-to-thickness ratio or depth-to-thickness ratio. However, the local instability of the web of a general compression member does not mean that the entire member loses its load-bearing capacity. Under certain conditions, the post-buckling strength of the web can also be used for design.

(5) The axial compression columns are divided into solid web columns and laced columns. Under the same cross-sectional area condition, the laced column can be adjusted by adjusting the distance of multiple split limbs to obtain a larger sectional moment of inertia. However, the shear deformation of the laced column is relatively large when the overall stability is lost. Therefore, in order to consider the influence of shear deformation on the overall stability and bearing capacity of the column, the overall stability of the laced column needs to use the equivalent slenderness ratio.

## Questions

5.1 What is the axially loaded member? What are the similarities and differences between the axially loaded tension member and the axially loaded compression member in the ultimate bearing capacity and the normal service limit state?

5.2 What is the difference in the calculation method for axially loaded tension members and axially loaded compression members when there are holes in the section? Why?

5.3 Why can the transmitted force of the hole be considered when the axial tension member with high-strength bolts is designed according to the friction-type connection? Can the axial tension member of the bearing-type connection consider the forward force of the hole? Why?

5.4 What is the measure of the stiffness of the axially loaded member? Why should the slenderness ratio of the axial tension member be restricted?

5.5 What are the overall instability forms of axially compression steel members?

5.6 What are the three forms of overall buckling of an ideal pin-ended axial compression column? What are the differences and characteristics?

5.7 Why does the residual stress affect the stable bearing capacity of the axially compressed column but not the strength?

5.8 What are the initial defects that affect the overall stable bearing capacity of axial compression members? How to think about it in theory?

5.9 What is the column strength curve of the axially loaded compression member?

5.10 What are the factors that affect the value of the overall stability factor of the axially loaded compression member?

5.11 What is the difference between the overall stability and the local stability of the axially loaded compression member? Why use equal stability conditions to design?

5.12 Why should the width-to-thickness ratio or the depth-thickness ratio be restricted for the component plates of the axially loaded compression member?

5.13 What is the post-buckling strength of the web? Why can the post-buckling strength be used in some cases?

5.14 How does the cross-sectional shear of a laced column be generated when the overall instability of the column? How to consider it in the design?

5.15 Why is the equivalent slenderness ratio used to calculate the overall stability of the laced column to the virtual axis? What factors are mainly considered in deriving the equivalent slenderness ratio calculation formula in the design standard GB 50017?

5.16 Why do laced columns and solid web columns need to be provided with transverse diaphragms in some cases? What is the function of the diaphragm?

## Exercises

5.1 Check the net section strength of steel plates connected by friction-type high-strength bolts, as shown in Figure 5.28. The bolt diameter is 20 mm, the hole diameter is 22 mm, the steel is Q355B, and the design value of bearing axial tension is $N=700$ kN.

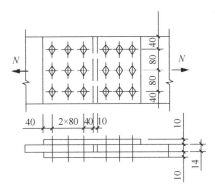

Figure 5.28 A friction-type high-strength bolt connection

5.2 As shown in Figure 5.29, check the strength and slenderness ratio of the horizontally placed axially loaded tension member composed of $2 \llcorner 63 \times 5$. The design value of the axial tension is 300 kN which only bears the action of static force, and the effective length is 4 m. The end has a row of holes with a diameter of 20 mm, and the steel is Q355B.

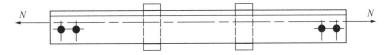

Figure 5.29 A axially loaded tension member

5.3 A horizontally placed axially loaded tension member, hinged at both ends, made of Q355B steel, 9 m long, and a T-shaped section composed of $2 \llcorner 90 \times 6$ with the tip of the limb downward. Can it withstand the tension force with a design value of 800 kN?

5.4 The two cross-sections (flame cutting edges) shown in Figures 5.30a and 5.30b have the same cross-sectional area, and the steels are all Q355C. When used as a hinged axially loaded compression column with a length of 9 m, can it safely withstand the design load of 3,500 kN?

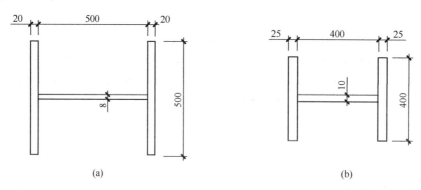

Figure 5.30  Axially loaded compression columns

5.5  One axially loaded compression column, the column height is 7.5 m, the two ends are hinged, and the axially compressive load is 1,500 kN (design value), the steel is grade Q355B. Holes do not weaken the cross-section. Try to design a lacing bar column and a batten plate column with a two-Channel limb, respectively.

# Chapter 6  FLEXURAL MEMBER (BEAM)

**GUIDE TO THIS CHAPTER:**

➤ *Knowledge points*

*The type and application of the flexural member (beam); the strength of the beam (bending strength, shear strength, local compressive strength, reduced strength); the stiffness (deflection) of the beam; the global stability of the beam; the local stability of the beam and the design of web stiffeners, section design of steel beams (hot-rolled steel beams and welded composite beams).*

➤ *Emphasis*

*The strength and stiffness of the beam are calculated. In addition, the global stability and local stability of the beam are calculated.*

➤ *Nodus*

*The local stability of the beam and the design of the web stiffener.*

## 6.1 Introduction

The flexural members that sometimes refer to the beams are among the most common members that one can find in structures. They are structural members that carry loads applied at right angles to the member's longitudinal axis. If the load is applied in one main plane, the beam is called the unidirectional bending member. If the load is applied in two main planes simultaneously, the beam is called a bidirectional bending member. The bending strength and shear strength of the steel structure must be ensured. In addition, the member's global stability and the local stability requirements of the compression flange plate must be ensured. For members that do not use the strength of the web after buckling, the local stability requirements of the web must be met. These belong to the first limit state problem of member design, namely the limit bearing capacity problem. The bending member must have sufficient rigidity to ensure that the member's deformation does not affect the normal use requirements. It is the second limit state problem of the member design: the service limit state problem. This chapter mainly introduces the basic concepts and related calculation methods of the strength, stiffness, global stability, local stability, and web buckling strength of solid web flexural members. Some applications of steel beams are shown in Figure 6.1.

When visualizing a beam for analysis or design purposes, it is convenient to think of

# Chapter 6  FLEXURAL MEMBER (BEAM)

Figure 6.1  Applications of beams
(a) the application in a steel building; (b) the application in a bridge

the member in some idealized form. This idealized form represents as closely as possible the actual structural member, but it has the advantage that it can be dealt with mathematically. For example, in Figure 6.2a, the beam is shown with simple supports. These supports, a pin (knife-edge or hinge) on the left and a roller on the right, create conditions that are easily treated mathematically when it becomes necessary to find beam reactions, shear, moments, and deflections. In buildings, each support is generally capable of furnishing vertical and horizontal reactions. The beam, however, is still considered to be simply supported since the requirement of a simple support is to permit freedom of rotation. For example, in Figure 6.2b, the cantilever beam has a fixed support on the left side. This type of support provides vertical and horizontal reactions as well as resistance (or a reaction) to rotation. Although the idealized conditions generally will not exist in the actual structure, they are close enough to allow for an appropriate analysis or design.

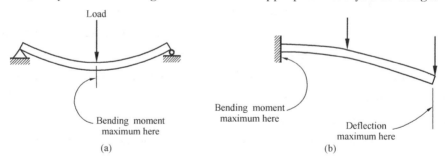

Figure 6.2  Beam types
(a) simply supported beam; (b) cantilever beam

Beams are sometimes called by other names that are indicative of some specialized functions (s):

Girder: a major, or deep, beam that often provides support for other beams;

Stringer: a main longitudinal beam, usually in bridge floors;

Floor beam: a transverse beam in bridge floors;

Joist: a light beam that supports a floor;

Lintel: a beam spanning an opening (a door or a window), usually in masonry construction;

Spandrel: a beam on the outside perimeter of a building that supports, among other loads, the exterior wall;

Purlin: a beam that supports a roof and frames between or over supports, such as roof trusses and rigid frames;

Girt: generally, a light beam that supports only the lightweight exterior sides of a building (typical in pre-engineered metal buildings).

## 6.2 The Strength and Stiffness of the Beam

In the structure, the beam's main function is to bear the lateral load from the floor and other members, and it also bears the horizontal force in the frame structure. These loads may generate bending moment and shear in the bending member. However, if the shear does not act on the member section's shear center, the member will also generate torque and bending deformation. This section describes the bending member's strength and stiffness under bending moment and shear. The torsion problem is introduced in Section 6.3.

### 6.2.1 Bending strength

It is known from material mechanics: in the elastic stage, when the member section acts on the bending moment around the $x$-axis of the principal centroid axis (Figure 6.3), the maximum normal stress at the edge of the member section is:

$$\sigma = \frac{M_x}{W_x}, \text{ or } \sigma = \frac{M_x}{W_{nx}} \tag{6-1}$$

where  $M_x$ = the maximum applied moment about $x$-axis;

$W_x$ = the elastic section modulus of the cross-section about the neutral axis, $x$-axis;

$W_{nx}$ = the net section modulus of the cross-section about the neutral axis, $x$-axis.

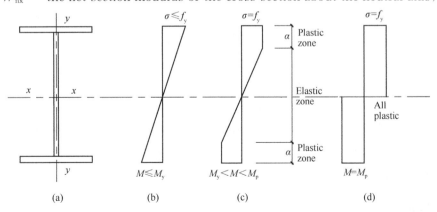

Figure 6.3  The normal stress distribution on the beam section at each load stage

When $\sigma$ reaches the steel yield point $f_y$, the member section is in the ultimate elastic state (Figure 6.3b), and the bending moment acting on it is the yield bending moment $M_{ex} = W_x f_y$. With the bending moment increasing, the member section begins to develop plasticity inward. Enter the elastoplastic state, and the stress state is shown in Figure 6.3(c). As shown in Figure 6.3(d), when the entire member section is completely plastic, the section reaches the maximum flexural bearing capacity called the plastic bending moment $M_p = W_p f_y$. At this time, the section forms a plastic hinge and reaches the ultimate plastic state. $W_p$ is the section plastic modulus of the section to the $x$-axis. Usually, $\gamma_{sp} = M_p/M_{ex}$ is defined as the sectional plastic coefficient around the $x$-axis. In the design of steel beams, if the plastic hinge is formed according to the section, although steel can be saved, the deformation is relatively large, which sometimes affects normal use. Therefore, GB 50017 stipulates that plasticity can be used to a limited extent by restricting the plastic development zone. Generally, the $a$ is between $h/8$ and $h/4$, which is the plastic zone length in Figure 6.2(c), and the plastic development coefficient is determined according to this working stage. Table 6.1 shows the plastic development coefficients of common sections. For example, for a biaxial symmetric I-shaped section $\gamma_x = 1.05$, when bending around the $y$-axis, $\gamma_y = 1.2$, for box section, $\gamma_x = \gamma_y = 1.05$. At this time, the bending strength of the beam should meet:

$$\frac{M_x}{\gamma_x W_{nx}} \leqslant f_y/\gamma_n = f \tag{6-2}$$

where  $\gamma_x =$ the plastic development coefficient about $x$-axis;

$\gamma_n =$ the partial coefficient of material resistance, which is 1.087 for Q235 steel and 1.111 for Q345, Q390, and Q420 steels.

In the same way, for a beam subjected to bidirectional bending, its strength should satisfy:

$$\frac{M_x}{\gamma_x W_{nx}} + \frac{M_y}{\gamma_y W_{ny}} \leqslant f_y/\gamma_n = f \tag{6-3}$$

where  $M_y =$ the maximum applied moment about $y$-axis;

$W_{ny} =$ the net section modulus of the cross-section about the neutral axis, $y$-axis;

$\gamma_y =$ the plastic development coefficient about $y$-axis.

For beams that need to calculate fatigue, the development of plasticity should not be considered. At this time, $\gamma_x$ and $\gamma_y$ take 1.0 in Equations (6-2) and (6-3).

In order to ensure that the compression flange of the beam will not locally buckle before the strength failure occurs when the ratio of the free outer elongation $b_1$ of the beam flange to the thickness $t$ is greater than $13\sqrt{235/f_y}$ but not more than $15\sqrt{235/f_y}$, $\gamma_x = \gamma_y = 1.0$. $f_y$ is the yield strength of the steel used in the beam.

## 6.2.2 Shear strength

(1) Shear center

A point can be found in the plane where the section of the member is located. When the shear force generated by the external force acts on this point, the member only produces linear displacement without twisting. This point is called the shear center of the member.

**The corresponding plastic development coefficient of common sections**      Table 6.1

| Item number | Section form | $\gamma_x$ | $\gamma_y$ |
|---|---|---|---|
| 1 | | | 1.2 |
| 2 | | 1.05 | 1.05 |
| 3 | | $\gamma_{x1}=1.05$ $\gamma_{x2}=1.05$ | 1.2 |
| 4 | | | 1.05 |
| 5 | | 1.2 | 1.2 |
| 6 | | 1.12 | 1.15 |
| 7 | | 1.0 | 1.05 |
| 8 | | | 1.0 |

According to the shear flow distribution law of the section, the shear center must be located on the axis of symmetry. For a section composed of several long and narrow rectangular sections with their center lines intersecting at one point, the shear center is at this point. The course of *Materials Mechanics* has given a clear method for the shear center position's determination, which will not be repeated in this book.

Figure 6.4 shows the location of the shear center of common sections. When the shear force generated by the external load acts on other positions other than the shear center, it can be moved to the shear center. At this time, not only is the shear applied to the shear center, but also the torque generated by the translational shear is applied. Torque causes the entire section to rotate around the shear center. The stress generated by the torque will be introduced in Section 6.3. This section describes the calculation of shear stress caused by bending.

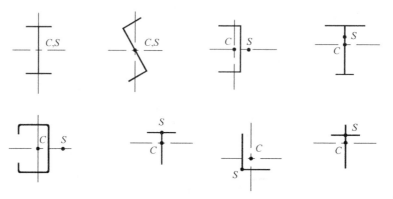

Figure 6.4 Schematic diagram of the shear center of the open section

(2) Calculation of shear stress caused by bending moment

According to the knowledge of material mechanics, the shear stress on the section of a solid web beam (Figure 6.5) is:

$$\tau = \frac{VS}{It} \tag{6-4}$$

where  $\tau$ = shear stress on a horizontal plane located with reference to the neutral axis;

$V$ = vertical shear force at that particular section;

$S$ = statical moment of the area between the plane under consideration and outside of the section, about the neutral axis;

$I$ = moment of inertia of the cross-section about the neutral axis;

$t$ = thickness of the section at the plane being considered.

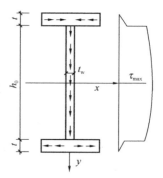

Figure 6.5 Distribution of shear stress on beam section

When the member acts on the shear force in the two main axis directions, the cross-section is shown in Figure 6.5, and the shear stress is calculated as follows:

$$\tau = \frac{V_y S_x}{I_x t} + \frac{V_x S_y}{I_y t} \tag{6-5}$$

where   $V_x$ = horizontal shear force at that particular section in the $y$-axis direction;
  $V_y$ = vertical shear force at that particular section in the $x$-axis direction;
  $S_x$ = statical moment of the area between the plane under consideration and outside of the section, about the $x$-axis;
  $S_y$ = statical moment of the area between the plane under consideration and outside of the section, about the $y$-axis;
  $I_x$ = moment of inertia of the cross-section about the $x$-axis;
  $I_y$ = moment of inertia of the cross-section about the $y$-axis;
  $t$ = thickness of the section at the plane being considered.

According to the elastic design, the maximum shear stress of the section reaches the limit state when the steel's shear yield point is reached. Therefore, the design should satisfy the following formula.

$$\tau_{max} \leqslant f_v \tag{6-6}$$

where   $f_v$ = design value of steel shear strength.

### 6.2.3 Local compressive strength

As shown in Figure 6.6, without supporting stiffener at the place where is fixed concentrated load, including the reaction force of the supporting or the concentrated moving load (such as crane wheel pressure), the beam web will bear the local compressive stress caused by the concentrated load. The local compressive stress is the largest at the junction of the beam web and the upper flange and decreases to zero at the lower flange, as shown in Figure 6.6(b).

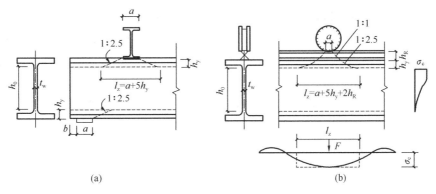

Figure 6.6   Local stress distribution at the edge of the web

In the calculation, it is assumed that the local compressive stress spreads at an angle of 45° within the height range $h_y$ (sometimes is the height of the crane track) below the load point, spreads at a ratio of 1 : 2.5 within the $h_y$ height range, and spreads to the junction of the web and the flange. Thus, the local compressive stress is not uniformly distributed along the longitudinal axis of the beam. Still, it is assumed that the local compressive stress is uniformly distributed in the range of $l_z$ to simplify the calculation. Therefore, the local compressive stress at the edge of the web is calculated as follows.

$$\sigma_c = \frac{\psi F}{t_w l_z} \leqslant f \tag{6-7}$$

where   $F$ = the concentrated load, and the dynamic coefficient should be considered for the dynamic load;

$\psi$ = the concentrated load amplification factor, for heavy-duty crane beams, $\psi$ = 1.35; for other beams, $\psi$ = 1.0; for the support reaction force, $\psi$ = 1.0;

$l_z$ = the assumed distribution length of the upper edge of the concentrated load on the calculated height of the web, calculated as follows.

Interior concentrated load (defined as applied at a distance from the end of the beam):

$$l_z = a + 5h_y + 2h_R \tag{6-8}$$

End reactions:

$$l_z = a + 2.5h_y + b \tag{6-9}$$

where   $a$ = the supporting length of the concentrated load along the longitudinal axis of the beam, taken as 50 mm;

$h_y$ = the distance from the beam's top surface to the upper edge of the calculated height of the web;

$h_R$ = the height of the crane track, for beams without tracks on the top flange, $h_R$ = 0;

$b$ = the distance from the beam end to the outer edge of the support plate; if $b$ is greater than 2.5 $h_y$, take 2.5 $h_y$;

$f$ = the design value of the compressive strength of the steel.

About the calculated height $h_0$ of the web: for rolled steel beams, it is the distance between the starting points of the two inner arcs where the webs connect with the upper and lower flanges; for welded composite beams, it is the height of the web; for riveted or high-strength bolt composite beam, it is the shortest distance between the rivets (or high-strength bolts) connecting the upper and lower flanges and the web.

When the calculation fails to meet the requirements for concentrated loads (including support reaction forces), support stiffeners should be used. For the concentrated moving load, such as crane loads, only increasing the web thickness can be used.

### 6.2.4 Reduced stress

Generally, there is shear and bending moment on the beam simultaneously, and sometimes there are concentrated local forces. When checking the beam design, strength is not only the maximum shear stress, maximum normal stress, and the local compressive stress to meet the requirements. If in some edge points of calculation hight have larger normal stress, shear stress, and local compressive stress, or quite normal stress and shear stress at the same time, such as continuous beam central support or beam flange section changing, etc. In these areas, despite the normal stress and shear stress is not the biggest, but there may be more dangerous under the effect of them at the same time. It needs to check these parts when designing a steel beam. According to the *Fourth Strength Theory*, in a complex stress state, if the reduced stress of a certain point reaches the yield strength in uniaxial tension, the point enters the plastic state. The reduced stress $\sigma_z$ at the dangerous point in the design should satisfy:

$$\sigma_z = \sqrt{\sigma^2 + \sigma_c^2 - \sigma\sigma_c + 3\tau^2} \leqslant \beta_1 f \qquad (6\text{-}10)$$

where $\quad \sigma, \tau, \sigma_c =$ normal stress, shear stress, and local compressive stress generated at the same point on the edge of the web's calculated height at the same time; $\sigma$ and $\sigma_c$ are positive for tensile stress and negative for compression. $\tau$ and $\sigma_c$ is calculated according to Equations (6-4) and (6-7), $\sigma$ is calculated as follows:

$$\sigma = \frac{M_x}{I_n} y_1 \qquad (6\text{-}11)$$

$M_x =$ the maximum applied moment about $x$-axis;

$I_n =$ the moment of inertia of net cross-section of the beam;

$y_1 =$ the distance from the calculated point to the neutral axis of the beam;

$\beta_1 =$ the strength design value increase factor. Consider the reduced stress at the web edge of a certain beam section that reaches yield, the web edge is limited to a local area, so the design strength is improved. Simultaneously, considering that the different sign stress fields will increase the plastic performance of the steel, $\beta_1$ can be larger. Therefore, when $\sigma$ and $\sigma_c$ have different signs, take $\beta_1 = 1.2$. When $\sigma$ and $\sigma_c$ have the same sign or $\sigma_c = 0$, take $\beta_1 = 1.1$.

### 6.2.5 The stiffness of the beam

The deflection measures the stiffness of the beam under the standard value of the applied loads. Insufficient beam stiffness will affect normal use or appearance. The so-called normal use refers to the equipment's normal operation, the protection of decorations and non-structural members, and people's comfort. Generally, the vibration of a beam under the influence of dynamic force can also be controlled by limiting the beam's deformation. Therefore, the stiffness of the beam can be checked as follows:

$$v \leqslant [v] \qquad (6\text{-}12)$$

where $\quad v =$ the maximum deflection in the beam caused by the standard value of the applied loads, without considering the sub-factor and dynamic coefficient of loads;

## Chapter 6  FLEXURAL MEMBER (BEAM)

$[v]$ = the allowable deflection value of the beam. It is adapted according to the design standard GB 50017, shown in Table 6.2, including $[v_T]$ and $[v_Q]$. When there is a practical experience or special requirements, it can be adjusted appropriately according to the principle of not affecting the normal use and appearance.

The deflection $v$ of the beam can be calculated according to material mechanics and structural mechanics or checked from the manual of static structural calculation. For simply supported beams that bear multiple (4 or more) concentrated loads distributed at equal intervals along the beam span direction, the accurate calculation of the deflection is more complicated. However, its deflection is close to the deflection caused by the same maximum bending moment under the uniformly distributed load. Therefore, the deflection of the beam can be checked according to the following approximate equations.

Simply supported beams with equal cross-section:

$$\frac{v}{l} = \frac{5}{384} \frac{q_k l}{EI_x} = \frac{5}{48} \cdot \frac{q_k l^2 \cdot l}{8EI_x} \approx \frac{M_k l}{10EI_x} \leqslant \frac{[v]}{l} \qquad (6\text{-}13\text{a})$$

Simply supported beams with variable flange width:

$$\frac{v}{l} = \frac{M_k l}{10EI_x}\left(1 + \frac{3}{25} \cdot \frac{I_x - I_{x1}}{I_x}\right) \leqslant \frac{[v]}{l} \qquad (6\text{-}13\text{b})$$

where  $q_k$ = the standard value of uniformly distributed loading;

$M_k$ = the maximum bending moment produced by the load's standard value;

$I_x$ = the moment of inertia of the gross cross-section at mid-span;

$I_{x1}$ = the gross cross-sectional moment of inertia near the support.

Since the deflection is the overall mechanical behavior of the flexural member, it is not necessary to use the net section parameters like the strength calculation, and only the gross cross-section parameters are used for calculation. The following stability calculation part also uses the gross cross-section for the same reason as the stiffness calculation.

It should be noted that for floor beams and working platform beams, the deflection generated by all load standard values and the deflection generated by only the variable load standard values should be checked separately. That is to say, for these two types of beams, there are two allowable deflection values, $[v_T]$ is the allowable value of the deflection generated by the standard value of all loads (if arching should be subtracted), and $[v_Q]$ is the allowable value of the deflection produced by the variable load standard value.

In Table 6.2, $l$ is the span of the bending member, and for cantilever beams and outrigger beams, it is twice the length of the cantilever. When the crane girder or crane truss span is greater than 12 m, the allowable deflection value $[v_T]$ should be multiplied by a factor of 0.9. When the wall adopts malleable material or the structure adopts flexible connection, the wall pillar's allowable value of the horizontal displacement can be taken as $l/300$. The allowable value of the horizontal displacement of the wind-resistant truss (when used as the support of continuous pillars) can be taken as $l/800$.

**Allowable value of deflection of bending member**  Table 6.2

| Item | Member category | Allowable deflection $[v_T]$ | Allowable deflection $[v_Q]$ |
|---|---|---|---|
| 1 | Crane beam and crane truss (calculate the deflection according to its weight and the largest crane) | | — |
| 1 | (1) Manual cranes and single beam cranes (including suspension cranes) | $l/500$ | |
| 1 | (2) Light duty bridge crane | $l/750$ | |
| 1 | (3) Intermediate duty bridge crane | $l/900$ | |
| 1 | (4) Heavy duty bridge crane | $l/1000$ | |
| 2 | Rack beam of manual or electric hoist | $l/400$ | — |
| 3 | Work platform beam with heavy rail (weight equal to or greater than 38 kg/m) | $l/600$ | — |
| 3 | Work platform beam with light rail (weight equal to or less than 24 kg/m) rail | $l/400$ | |
| 4 | Floor (roof) beams or trusses, working platform beams (except item 3), and platform slabs | | |
| 4 | (1) Main beam or truss (including beams and trusses with suspended lifting equipment) | $l/400$ | $l/500$ |
| 4 | (2) Supporting only profiled sheeting roofs and cold-formed steel purlins | $l/180$ | |
| 4 | (3) Supporting profiled metal sheet roofs and cold-formed steel purlins, and there are also suspended ceilings | $l/240$ | |
| 4 | (4) Secondary beam of plastered ceiling | $l/250$ | $l/350$ |
| 4 | (5) Beams other than clauses (1)-(4), (including stair beams) | $l/250$ | |
| 4 | (6) Roof purlin | | $l/300$ |
| 4 | Supporting the profiledsheeting roof | $l/150$ | |
| 4 | Support other roofing materials | $l/200$ | |
| 4 | With suspended ceiling | $l/240$ | |
| 4 | (7) Platform slabs | $l/150$ | |
| 5 | Wall frame parts (wind load does not consider the vibration wind coefficient) | | |
| 5 | (1) Pillar (horizontal direction) | — | $l/400$ |
| 5 | (2) Wind-resistant truss (horizontal displacement when used as a support for continuous pillars) | — | $l/1000$ |
| 5 | (3) Beam of masonry wall (horizontal direction) | — | $l/300$ |
| 5 | (4) Beam supporting profiled metal plate (horizontal direction) | — | $l/100$ |
| 5 | (5) Beam supporting other wall materials (horizontal direction) | — | $l/200$ |
| 5 | (6) Beams with glass windows (vertical and horizontal directions) | $l/200$ | $l/200$ |

## 6.3 Torsion of the Beam

### 6.3.1 Free torsion

When the shear force acting on the beam does not pass through the shear center, the beam will bend and deform and twist around the shear center. When torsion occurs, the cross-section of members with circular cross-sections remains plane. The other cross-sections of members will be uneven due to the longitudinal displacement of the cross-section fibers. The cross-section will no longer remain plane, resulting in warping deformation. If the displacement of each fiber along the longitudinal axis of the member is not restricted, it is pure torsion or free torsion.

Figure 6.7 Free torsion of I-shaped members

Figure 6.7 shows that the I-shaped member of the uniform cross-section is under the same torque at both ends, and no special structural measures are added to the end. Therefore, the cross-section can be freely warped, and the member is pure torsion. The shear flow generated by pure torsion on the open section member is shown in Figure 6.8. The shear flow direction is parallel to the wall thickness centerline and linearly changes along the wall thickness direction. The shear stress in the middle centerline of the wall thickness equals zero, and the shear stress equals zero at the two wall surfaces.

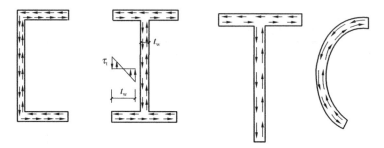

Figure 6.8 Free torsion shear stress

When the maximum shear stress value, $\tau_t$, is reached, the size of $\tau_t$ is directly proportional to the change rate $\varphi'$ of the member's torsion angle. $\varphi'$ is called the torsion rate. The resultant moment of the shear flow against the external torque is $GI_t\varphi'$, then the free torque $M_t$ acting on the member is:

$$M_t = GI_t\varphi' \tag{6-14}$$

where  $G =$ the shear modulus of the material;

$\varphi' =$ the torsion angle of the cross-section and its sign is determined by the right-hand spiral law like $M_t$;

$I_t$ = the uniform torsional constant, also called the torsional moment of inertia. For an open thin-walled section composed of multiple long and narrow plates, $I_t$ is calculated by the following equation:

$$I_t = \frac{k}{3} \sum_{i=1}^{n} b_i t_i^3 \qquad (6\text{-}15)$$

where  $b_i$ = the width and thickness of the $i$-th plate;

$k$ = a factor that considers the beneficial influence of the hot-rolled shapes at the junction corner, and experiments determine its value. It is 1.0 for angles, 1.15 for T-shapes, 1.12 for the Channels, and 1.25 for the I-shapes.

The relationship between the maximum shear stress $\tau_t$ and torque $M_t$ is:

$$\tau_t = \frac{M_t t}{I_t} \qquad (6\text{-}16)$$

For the closed section, the shear flow distribution shown in Figure 6.9 is closed along the member section.

For a thin-walled closed section, it can be considered that the shear stress $\tau$ is uniformly distributed along with the wall thickness, the direction is tangent to the centerline of the section, and $\tau_t$ is constant at any place along the section of the member. Therefore, torque $M_t$ is calculated by the following equation:

$$M_t = \oint \rho \tau t \, \mathrm{d}s = \tau t \oint \rho \, \mathrm{d}s \qquad (6\text{-}17)$$

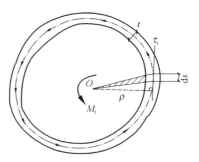

Figure 6.9  Free torsion of the closed cross-section

where  $\rho$ = the distance from the center of shear force to the centerline of the micro-element segment $\mathrm{d}s$.

So, $\oint \rho \, \mathrm{d}s$ is twice the area $A_0$ enclosed by the centerline of the sectional thickness, and rewrite Equation (6-17) as Equation (6-18).

$$M_t = 2\tau t A_0, \text{ or } \tau = \frac{M_t}{2tA_0} \qquad (6\text{-}18)$$

where  $t$ = the wall thickness at the calculated section;

$A_0$ = the area enclosed by the centerline of the sectional thickness.

For a closed thin-walled section, $I_t$ is calculated by the following equation:

$$I_t = \frac{4A_0^2}{\oint \dfrac{\mathrm{d}s}{t}} \qquad (6\text{-}19)$$

It should be pointed out that the integral is the closed-circuit integral in the thickness of each plate in the section.

From the above description, it can be seen that the closed section has stronger resistance to torsion than the open section.

## 6.3.2 Constrained torsion of open section members

As shown in Figure 6.10, the cantilever I-shaped member is twisted under the action of the torque $M_t$. Although the free end is not constrained, the section at the fixed end cannot be warped, so the middle sections are constrained to different degrees. After the sectional warping deformation is restrained, the longitudinal warping normal stress is generated, and the warping shear stress is also generated. The warping shear stress forms the ability to resist the warping torque $M_\omega$ around the sectional shear center. The member that undergoes the torsion produces not only free torsion but also constrained warping torsion. The total torque is divided into two parts. One is the free torque $M_t$, and the other is the warping torque $M_\omega$. The member torsion balance equation is:

$$M_z = M_t + M_\omega \tag{6-20}$$

In Equation (6-20), $M_t$ is calculated by Equation (6-14) for the opening section; warping restraint (or non-uniform) torsion $M_\omega$ can be calculated by the following equation:

$$M_\omega = -EI_\omega \varphi''' \tag{6-21}$$

Substituting Equations (6-14) and (6-21) into Equation (6-20) to obtain the torque balance equation:

$$M_z = GI_t \varphi' - EI_\omega \varphi''' \tag{6-22}$$

where  $I_\omega$ = the warping constant and is also known as the sector moment of inertia, and ahe dimension is $L^6$.

The general calculation formula of $I_\omega$ is:

$$I_\omega = \int_0^s \omega^2 t \, ds = \int_A \omega^2 \, dA \tag{6-23}$$

where  $\omega$ = the main sector coordinate. And its dimension is $L^2$.

The calculation method of $\omega$ is described below.

As shown in Figure 6.11, the length along the sectional centerline with $O_1$ as the starting point is defined as the curve coordinate $s$. The sectoral coordinates of any point $P$ on the midline of the section are twice the area enclosed by the arc between points $O_1$ and $P$

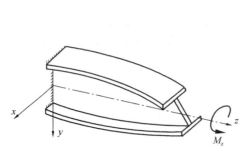

Figure 6.10  The restrained torsion of the cantilever I-shaped member

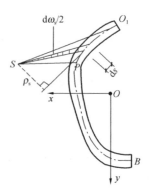

Figure 6.11  Sector coordinate calculation

and the shear center $S$. Choose any micro-element segment $ds$ between $O_1$ and $P$, the vertical distance between $S$ and $ds$ is $\rho$, the sector area of this micro-segment, $V\dfrac{d\omega_s}{2}=\dfrac{\rho_s ds}{2}$, and the sector coordinate $\omega_s$ of point $P$ is:

$$\omega_s = \int_0^s \rho_s ds \tag{6-24}$$

The $O_1$ point is the zero point of the sector coordinate, and the sector coordinate $\omega_s$ obtained by rotating when the vector $SP$ in the counterclockwise direction is positive. Any point on the cross-section can be selected as point $O_1$. Of course, $\omega_s$ changes with the change of point $O_1$. After obtaining the sectoral coordinates, the main sectoral coordinates $\omega_s$ can be calculated as follows:

$$\omega = \omega_s = -\dfrac{\int_A \omega_s dA}{A} \tag{6-25}$$

where $A =$ the cross-sectional area.

If the selected point $O_1$ happens to make $\omega = \omega_s = 0$, that is, $\int_A \omega_s dA = 0$, then $\omega_s$ is the main sector coordinate.

For a certain section, the main sector coordinate $\omega$ of each point on the section is a certain value. Now take the torsion cantilever I-shaped member shown in Figure 6.10 as an example to illustrate the calculation method of restrained torsion.

Suppose the torsion angle of any section from the coordinate origin $Z$ is $\varphi$, then as shown in Figure 6.12(a), the horizontal displacement of the upper and lower flanges is:

$$u = \dfrac{h}{2}\varphi \tag{6-26}$$

Considering each flange as a single bending member, there are:

$$M_f = -EI_f u'' = -EI_f \dfrac{h}{2}\varphi'' \tag{6-27}$$

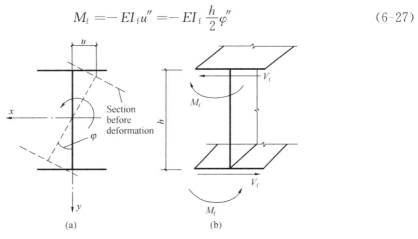

Figure 6.12 Deformation and internal force of the restrained torsion

where  $I_f$ = the moment of inertia of one flange about the y-axis, $I_f = \frac{1}{2}I_y$;

$M_f$ = the bending moment in the plane of one flange.

According to the relationship between the bending moment and the shear force, the shear force in the flange is:

$$V_f = -\frac{dM_f}{dz} = -EI_f \frac{h}{2}\varphi''' \qquad (6-28)$$

Then the warping torque formed by the two shear forces in opposite directions of the upper and lower flanges (Figure 6.12b) is:

$$M_\omega = -V_f h = -EI_f \frac{h^2}{2}\varphi''' \qquad (6-29)$$

Define the product of the two opposite bending moments $M_f$ and the distance $h$ between the upper and lower flanges in Figure 6.12(b) as *Bimoment B*:

$$B = M_f h = -EI_f \frac{h^2}{2}\varphi'' \qquad (6-30)$$

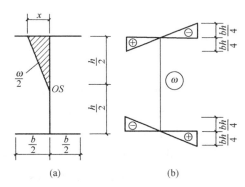

Figure 6.13  The sectoral coordinates of the I-shaped section

To convert the above equations to a more general expression, study the I-shaped sectoral coordinates below. It can be proved that the main sector coordinate of the centroid (also the shear center) of the biaxial symmetric cross-section is zero, and the sector coordinate of each point of the cross-section can be obtained by Equation (6-24). Note that, $\rho = h/2$, is a constant shown in Figure 6.13a, then the sectoral coordinates of the I-shaped section (only the flange has a non-zero value) can be expressed as:

$$\omega = \frac{h}{2}x \qquad (6-31)$$

The sectoral coordinates of each point of the I-shaped section are shown in Figure 6.13(b). It can be proved by Equation (6-31).

$$I_f \frac{h}{2} = \left(\frac{h}{2}\right)^2 (2I_f) = \left(\frac{h}{2}\right)^2 I_y = \left(\frac{h}{2}\right)^2 \int_A x^2 dA = \int_A \omega^2 dA = I_\omega \qquad (6-32)$$

Substituting Equation (6-32) into Equations (6-29) and (6-30), the general expressions of warping torque [Equation (6-21)] and *Bimoment B* are:

$$M_\omega = -EI_\omega \varphi''', B = -EI_\omega \varphi'' \qquad (6-33)$$

The warping normal stress and warping shear stress generated by constrained torsion are:

$$\sigma_\omega = -EI_\omega \varphi'' = \frac{B}{I_\omega}\omega = \frac{B}{W_\omega} \qquad (6-34a)$$

$$\tau_\omega = \frac{M_\omega S_\omega}{I_\omega t} \qquad (6\text{-}34b)$$

where  $W_\omega$ = the section sector modulus at a certain point in the section, $W_\omega = I_\omega/\omega$;

$S_\omega$ = statical moment of the area below a certain calculated point $P$ on the section (see Figure 6.11), which is a function of the curve coordinate $S$, and the dimension is $L^3$. The calculation equation of $S_\omega$ is:

$$S_\omega = \int_P^B \omega t \, ds \qquad (6\text{-}35)$$

## 6.4 The Overall Stability of the Beam

### 6.4.1 Several concepts

(1) The overall instability of the beam

When the load on the beam increases to a certain value, the beam suddenly leaves the bending plane, exhibits significant lateral bending and torsion, and immediately loses its bearing capacity. That is the overall instability of the beam (Figure 6.14).

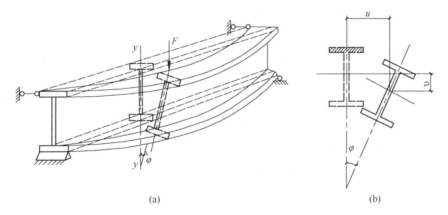

Figure 6.14  The overall instability of simply supported beams

To improve its bending strength and stiffness, most steel beams adopt high and narrow I-sections, and the moments of inertia of the two main axes are very different, that is, the $x$-axis is the strong axis and the $y$-axis is the weak axis. Therefore, when the beam is subjected to a not too large lateral load $F$ in its maximum stiffness plane (Figure 6.14a), and since the load action line passes through the shear center, the beam will only be bent and deformed in the maximum stiffness plane, without a twist. However, when the lateral load $F$ gradually increases to a certain value, the beam suddenly appears lateral bending and torsion due to the small lateral bending stiffness and immediately loses its bearing capacity.

For large-span beams without lateral support between spans, their bearing capacity when they lose overall stability is generally lower than the bearing capacity determined by their strength. Therefore, the determination of the cross-sectional dimensions of these beams is often controlled by the overall stability.

The overall instability failure of the beam occurs suddenly, without obvious warning beforehand, and it is more dangerous than the strength failure of the beam, so special attention should be paid to it.

(2) The overall instability form of the beam is lateral-torsional buckling

The reason for the lateral-torsional buckling when the beam is overall instability can be understood as follows: the upper flange of the beam is a compression plate. If the web provides no continuous support, there will be a flange along the direction of less rigidity— the possibility of buckling in directions outside the plane of the plate. However, due to the restriction of the web, the actual rigidity in this direction is greatly improved. Therefore, the upper flange under compression can only buckle in the plane of the flange. Furthermore, the compression flange and the web of the compression zone of the beam are not the same as the axial compression member. Instead, they are directly connected to the tension flange and the web of the tension zone of the beam. Therefore, when the compression flange buckles, it is restrained by the tension part, which results in a serious slope of the compression flange and a small slope of the tension part. Therefore, the form of the overall instability of the beam must be lateral bending torsional buckling (Figure 6.14).

(3) Critical bending moment and critical stress of the beam

The maximum bending moment that the beam can withstand before it loses its overall stability is called the critical bending moment $M_{cr}$. The maximum bending compressive stress that the beam can withstand before it loses its overall stability is critical stress $\sigma_{cr}$. $\sigma_{cr} = M_{cr}/W_x$, where $W_x$ is the elastic section modulus of the cross-section about the neutral axis, $x$-axis.

### 6.4.2 Calculation of critical bending moment of beams

(1) Critical bending moment of a biaxially symmetric I-section simply supported beam under pure bending

As shown in Figure 6.15, a biaxial symmetrical I-section simply supported beam reaches a critical state under pure bending and slightly lateral bending and torsion. The simple support at the end of the beam is essentially clamping support, which is referred to in Figure 6.18, and the displacement of the support section in the $x$-axis and $y$-axis directions is constrained, and the torsion around the $z$-axis is also constrained. But in the support, the cross-section can be freely warped and rotate freely around the $x$-axis and $y$-axis. Under the conditions mentioned above, when the beam occurs overall instability, its critical bending moment expression is shown below.

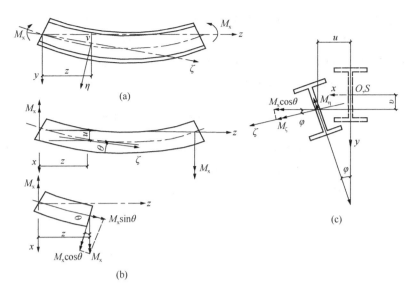

Figure 6.15 Minor deformation state of a biaxially symmetric I-section simply supported beam under pure bending

$$M_{cr} = \frac{\pi^2 EI_y}{l^2}\sqrt{\frac{I_\omega}{I_y}\left(1+\frac{GI_t l^2}{\pi^2 EI_\omega}\right)} \qquad (6\text{-}36)$$

In Equation (6-36), the $\pi^2 EI_y/l^2$ before the root sign is the Euler critical load of the axially loaded compression member about the y-axis. Thus, from Equation (6-36), it can be seen that the factors affecting the critical bending moment of a biaxial symmetric I-section simply supported beam under pure bending include the lateral bending stiffness of the beam $EI_y$, the torsional rigidity $GI_t$, the warping rigidity $EI_\omega$, and the lateral unsupported span $l$ of the beam.

(2) Critical bending moment of a uniaxial symmetrical I-section beam under lateral load

In the single axial symmetrical I-section (Figures 6.16a and 6.16c), the shear center $S$ does not coincide with the centroid $O$. When the beam subjected to lateral load is in a state of equilibrium with slight lateral-torsional deformation, its lateral-torsional buckling

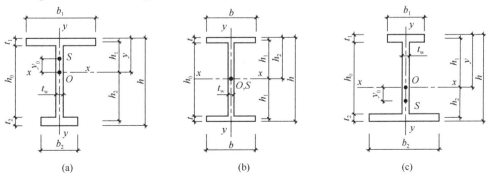

Figure 6.16 Welded I-shapes

differential equations are no longer constant coefficient differential equations. So it is impossible to get accurate analytical solutions, only numerical solutions and approximate solutions. The following is an approximate solution of the critical bending moment obtained by the energy method under different loads.

$$M_{cr} = \beta_2 a + \beta_3 B_y + \sqrt{(\beta_2 a + \beta_3 B_y)^2 + \frac{I_\omega}{I_y}\left(1 + \frac{l_1^2 G I_t}{\pi^2 E I_\omega}\right)} \tag{6-37}$$

where $\beta_1$, $\beta_2$, and $\beta_3$ is the coefficient related to the load type and referred to Table 6.3; $l_1$ is the unsupported lateral span of the beam; $a$ is the distance from the point of lateral load to the shear center of the section. When its direction is consistent with the deflection direction, it is negative; otherwise, it is positive.

$B_y$ is a parameter reflecting the degree of cross-section asymmetry. When the cross-section is biaxially symmetric, $B_y = 0$; when the cross-section is asymmetric:

$$B_y = \frac{1}{2I_x}\int_A y(x^2 + y^2)dA - y_0 \tag{6-38}$$

where $y_0$ is the distance from the shear center $S$ to the centroid $O$. When its direction is consistent with the deflection direction, and it is negative; otherwise, it is positive:

$$y_0 = \frac{I_2 h_2 - I_1 h_1}{I_y} \tag{6-39}$$

where $I_1$ and $I_2$ are the moments of inertia of the compression flange and the tension flange to the y-axis; $h_1$ and $h_2$ are the distance from the compression flange centroid and the centroid of the tension flange to the centroid of the entire section, respectively.

**Coefficients in Equation (6-37) for lateral-torsional buckling of a beam with two simply supported ends**　　　　Table 6.3

| Load category | $\beta_1$ | $\beta_2$ | $\beta_3$ |
|---|---|---|---|
| Concentrated load at mid-span | 1.35 | 0.55 | 0.40 |
| Uniformed distributed load across span | 1.13 | 0.46 | 0.53 |
| Pure bending | 1.00 | 0 | 1.00 |

(3) Critical bending moment of a beam in elastoplastic stage

Equations (6-36) and (6-37) are only suitable for solving the critical bending moment of elastic bending-torsion buckling steel beams, that is, the critical stress $\sigma_{cr} \leqslant f_p$ (proportional limit) when the beam is instability. These beams are often slender and have no lateral support in the middle of the span, and their critical stress is smaller.

Steel beams that are not slender or have enough lateral support may undergo elastoplastic buckling; that is, the critical stress of the overall instability of the beam is $\sigma_{cr} > f_p$. Simultaneously, the elastic modulus $E$ and shear modulus $G$ of the steel is no longer constant but gradually decrease as the critical stress increases.

There are residual stresses in steel beams in actual engineering. Therefore, when determining whether the steel beam material enters the elastoplastic working stage, the residual stress must be added to the stress caused by the structural load.

For simply supported beams under pure bending and whose section is symmetrical to the bending moment action plane, an analytical formula for the critical bending moment of elastic-plastic lateral-torsional buckling expressed by the tangent modulus $E_t$ can also be written. For impure bending beams, because of the different distribution of elastic zone and plastic zone in each section, the effective stiffness distribution of each section is different. Thus, it becomes a variable stiffness beam. The calculation of the critical bending moment of elastoplastic flexural torsional buckling will become very complicated. Generally, no analytical solution can be obtained.

### 6.4.3 Main factors affecting the overall stability of steel beams

It can be seen from Equation (6-37) that the main factors affecting the critical bending moment of steel beams are:

(1) The lateral bending stiffness of the section $EI_y$, the greater the torsional stiffness $GI_t$ and the warping stiffness $GI_\omega$, the greater the critical bending moment and the better the overall stability of the beam. In addition, by widening the compression flange of the beam, the critical bending moment increases.

(2) The smaller the unsupported lateral length of the beam or the smaller the distance between the lateral support points of the compression flange, the greater the critical bending moment and the better the overall stability of the beam.

(3) Load type. The bending moment diagram formed by the load acting on the beam is more evenly distributed along the span, and the critical bending moment is smaller. When the bending moment diagram is rectangular under pure bending, the $M_{cr}$ is the smallest. When the bending moment diagram is parabolic under uniformly distributed, and the $M_{cr}$ is slightly larger. For a concentrated load in the mid-span, the bending moment diagram is triangular, and the $M_{cr}$ is the largest.

(4) The higher position of the load point and the depth of the beam section, the smaller the $M_{cr}$. When the load is applied to the upper flange, the value of $a$ in Equation (6-37) is negative, and the critical bending moment is small; when the load is applied to the lower flange, the value of $a$ is positive, and the critical bending moment is larger. It can also be seen from Figure 6.17 that when the beam produces slight lateral-torsional deformation, the load acting on the upper flange produces an unfavorable additional bending moment on the shear center $S$, which intensifies the beam torsion (Figure 6.17a). The load acting on the lower flange (Figure

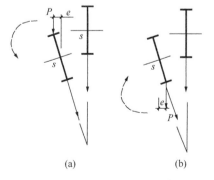

Figure 6.17 The influence of the position of the load application point on the stability of the beam

6.17b) prevents the additional bending moment generated by the shear center $S$ from rotating the beam section and delays the overall instability of the beam.

(5) The beam end support restrains the cross-section, especially the greater the degree of restraint on the rotation around the $y$-axis and the larger the critical bending moment $M_{cr}$.

### 6.4.4 The overall stability factor of the beam

To ensure that the beam does not lose its overall stability, the maximum compressive fiber bending normal stress should not exceed the critical stress of the overall stability of the beam divided by the partial coefficient of resistance:

$$\sigma = \frac{M_x}{W_x} \leqslant \frac{M_{cr}}{W_x} \cdot \frac{1}{\gamma_R} = \frac{\sigma_{cr}}{\gamma_R} = \frac{\sigma_{cr}}{f_y} \cdot \frac{f_y}{\gamma_R} = \varphi_b f \tag{6-40}$$

where $\varphi_b$ is the overall stability factor of the beam.

The overall stability factor $\varphi_b$ of welded I-section simply-supported beams with a constant cross-section (including hot-rolled H-shape)

$$\varphi_b = \beta_b \frac{4320}{\lambda_y^2} \cdot \frac{Ah}{W_x} \left[ \sqrt{1 + \left(\frac{\lambda_y t_1}{4.4h}\right)^2} + \eta_b \right] \varepsilon_k^2 \tag{6-41}$$

where  $\beta_b$ = the equivalent critical bending moment coefficient for the overall stability of the beam is adopted according to the Attached Table C.1;

$\lambda_y$ = the slenderness ratio of the beam between the lateral support points to the weak axis of the section ($y$-axis), $\lambda_y = l_1/i_y$;

$l_1$ = the distance between the lateral support points of the compression flange of the beam (mm); for the beam without a lateral support point in the middle of the span, it is the span (the beam support is treated as having lateral support);

$i_y$ = the radius of gyration of the beam section to the $y$ axis;

$A$ = gross cross-sectional area of a beam;

$W_x$ = the modulus of gross cross-section determined by the largest fiber under compression;

$h$ = the beam depth;

$t_1$ = thickness of beam compression flange;

$\eta_b$ = cross-section asymmetry influence coefficient. Take $\eta_b = 0$ for biaxial symmetry; $\eta_b = 0.8(2\alpha_b - 1)$ when strengthening the compression flange; $\eta_b = 2\alpha_b - 1$, when strengthening the tension flange. $\alpha_b = I_1/(I_1 + I_2)$, $I_1$ and $I_2$ are the moments of inertia of the compression flange and the tension flange to the $y$-axis, respectively.

There are not many cases where the beam is subjected to pure bending. However, when the beam is subjected to any lateral load or the beam has a single axis symmetrical section, Equation (6-41) should be corrected. Please refer to Appendix C for the requirements of the design standard GB 50017 on the overall stability factor of the beam.

When $\varphi_b$ calculated according to the method in Appendix C or found is greater than 0.6, it indicates that the beam has entered the elastoplastic stage. At this time, $\varphi'_b$ is used instead and $\varphi'_b$ is calculated as follows:

$$\varphi'_b = 1.07 - 0.282/\varphi_b \leqslant 1.0 \qquad (6\text{-}42)$$

### 6.4.5 Measures to ensure the overall stability of beams and calculation methods

(1) Cannot calculate the overall stability of the beam

In practical applications, beams are often connected to other components, which helps prevent beams from losing overall stability. When one of the following conditions is met, the overall stability of the beam may not be calculated.

① When a plank (various reinforced concrete slabs and steel plates) is densely laid on the compression flange of the beam and firmly connected to it, it can prevent the lateral displacement of the compression flange of the beam.

It should be emphasized that the theoretical basis for the calculation of the overall stability of steel beams is based on the premise that no torsional deformation occurs at the support of the beam, and the torsion angle of the section must be equal to zero at the support of the beam. Therefore, it should be considered in the structure to provide reliable lateral support at the upper flange of the beam support to avoid the beam from twisting here. Generally, two structural measures can improve the torsion resistance of simply supported end steel beams. The first measure uses steel plates to connect the top beam flange to the supporting member to prevent the beam's torsion (Figure 6.18a), and its effect is better than the second measure. The second measure (Figure 6.18b) is to install stiffeners at the beam end to form a rigid section and use the bolts connecting the lower flange to the support, providing a certain degree of torsion resistance. Figure 6.18(c) is a schematic diagram of the torsional deformation of the beam end section when the above two measures are not adopted, and the requirement of $\varphi = u = 0$ is not met here. When the simply sup-

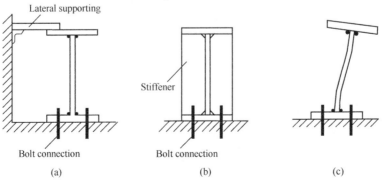

Figure 6.18 The diagram of the torsion resistance structure measures at the simply supported end of the steel beam

ported beam is connected with adjacent members only by the web, the distance between the lateral support points should be taken as 1.2 times the actual distance when calculating the stability of the steel beam.

② Box-shaped section simply supported beam (Figure 6.19), its section size satisfies $h/b_0 \leqslant 6$ and $l_1/b_0 \leqslant 95\varepsilon_k^2$, the overall stability of the beam may not be calculated, and $l_1$ is the distance between the lateral support points of the compression flange (the beam's support is considered to have lateral support). However, since the box-shaped section's lateral bending rigidity and torsion rigidity are far greater than that of the I-section, and the overall stability is very strong, the limits of $h/b_0$ and $l_1/b_0$ specified in this article can easily be met.

(2) The calculation formula for overall beam stability

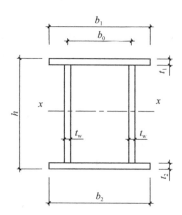

Figure 6.19 Box section

For beams that do not meet any of the conditions described above, the overall stability calculation should be carried out.

① For members that are bent in the principal plane of maximum stiffness, their overall stability shall be calculated as follows:

$$\frac{M_x}{\varphi_b W_x f} \leqslant 1.0 \qquad (6\text{-}43)$$

where $M_x$ = the design value of the maximum bending moment acting around the strong axis of the section ($x$-axis);

$W_x$ = the modulus of gross cross-section determined by the largest fiber under compression. When the width-to-thickness ratio grade of the sectional plate is S1, S2, S3, or S4, the full section modulus shall be adopted. When the width-to-thickness ratio class of the section plate is S5, the effective section modulus shall be taken, and the effective extension width of the uniformly compressed flange may be $15\varepsilon_k$; the effective section of the web may be following Article 8.4.2 of the design standard GB 50017 are adopted.

$\varphi_b$ = the overall stability factor of the beam and is calculated according to the method in Appendix C.

② For members with cross-section such as I-section or hot-rolled H-shapes that are bent on two main planes, their overall stability shall be calculated as follows:

$$\frac{M_x}{\varphi_b W_x f} + \frac{M_y}{\gamma_y W_y f} \leqslant 1.0 \qquad (6\text{-}44)$$

where $W_x$ and $W_y$ = the gross sectional modulus to the $x$-axis (strong axis) and the $y$-axis determined according to the maximum fiber under compression;

$\varphi_x$ = the overall stability factor of the beam determined by bending to the strong axis;

$\gamma_y$ = the plastic development coefficient of the section bent around the y-axis can be determined by referring to Table 6.1.

Equation (6-44) is an empirical formula in which the second term expresses the effect of bending around the weak axis. The denominator only reduces this effect appropriately and does not mean that the section allows the development of plasticity.

## 6.5 Local Stability of Steel Plate Composite Beam

In order to increase the moment of inertia of the beam section, it is often necessary to increase the width-to-thickness ratio or depth-to-thickness ratio of each plate in the section when choosing the size of the beam section. When determining the cross-sectional area of the flange plate of the required I-shaped section, $A_f = bt$, when $b$ and $t$ are specifically selected, if the ratio of $b/t$ is larger, the obtained section $I_y$ is also larger.

Increasing the ratio of depth (width)-to-the thickness of the beam can obtain a more economical beam section, but at the same time, it brings another problem: each plate may lose its local stability first. However, due to the limitation of rolling conditions, the flange and web thickness of rolled steel beams are large, so there is no local stability problem, which must be considered in the steel plate beam design. Therefore, the consequences of the loss of local stability are not as serious as the immediate loss of bearing capacity of the beam due to the loss of global stability. Still, the loss of local stability will change the stress condition of the beam and reduce the global stability and stiffness of the beam because the problem of local stability must still be taken seriously.

The following three methods are adopted to deal with the local stability of the I-section welded steel plate composite beam.

(1) For the flange, the width-to-thickness ratio is limited to avoid local instability of the flange.

(2) For crane beams directly bearing dynamic load or other plate beams without considering the strength of the web after buckling, the web is provided with a stiffener. The web is divided into several compartments, and the stability of each compartment is calculated to avoid local instability. After buckling, the web strength is not considered for the crane beam to prevent the web fatigue crack caused by repeated buckling.

(3) If considering the post-buckling strength of the web, the flexural and shear capacity of the beam section of the partially buckling web is calculated.

### 6.5.1 Local stability of the compression flange

The compression flange of the I-section composite beam can be seen as two rectangular slats supported on three sides and free on one side under uniform compression on the plane of the plate. One of its longitudinal edges is connected to the web. Because the thickness of the web is usually less than that of the flange, the web has little rotation constraint on the flange, and the edge is a simple support edge. The two lateral supporting edges can be regarded as simply supported on the top of the supporting stiffener or the transverse stiffener. The plane size of the slats is $a \times b_1$, $a$ is the spacing of the transverse stiffeners of the web, and $b_1$ is the free overhang width of the compression flange, as shown in Figure 6.20 below. The welded components are taken from the design standard GB 50017 in China.

$$b_1 = \frac{b - t_w}{2} \tag{6-45}$$

That is, $b_1$ is the distance between the web surface and the edge of the flange. In the specific design, it is usually safe to take $b_1$ as half of the flange width $b$.

From elastic stability theory, the elastic stable critical stress formula of a thin rectangular plate is obtained.

$$\sigma_{cr} = \frac{\chi K \pi^2 E}{12(1-\nu^2)} \left(\frac{t}{b}\right)^2 \tag{6-46}$$

where $\chi$ is the elastic fixed coefficient, and $\chi = 1.0$ is taken as the simply-supported edge; $b$ is the width of the rectangular plate, where $b$ should be changed to $b_1$. When one of the three simply supported sides is free under longitudinal uniform compression, the buckling coefficient $K$ is:

$$K = 0.425 + \left(\frac{b_1}{a}\right)^2 \tag{6-47}$$

Because of $\frac{b_1}{a} \ll 1$, $K = 0.425$ is often approximated. In order to obtain the maximum width-to-thickness ratio of the free overhanging portion of the I-section flange at the elastic stage, the critical stress should be reached to the maximum possible stress of the plate. When bending in the elastic stage, the maximum edge fiber stress of the beam is $f_y$. If the change of stress on the flange thickness is not taken into account, approximate $\sigma_{cr} = f_y$, $E = 206 \times 10^3 \text{ N/mm}^2$, and $\nu = 0.3$ can be obtained.

$$\frac{b_1}{t} \leqslant \sqrt{\frac{0.425 \times \pi^2 \times 206 \times 10^3}{12(1-0.3^2) \times 235}} \cdot \sqrt{\frac{235}{f_y}} = 18.6 \sqrt{\frac{235}{f_y}} = 18.6 \varepsilon_k \tag{6-48}$$

For beams that do not require fatigue checking, considering the influence of different degrees of beam plasticity, the compression flange width-to-thickness ratios of S1, S2, S3, S4, and S5 classifications are respectively on the right side of the Equation (6-46). Therefore, 0.5, 0.6, 0.7, 0.8, and 1.1 times of 18.6 can be used according to Table 3.1 after rounding the integers.

The section width-to-thickness ratio grade S1 or S2 is a plastic section. Because the frame beam is often designed as a plastic energy dissipation zone in the seismic performance design of civil buildings, it must form a plastic hinge under the seismic action of fortification intensity. Therefore, the design standards have stricter restrictions on the width-to-thickness ratio.

The section width-to-thickness ratio grade S3 is an elastoplastic section. When considering the development of plastic deformation in the part of the section, plastic and elastic zones are formed on the section. The stress on the entire thickness of the flange can reach the yield strength $f_y$, but it is still elastic in the direction perpendicular to the compressive stress. This situation is an orthotropic plate, and the accurate calculation of the critical stress is more complicated. Generally, $\sqrt{\eta} E$ can replace the elastic modulus $E$ to consider this effect (the coefficient $\eta \leqslant 1$, and $\eta$ is the tangent elastic modulus $E_t$ ratio to the elastic modulus E). If $\eta$ is equal to 0.25,

$$\frac{b_1}{t} \leqslant \sqrt{\frac{3}{4}} \cdot \left(15\sqrt{\frac{235}{f_y}}\right) = 13\sqrt{\frac{235}{f_y}} = 13\varepsilon_k \qquad (6\text{-}49)$$

The section width-to-thickness ratio grade S4 is the elastic limit. Still, considering the influence of residual stress, the longitudinal stress in some areas of the flange has exceeded the effective ratio limit and has entered the elastoplastic stage. Therefore, if $\eta$ is greater than or equal to 0.5, the Equation (6-46) $\sigma_{cr}$ is greater than or equal to $f_y$. The condition that local instability does not yield before the maximum stress of the compressed edge is satisfied. Then, the following equation will be gotten.

$$\frac{b_1}{t} \leqslant \sqrt{\frac{2}{3}} \times 18.6\sqrt{\frac{235}{f_y}} = 15\sqrt{\frac{235}{f_y}} = 15\varepsilon_k \qquad (6\text{-}50)$$

The section width-to-thickness ratio grade S5 is called a thin-walled section. That is because plates with free edges may change the center of section stiffness after partial removal, thereby changing the force of the member. Therefore, even the S5 grade can be calculated by the effective section method. However, bearing capacity, the design standard GB 50017 still imposes restrictions on the width-to-thickness ratio of the plate.

When the plastic deformation of the section is considered, the plastic zone and the elastic zone are formed on the section, and the stress on the entire thickness of the flange plate can reach the yield strength $f_y$. In the design standard GB 50017, the plastic development coefficient $\gamma_x = 1.05$, equivalent to limiting each side's plastic deformation development depth to 1/8 of the beam section depth. As shown in Figure 6.20, the strain of edge fiber at this time is $\frac{4}{3}\varepsilon_y$. Considering that the section enters into the elastic-plastic work, the elastic modulus $E$ in the critical stress formula is replaced by the secant modulus $E_{sec}$, equivalent to the edge strain is $\frac{4}{3}\varepsilon_y$.

Therefore, considering the development of plastic deformation, the approximate limit value of the ratio of free outstanding width $b_1$ and thickness $t$ of the compression flange is as follows:

$$E_{sec} = \frac{f_y}{\frac{4}{3}\varepsilon_y} = \frac{3}{4}E \qquad (6\text{-}51)$$

It is obtained that

$$\frac{b_1}{t} \leqslant \sqrt{\frac{3}{4}} \cdot \left(15\sqrt{\frac{235}{f_y}}\right) = 13\sqrt{\frac{235}{f_y}} = 13\varepsilon_k \qquad (6\text{-}52)$$

It is another regulation in the design standard GB 50017 when the design considers the development of plastic deformation.

Equation (6-48) is changed to Equation (6-50) to consider the effect of defects such as residual stress on elastic stability. However, changing from Equation (6-50) to Equation (6-52) to consider the influence of plastic deformation on the section is not the same concept.

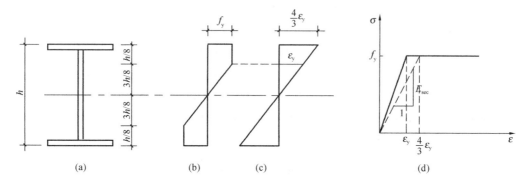

Figure 6.20 Stress and strain diagrams of I-section beams in elastoplastic working stage
(a) cross-section shape; (b) stress diagram; (c) strain figure; (d) stress-strain relationship

For box-shaped beams under compression, the part of the flange between the two webs can be regarded as simply supported on four sides and uniformly compressed in the longitudinal direction. Therefore, the buckling coefficient, $K=4.0$, elastic restraint coefficient, $\chi=1.0$, according to Equation (6-46), let $\sigma_{cr}$ is greater than or equal to $0.9f_y$, then the plate width-to-thickness ratio is.

$$\frac{b_0}{t} \leqslant 42\sqrt{\frac{235}{f_y}} = 42\varepsilon_k \qquad (6\text{-}53)$$

where  $b_0$ = the width of the compression flange between the two weds;
$t$ = the thickness of the compression flange.

### 6.5.2 Local stability of the web

For steel plate composite beams that bear static loads and indirect dynamic loads, generally consider the strength of the web after buckling, arrange the stiffeners according to the requirements of the design standard GB 50017, and calculate its bending and shear resistance. For crane beams and similar members that directly bear dynamic loads, stiffeners are configured according to the following regulations, and the stability of each plate section is calculated.

The most commonly used method for setting stiffeners is to use two rectangular steel plates to be welded on both sides of the web, respectively, as shown in Figure 6.21(c). Then, according to the size of the depth-to-thickness ratio, $h_0/t_w$ ($h_0$ is the calculated height of the web), and the load. Web stiffeners are of the four types shown in Figure 6.21: bearing stiffeners, transverse stiffeners, longitudinal stiffeners, and short stiffeners.

(1) When $h_0/t_w \leqslant 80\sqrt{235/f_y} = 80\varepsilon_k$, for beams with local compressive stress ($\sigma_c \neq 0$), transverse stiffeners should be arranged according to the structure. When the local compressive stress is small, transverse stiffeners may not be arranged, as shown in Figure 6.21.

(2) When $h_0/t_w > 80\sqrt{235/f_y} = 80\varepsilon_k$ transverse stiffeners should be arranged, as shown in Figure 6.21.

(3) When $h_0/t_w > 170\sqrt{235/f_y} = 170\varepsilon_k$ (the torsion of the compression flange is constrained, such as it is connected with rigid pavement, brake plate, or welded rail) or (the torsion of the compression flange is not constrained) or as required by the calculation, it should be in the larger bending moment area add longitudinal stiffeners in the compression zone, as shown in Figure 6.21. For beams with large local compressive stress, short stiffeners should be arranged in the compression zone when necessary, as shown in Figure 6.21.

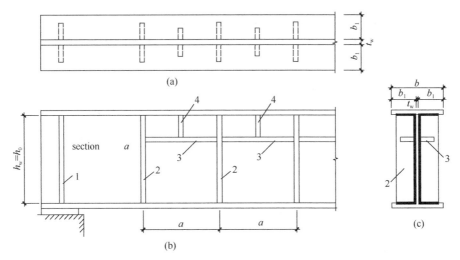

Figure 6.21 Weld stiffeners for I-section beams
1—bearing stiffener; 2—transverse stiffener; 3—longitudinal stiffener; 4—short stiffener

Under any condition, $h_0/t_w < 250\varepsilon_k$.

In the above description, $h_0$ is the calculated height of the web. Still, the welded beam

$h_0$ is equal to the height of the web, and for the rolled steel beam, $h_0$ is the distance between the starting points of the two inner arcs at the intersection of the web and the upper and lower flanges. Thus, for uniaxially symmetric beams, $h_0$ in the third clause above should be two times the height $h_c$ of the compression zone of the web.

Because of the different uses of the beam and the different positions of the webs divided by stiffeners, loads of the webs are different. Therefore, to check each web cell's local stability, the critical stress of maintaining stability in each grid under various individual loads should be obtained first. Then the local stability of each grid should be checked by using the critical conditions under the simultaneous action of various stresses. These are described below.

(1) Critical stress of the webzone under the single action of bending stress

Figure 6.22(a) shows the buckling of a simple four-sided supported plate under pure bending. In the transverse direction of the plate, it flexes into a half-wave. Depending on the aspect ratio $a/h_0$ of the plate, the plate's longitudinal direction may buckle into a half-wave or more.

$$\sigma_{cr} = \frac{\chi K \pi^2 E}{12(1-v^2)} \left(\frac{t_w}{h_0}\right)^2 = 18.6\chi K \left(\frac{100 t_w}{h_0}\right)^2 \quad (6-54)$$

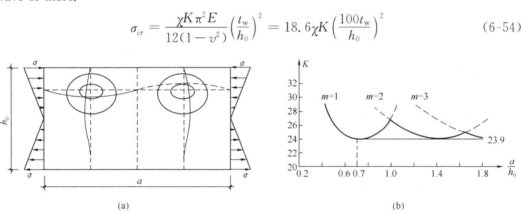

(a)  (b)

Figure 6.22 Elastic buckling of four-sided simply supported plate under pure bending

The buckling coefficient $K$, the lowest value 23.9, applies to the case where the quadratically simply supported $a/h_0 > 0.7$ bends into one or several half-waves, as shown in Figure 6.22(b). When the two loaded edges of the plate are simply supported and the top and bottom are fixed, the buckling coefficient increases to $K = 39.6$, equivalent to introducing the elastic fixation coefficient $\chi = 39.6/23.9 = 1.66$. To the web of the I-section beam, the lower edge is connected with the tensile flange so that it is close to the fixed edge; the upper edge of the beam is connected to the compression flange, so it cannot be regarded as completely fixed. Generally, it can be considered in two cases. When the compression flange of the beam is connected with rigid pavement, brake plate, or welded rail, the torsion deformation of the compression flange is constrained, and the upper edge can be regarded as completely fixed and $\chi = 1.66$. The other is that the torsion of the compression flange is not constrained. At this time, the upper edge can only be regarded as simply supported and $\chi = 1.0$.

The definition of web's general depth-to-thickness ratio is the yield strength of steel subjected to bending, shear, or compression divided by the square root of the critical stress of elastic buckling of the corresponding web region of bending, shear, or local pressure. For example, the general depth-to-thickness ratio is:

$$\lambda_b = \sqrt{\frac{f_y}{\sigma_{cr}}} = \frac{h_0/t_w}{100}\sqrt{\frac{f_y}{18.6\chi K}} = \frac{h_0/t_w}{28.1\sqrt{\chi K}}\sqrt{\frac{f_y}{235}} \quad (6\text{-}55)$$

When the torsion of the compression flange of the beam is constrained, take $\chi=1.66$ and $K=23.9$, get:

$$\lambda_b = \frac{h_0/t_w}{177}\sqrt{\frac{f_y}{235}} \quad (6\text{-}56a)$$

When the torsion of the compression flange of the beam is not restrained, take $\chi=1.0$ and $K=23.9$, get:

$$\lambda_b = \frac{h_0/t_w}{138}\sqrt{\frac{f_y}{235}} \quad (6\text{-}56b)$$

The subscript b of $\lambda_b$ stands for bending stress. According to the definition of general depth-to-thickness ratio, the relationship between critical stress $\sigma_{cr}$ and $\lambda_b$ in the elastic stage must be:

$$\sigma_{cr} = \frac{f_y}{\lambda_b^2} \text{ or } \frac{\sigma_{cr}}{f_y} = \frac{1}{\lambda_b^2} \quad (6\text{-}57)$$

Its curve is shown in the curve $ABEG$ in Figure 6.23, which intersects with the horizontal dash line of $\sigma_{cr} = f_y$ at point $E$, corresponding to $\lambda_b = 1.0$. The line $ABEF$ is an ideal curve. However, in practice, there must be defects. Therefore, in the design standard GB 50017, the critical stress curve of the web cell under pure bending is adopted, curve $ABCD$ as shown in Figure 6.23. The curve consists of three sections: curve $AB$ represents the critical stress dur-

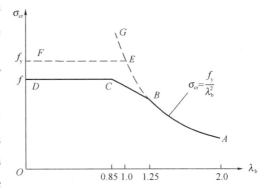

Figure 6.23 Critical stress curve of rectangular web cells under pure bending

ing elastic state; horizontal line $CD$ represents $\sigma_{cr}$ is equal to $f$, which is the design strength of the steel; tilt straight line $BC$, which is the critical stress curve from the elastic stage to the critical stress equal to $f$.

Corresponding to the three-section curve in Figure 6.23, the calculation formula of $\sigma_{cr}$ in the design standard GB 50017 is:

When $\quad\quad\quad\quad\quad\quad\quad\quad \lambda_{n,b} \leqslant 0.85 , \sigma_{cr} = f \quad (6\text{-}58a)$

When $\quad 0.85 < \lambda_{n,b} \leqslant 1.25$, $\sigma_{cr} = [1 - 0.75(\lambda_{n,b} - 0.85)]f \qquad (6\text{-}58b)$

When $\quad \lambda_{n,b} > 1.25$, $\sigma_{cr} = 1.1f/\lambda_{n,b}^2 \qquad (6\text{-}58c)$

where $\lambda_{n,b}$ = the regularized width-to-thickness ratio for calculation of beam webs under pure bending.

When the torsion of the compression flange of the beam is constrained, it can get

$$\lambda_{n,b} = \frac{2h_c/t_w}{177}\sqrt{\frac{f_y}{235}} = \frac{2h_c/t_w}{177} \cdot \frac{1}{\varepsilon_k} \qquad (6\text{-}59a)$$

When the torsion of the compression flange of the beam is unconstrained, it can get

$$\lambda_{n,b} = \frac{2h_c/t_w}{138}\sqrt{\frac{f_y}{235}} = \frac{2h_c/t_w}{138} \cdot \frac{1}{\varepsilon_k} \qquad (6\text{-}59b)$$

$h_c$ is the height of the compression zone of the web when the beam is subject to pure bending. When the beam section is biaxial symmetry, $h_0 = 2h_c$. When it is the single axisymmetric section of the strengthened upper flange, $h_0 > 2h_c$. $2h_c$ replaces $h_0$ to improve the critical buckling stress of the web. It is appropriate to use $2h_c$ instead of $h_0$ because the bending compression zone is the main factor that makes the pattern part of the web unstable. In Figure 6.23, line $BCD$ replaces the theoretical line $BEF$ to consider both inelastic work and the impact of defects.

It should be noted that although the three Equations (6-58a)~(6-58c) of the critical stress $\sigma_{cr}$ are based on the steel strength design value $f$, the critical stress (6-58c) is expressed in the elastic phase. After $f$ multiplied by 1.1, it is equivalent to $f_y$, which means the partial coefficient of resistance is not calculated. The elastic and inelastic ranges are treated differently because when the plate is in the elastic range, there is a larger post-buckling strength, and the safety factor can be smaller. In the subsequent expression of the elastic critical shear stress $\tau_{cr}$ and the expression of the critical normal stress $\sigma_{c,cr}$ in the elastic phase, $f$ is multiplied by 1.1 for the same reason.

(2) The critical stress of the web zone under pure shear

Figure 6.24 shows the elastic buckling of a four-side simply supported plate under the action of uniformly distributed shear stress. During buckling, the plate surface is concave and convex along a direction of approximately 45°, close to the principal stress direction under pure shear. The equation for the critical shear stress is:

$$\tau_{cr} = \frac{\chi K \pi^2 E}{12(1-\nu^2)}\left(\frac{t_w}{h_0}\right)^2 \qquad (6\text{-}60)$$

The value of the embedding coefficient $\chi$ does not distinguish whether the torsion of the compression flange of the beam is constrained or not, and $\chi = 1.23$ is unified. Take $\chi = 1.23$, $E = 2.06 \times 10^5 \text{N/mm}^2$, and $\nu = 0.3$ into Equation (6-60), then the critical shear stress of web under pure shear is:

$$\tau_{cr} = 22.9K\left(\frac{100t_w}{h_0}\right)^2 \qquad (6\text{-}61)$$

The general depth-to-thickness ratio is:

## 6.5 Local Stability of Steel Plate Composite Beam

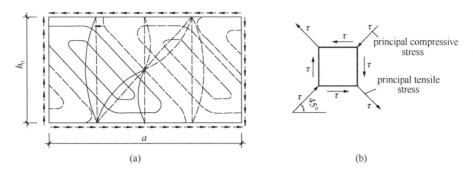

Figure 6.24  Elastic buckling of four-side simply supported plates under uniform shear stress

$$\lambda_s = \sqrt{\frac{f_{vy}}{\tau_{cr}}} = \sqrt{\frac{f_y}{\sqrt{3}\tau_{cr}}} = \frac{h_0/t_w}{41\sqrt{K}}\sqrt{\frac{f_y}{235}} = \frac{h_0/t_w}{41\sqrt{K}} \cdot \frac{1}{\varepsilon_k} \quad (6\text{-}62)$$

According to the elastic stability theory, the buckling coefficient $K$ in pure shear is:

$$\left.\begin{array}{l} \text{When } a/h_0 \leqslant 1.0 \quad K = 4 + 5.34\left(\dfrac{h_0}{a}\right)^2 \\[6pt] \text{When } a/h_0 > 1.0 \quad K = 5.34 + 4\left(\dfrac{h_0}{a}\right)^2 \end{array}\right\} \quad (6\text{-}63)$$

In the design standard GB 50017 of China, take Equation (6-63) into Equation (6-62) to obtain that

$$\text{When } a/h_0 \leqslant 1.0, \ \lambda_{n,s} = \frac{h_0/t_w}{37\eta\sqrt{4 + 5.34\,(h_0/a)^2}} \cdot \frac{1}{\varepsilon_k} \quad (6\text{-}64a)$$

$$\text{When } a/h_0 > 1.0, \ \lambda_{n,s} = \frac{h_0/t_w}{37\eta\sqrt{5.34 + 4\,(h_0/a)^2}} \cdot \frac{1}{\varepsilon_k} \quad (6\text{-}64b)$$

where  $\lambda_{n,s}$ = the regularized depth-to-thickness ratio for the calculation of beam webs under pure shear;

$a$ = the spacing between transverse stiffeners, when there is no middle transverse stiffener in the beam span, use $a/h_0 = \infty$ in Equation (6-64b);

$\eta$ = 1.11 for the simply supported beam and takes 1.0 for the maximum stress area at the beam end of the frame beam.

The critical stress formula is also divided into three sections; the calculation formula of $\tau_{cr}$ in the design standard GB 50017 is:

When $\quad\quad\quad\quad\quad\quad\quad \lambda_{n,s} \leqslant 0.8, \ \tau_{cr} = f_v \quad\quad\quad\quad\quad\quad\quad (6\text{-}65a)$

When $\quad\quad\quad 0.8 < \lambda_{n,s} \leqslant 1.2, \ \tau_{cr} = [1 - 0.59(\lambda_{n,s} - 0.8)]f_v \quad (6\text{-}65b)$

When $\quad\quad\quad\quad\quad\quad\quad \lambda_{n,s} > 1.2, \tau_{cr} = 1.1f_v/\lambda_{n,s}^2 \quad\quad\quad\quad (6\text{-}65c)$

Its curve is similar to Figure 6.23, except that the straight line's upper and lower dividing points in the middle transition section are different.

When the web is not provided with transverse stiffeners, $a/h_0 = \infty$ can be taken, then $K = 5.35$. If required, $\lambda_{n,s}$ should not exceed 0.8. At this time, the limit of web depth-to-thickness ratio can be obtained from Equation (6-64b):

191

$$\frac{h_0}{t_w} = 0.8 \times 41\sqrt{5.34}\varepsilon_k = 75.8\varepsilon_k \qquad (6\text{-}66)$$

It is generally considered that the steel shear ratio limit is equal to $0.8f_{vy}$, and the limit specified by the design standard is $0.8\varepsilon_k$.

(3) The critical force of web zone under single action of local compressive stress

Figure 6.25 shows the elastic buckling of the four-side simply supported web under local compressive stress. During buckling, only a half-wave appears in the longitudinal and transverse directions of the plate, and the critical stress is:

$$\sigma_{c,cr} = \frac{\chi K \pi^2 E}{12(1-v^2)}\left(\frac{t_w}{h_0}\right)^2 \qquad (6\text{-}67)$$

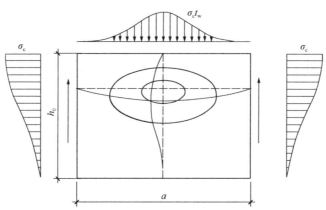

Figure 6.25 Elastic buckling of four-side simply supported plate under local compressive stress

The buckling coefficient $K$ varies with the ratio of $a/h_0$, and can be approximated as:

When
$$0.5 \leqslant a/h_0 \leqslant 1.5, \ K = \left(7.4 + 4.5\frac{h_0}{a}\right)\frac{h_0}{a} \qquad (6\text{-}68)$$

When
$$1.5 < a/h_0 \leqslant 2.0, \ K = \left(11 - 0.9\frac{h_0}{a}\right)\frac{h_0}{a} \qquad (6\text{-}69)$$

For web under local compressive stress, the design standard GB 50017 adopts elastic fixation coefficient:

$$\chi = 1.81 - 0.255\frac{h_0}{a} \qquad (6\text{-}70)$$

The general depth-to-thickness ratio under local pressure is:

$$\lambda_c = \sqrt{\frac{f_y}{\sigma_{c,cr}}} = \frac{h_0/t_w}{28.1\sqrt{\chi K}} \cdot \frac{1}{\varepsilon_k} \qquad (6\text{-}71)$$

$\lambda_c$ can be obtained by substituting the above Equations (6-68)~(6-70) into Equation (6-71), but it is not difficult to find that this equation will be extremely complicated. To simplify the equation, the design standard GB 50017 gives a simple equation of general depth-to-thickness ratio $\lambda_{n,c}$ as follows:

When $\quad 0.5 \leqslant a/h_0 \leqslant 1.5$, $\lambda_{n,c} = \dfrac{h_0/t_w}{28\sqrt{10.9+13.4(1.83-a/h_0)^3}} \cdot \dfrac{1}{\varepsilon_k}$ (6-72a)

When $\quad 1.5 < a/h_0 \leqslant 2.0$, $\lambda_{n,c} = \dfrac{h_0/t_w}{28\sqrt{18.9-5a/h_0}} \cdot \dfrac{1}{\varepsilon_k}$ (6-72b)

The curve of Equation (6-72) is similar to that in Figure 6.23, except that the straight line's upper and lower dividing points in the middle transition section are different.

The critical stress $\sigma_{c,cr}$ adopted in the design standard GB 50017 is similar to $\sigma_{cr}$ and $\tau_{cr}$, and is also divided into three sections.

When $\qquad \lambda_{n,c} \leqslant 0.9$, $\sigma_{c,cr} = f$ (6-73a)

When $\qquad 0.9 < \lambda_{n,c} \leqslant 1.2$, $\sigma_{c,cr} = [1-0.79(\lambda_{n,c}-0.9)]f$ (6-73b)

When $\qquad \lambda_{n,c} > 1.2$, $\sigma_{c,cr} = 1.1f/\lambda_{n,c}^2$ (6-73c)

(4) Checking calculation of the local stability of the web zone under the combined action of various stresses

The beam web zone is generally subjected to the combined action of two or more stresses, so its local stability checking must meet the critical conditions under the common conditions of multiple stresses.

The stiffeners on the beam web can be divided into two categories according to their different functions. One is to divide the web into smaller cells to improve the local stability of the web, called interval stiffeners. Interval stiffeners include transverse stiffeners, longitudinal stiffeners, and short stiffeners. Transverse stiffeners mainly help prevent web instability caused by shear stress. Longitudinal stiffeners mainly help prevent web instability caused by bending compressive stress. Finally, short stiffeners mainly help prevent possible instability of the web caused by local compressive stress. In addition to the functions mentioned above, there is also a function of supporting and transmitting fixed concentrated load or supporting reaction force, called supporting stiffener.

The stiffeners are arranged on the steel beam webs to meet the requirements of local stability. Generally, the stiffeners should be arranged on the webs according to the structural requirements (see Section 6.5.3), and then the various areas of the webs should be checked. Finally, make necessary adjustments when it does not meet the requirements.

① Web zone with transverse stiffeners only

In order to study its local stability, some simplifications are often made. For example, the stress on webs with only transverse stiffeners is often simplified, as shown in Figure 6.26.

According to the design standard GB 50017 of China, the checking conditions for the local stability of webs with only transverse stiffeners are as follows:

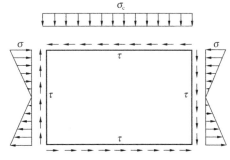

Figure 6.26　Web loading with transverse stiffeners only

$$\left(\frac{\sigma}{\sigma_{cr}}\right)^2 + \left(\frac{\tau}{\tau_{cr}}\right)^2 + \frac{\sigma_c}{\sigma_{c,cr}} \leqslant 1 \qquad (6\text{-}74)$$

where  $\sigma =$ the bending compressive stress (N/mm²) of the edge of the web depth generated by the average bending moment in the calculated grid. $\sigma = Mh_c/I$, $h_c$ is the depth of web zone subject to bending; for the biaxial symmetrical section (I-section, etc.), $h_c = h_0/2$ and $h_0$ is the calculation depth of the web;

$\tau =$ the average shear stress (N/mm²) generated by the average shear force in the calculated grid. $\tau = V/(h_w t_w)$ where $h_w$ is the depth of the web;

$\sigma_c =$ the local compressive stress (N/mm²) at the edge of the calculated depth of the web, calculated according to Equation (6-7), and the value of $\psi$ takes 1.0;

$\sigma_{cr} =$ the critical stress of the web zone under the single action of bending stress, calculated according to Equation (6-58);

$\tau_{cr} =$ the critical stress of the web zone under pure shear, calculated according to Equation (6-65);

$\sigma_{c,cr} =$ the critical force of web zone under single action of local compressive stress, calculated according to Equation (6-73).

② Web zone with both transverse stiffeners and longitudinal stiffeners

Figure 6.27 shows the web is equipped with both transverse and longitudinal stiffeners and its force diagram. The longitudinal stiffeners divide the web into two types of zones I and II, the heights of which are $h_1$ and $h_2$, respectively, and $h_1 + h_2 = h_0$. Local stability calculations should be made for these two divisions, respectively.

a. Zone I between the compression flange and the longitudinal stiffener

The force state of zone I is shown in Figure 6.27(b). The following equation calculates the local stability of the web in this zone.

$$\frac{\sigma}{\sigma_{cr1}} + \left(\frac{\tau}{\tau_{cr1}}\right)^2 + \left(\frac{\sigma_c}{\sigma_{c,cr1}}\right)^2 \leqslant 1.0 \qquad (6\text{-}75)$$

$\sigma_{cr1}$, $\tau_{cr1}$, and $\sigma_{c,cr1}$ is calculated by the following method, respectively.

- The calculation of $\sigma_{cr1}$ according to Equation (6-58), but $\lambda_{n,b1}$ replaces $\lambda_{n,b}$ in equations.

When the compressive flange is constrained,

$$\lambda_{n,b1} = \frac{h_1/t_w}{75\varepsilon_k} \qquad (6\text{-}76a)$$

When the compressive flange is unconstrained,

$$\lambda_{n,b1} = \frac{h_1/t_w}{64\varepsilon_k} \qquad (6\text{-}76b)$$

where $h_1$ is the distance from the longitudinal stiffener to the edge of the web depth (mm).

- The calculation of $\tau_{cr1}$ according to Equation (6-65), but $h_1$ replaces $h_0$ in equations.
- The calculation of $\sigma_{c,cr1}$ according to Equation (6-73), but $\lambda_{n,c1}$ replaces $\lambda_{n,b}$ in equations.

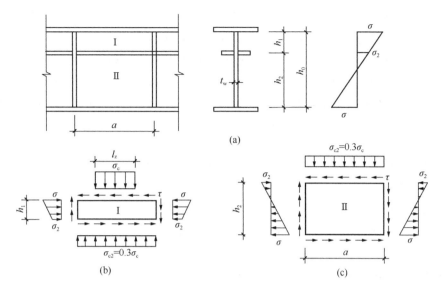

Figure 6.27  Both transverse stiffeners and longitudinal stiffeners strengthen the web

When the compressive flange is constrained,

$$\lambda_{n,c1} = \frac{h_1/t_w}{56\varepsilon_k} \quad (6\text{-}77\text{a})$$

When the compressive flange is unconstrained,

$$\lambda_{n,c1} = \frac{h_1/t_w}{40\varepsilon_k} \quad (6\text{-}77\text{b})$$

Zone I, as shown in Figure 6.27(b), is a long and narrow strip. When the upper part is under local pressure, the zone can be approximately regarded as a vertical central pressure strip. The width is approximated by the width of the middle section of the slab $l_z + h_1 \approx 2h_1$. When the torsion of the upper flange is restrained, the upper end of the strip is regarded as a fixed end, and the lower end is simply supported. When the upper flange torsion is not constrained, it is assumed that the upper and lower ends are simply supported. Therefore, according to Euler's formula, the critical stress in the two cases can be:

$$\sigma_{c,cr1} = \frac{\pi^2 E(2h_1 t_w^3)}{12(1-v^2)(0.7h_1)^2} \cdot \frac{1}{h_1 t_w} = \frac{4\pi^2 E}{12(1-v^2)}\left(\frac{t_w}{h_1}\right)^2 \quad (6\text{-}78)$$

and

$$\sigma_{c,cr1} = \frac{2\pi^2 E}{12(1-v^2)}\left(\frac{t_w}{h_1}\right)^2 \quad (6\text{-}79)$$

from $\lambda_{c1} = \sqrt{f_y/\sigma_{c,cr}}$, we get Equations (6-77a) and (6-77b).

b. Zone II between the tensile flange and the longitudinal stiffener

The force situation of zone II is shown in Figure 6.27(c). The following equation calculates the local stability of the web of this zone.

$$\left(\frac{\sigma_2}{\sigma_{cr2}}\right)^2 + \left(\frac{\tau}{\tau_{cr2}}\right)^2 + \frac{\sigma_{c2}}{\sigma_{c,cr2}} \leqslant 1 \quad (6\text{-}80)$$

$\sigma_2$ = the bending compressive stress (N/mm$^2$) of the web at the longitudinal stiffener generated by the average bending moment in the calculated grid.

$\sigma_{c2}$ = the transverse compressive stress of the web at the longitudinal stiffener.

$\sigma_{cr2}$, $\tau_{cr2}$, cnd $\sigma_{c,cr2}$ can be calculated by the following method, respectively.

- The calculation of $\sigma_{cr2}$ according to Equation (6-58), but $\lambda_{n,b2}$ replaces $\lambda_{n,b}$ in equations.

$$\lambda_{n,b2} = \frac{h_2/t_w}{194\varepsilon_k} \qquad (6\text{-}81)$$

Equation (6-81) is still derived from Equation (6-55) but take $\chi=1.0, K=47.6$, so $(28.1\sqrt{\chi K})=194$.

- The calculation of $\tau_{cr2}$ according to Equation (6-65), but $h_2$ replaces $h_0$ in equations.
- The calculation of $\sigma_{c,cr2}$ according to Equation (6-73), but $h_2$ replaces $h_0$ in equations. When $a/h_2 > 2$, take $a/h_2 = 2$.

③ Web zone with short stiffener between the compression flange and longitudinal stiffener

Figure 6.28 shows the force diagram of the web reinforced with transverse stiffeners, longitudinal stiffeners, and short stiffeners at the same time. The stability calculation of zone II is the same as that in the above ② and calculated by Equation (6-75). Here we only explain the stability calculation of zone I in Figure 6.28. Equation (6-75) is still used to check the stability of zone I. $\sigma_{cr1}$ is still calculated by Equations (6-76) or (6-77); $\tau_{cr1}$ is calculated by Equation (6-65), still we need to change $h_0$ and $a$ to $h_1$ and $a_1$ respectively, and $a_1$ is the distance between short stiffeners; $\sigma_{c,cr1}$ is still calculated by Equation (6-73), but $\lambda_{n,c1}$ replaces $\lambda_{n,c}$ in equations.

For zones with $a_1/h_1 \leqslant 1.2$, when the compressive flange is constrained,

$$\lambda_{n,c1} = \frac{a_1/t_w}{87\varepsilon_k} \qquad (6\text{-}82a)$$

When the compressive flange is unconstrained,

$$\lambda_{n,c1} = \frac{a_1/t_w}{73\varepsilon_k} \qquad (6\text{-}82b)$$

For $a_1/h_1 > 1.2$, the above Equation (6-82) is still used, but the right side of the equation should be multiplied by $1/\sqrt{0.4+0.5a_1/h_1}$.

The source of Equation (6-82) is still from Equation (6-55). Zone I is a four-sided support plate. When the upper and lower compressive stress are equal and $a_1/h_1 \leqslant 1.2$, the buckling coefficient $K$ is approximately equal to 4.0. Now the compressive stress on the top is $\sigma_c$, and the compressive stress on the bottom is $0.3\sigma_c$, and $K$ is approximately equal to 6.8; take $\chi = 1.4$ and $\chi = 1.0$ into Equation (6-55), Equations (6-82a) and (6-82b) are obtained, respectively. When $a_1/h_1 > 1.2$, because the buckling coefficient $K$ will increase linearly with the increase of $a_1/h_1$, it is stipulated that the right side of the Equation (6-82) should be multiplied by $1/\sqrt{0.4+0.5a_1/h_1}$.

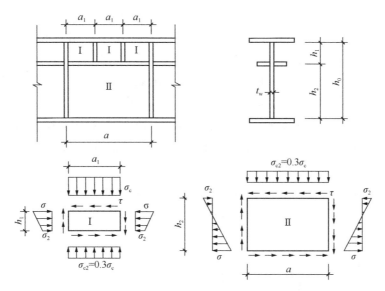

Figure 6.28 Webs reinforced with transverse stiffeners,
longitudinal stiffeners, and short stiffeners

### 6.5.3 The construction details and calculation of stiffener of the beam web

(1) The setting of web stiffeners of the welded beam

The main purpose of stiffeners on the beam web is to ensure the local stability of the web. Due to the direct bearing of dynamic loads in bridge engineering, it is generally not allowed to consider the post-buckling strength utilization of the webs. In housing construction projects, composite beams that bear static loads and indirect dynamic loads should be considered. The strength of the web after buckling is not considered in other cases. If the strength of the web after buckling is considered, the stiffeners can be arranged following the requirements of Section 6.6, and their flexural and shear bearing capacity can be calculated. If the strength after buckling is not considered, the webs of welded cross-section beams should be equipped with stiffeners following Section 6.5.2.

Afte rarranged stiffeners under Section 6.5.2, except for the stiffeners' configuration according to the structural requirements, the stability of each web cell shall be checked by Section 6.5.2. Suppose there is a situation that does not meet the requirements. In that case, it is necessary to make appropriate adjustments to the stiffeners' layout and then recheck the adjusted web cells until all cells meet the stability requirements. Generally, the spacing of the transverse stiffeners can be reduced to improve the local stability of the web cells under shear stress and local compressive stress. However, reducing the spacing cannot improve the local stability of the web cells under bending compressive stress. When the local stability of the web zone under the bending compressive stress does not meet the requirements, the problem can only be solved by setting up a longitudinal stiffening aid in the compression zone. However, setting a short stiffener will increase the manufacturing workload and complicate the structure, so it is generally only used in beams with large local compressive stress (such as crane beams with large wheel pressure).

(2) Structural requirements for web spacing stiffeners

① The position of the stiffener on the side of the web

The stiffeners should be arranged in pairs on both sides of the web (Figure 6.29a). When the web only bears static load or small dynamic load, to save steel or reduce manufacturing workload, the horizontal stiffener and longitudinal stiffeners can also be arranged on one side, as shown in Figure 6.29b. However, supporting stiffeners and stiffeners of heavy-duty crane beams should be arranged on two sides.

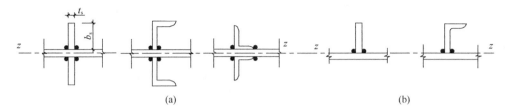

(a)                                                     (b)

Figure 6.29  The form of the beam web stiffener

② Section form and material of stiffener

The stiffener can be made of steel plate or steel shapes, and the welded beam is generally made of steel plate. Q235 steel is generally used for stiffeners because stiffeners mainly use their stiffness, and it is not economical to use high-strength steel as stiffeners.

③ Spacing and position of stiffener

The minimum spacing of transverse stiffeners is $0.5h_0$, and the maximum spacing is $2h_0$. For beams without local compressive stress: when $h_0/t_w \leqslant 100$, the maximum spacing can be $2.5h_0$. The distance between longitudinal stiffeners and the edge of the web shall be $h_c/2.5 - h_c/2$ range (for the single-axis symmetric cross-sections, that is, in the range of $h_0/5 - h_0/4$).

④ The stiffness requirements of the stiffener

The stiffener should have sufficient stiffness to be used as a reliable support to prevent lateral deflection and instability of the web. Therefore, the steel structure design code has the following requirements for the section size and section moment of inertia of the stiffener:

a. The transverse stiffeners of steel plates arranged in pairs on both sides of the web shall meet the following requirements:

Outstanding width:

$$b_s \geqslant \frac{h_0}{30} + 40 \text{ (mm)} \qquad (6\text{-}83)$$

Thickness:

Pressure stiffener $t_s \geqslant \dfrac{b_s}{15}$ (mm); unforced stiffener $t_s \geqslant \dfrac{b_s}{19}$ (mm). (6-84)

The transverse stiffener of the steel plate arranged on only one side of the web shall have an extension width greater than 1.2 times calculated according to Equation (6-83), and its thickness shall conform to the requirements of Equation (6-84). The 1.2 times here are based on the condition that the rigidity of the single-sided configuration and the double-sided configuration are the same.

b. In the web reinforced with transverse stiffeners and longitudinal stiffeners, at the intersection of longitudinal stiffeners and transverse stiffeners, the longitudinal stiffeners should be disconnected, and the transverse stiffeners should be continuous (Figure 6.30a). At this time, the cross-sectional dimensions of the transverse stiffeners should meet the requirements of a, and the cross-sectional moment of inertia $I_z$ should still meet the following requirements:

$$I_z \geqslant 3h_0 t_w^3 \qquad (6-85)$$

The moment of inertia $I_y$ of the longitudinal stiffener shall meet the following requirements:

$$\left. \begin{array}{l} \text{When } a/h_0 \leqslant 0.85, \quad I_y \geqslant 1.5 h_0 t_w^3 \\ \text{When } a/h_0 > 0.85, \quad I_y \geqslant \left(2.5 - 0.45 \dfrac{a}{h_0}\right)\left(\dfrac{a}{h_0}\right)^2 h_0 t_w^3 \end{array} \right\} \qquad (6-86)$$

where  $I_z$ = the moment of inertia of the transverse stiffener section with respect to the horizontal axis of the web ($z$-axis).

$I_y$ = the moment of inertia of the longitudinal stiffener section with respect to the vertical axis of the web ($y$-axis).

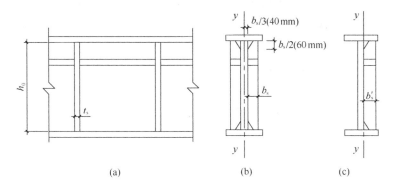

Figure 6.30  Structural requirements of stiffeners

For the $z$-axis and $y$-axis mentioned above, when the stiffeners are arranged on both sides, take them as the web axis (Figure 6.29a, Figure 6.30b). When the stiffeners are arranged on one side, take the web and the stiffener (Figure 6.29b, Figure 6.30c).

When the width-to-thickness ratio of the compression flange cannot meet the requirements of Equation (6-50) or Equation (6-53), stiffeners should be provided to increase the compressive bearing capacity of the compression flange. At the intersection of the longitudinal stiffener and the transverse stiffener, the transverse stiffener is often slotted to pass the longitudinal stiffener continuously.

c. The minimum spacing of short stiffeners is $0.75 h_1$. Therefore, the extension width of the short steel plate stiffener should be taken as 0.7-1.0 time the extension width of the transverse stiffener, and the thickness should not be less than 1/15 of the extension width of the short stiffener.

d. The section moment of inertia of the stiffener made of hot-rolled shapes (H-shape, I-section, Channel, and Angle welded to the web) shall not be less than the moment of inertia of the corresponding steel plate stiffener.

⑤ Stiffener cut corner

In order to avoid the intersection of the three-direction welds and reduce the residual welding stress, the transverse stiffeners of welded beams should be corner cut at the part connected to the flange. The width is about $b_s/3$ (but not more than 40 mm), the height is about $b_s/3$ (but not more than 60 mm) (Figure 6.30b). $b_s$ is the width of the stiffener. Therefore, the flange welds of the beam can pass continuously. When the cut corner is used as a welding process hole, the cut corner should adopt a 1/4 arc with a radius of $R=$ 30 mm. In addition, at the intersection of the longitudinal stiffener and the transverse stiffener, two oblique corners at both ends of the longitudinal stiffener should be cut off so that welds connecting the transverse stiffener and the web can pass continuously.

⑥ The structural requirement of the transverse stiffener of the crane beam

The width of the transverse stiffener of the crane beam should not be less than 90 mm. The upper end of the transverse stiffener of the crane beam should be flattened and topped tightly between the bottom surfaces of the upper flange. In the heavy-duty crane beam (A6-A8), the middle transverse stiffener should be arranged in pairs on both sides of the web, while the crane beams for light-duty and medium-duty (A1-A5) can be arranged on one side or staggered. In welded crane beams, transverse stiffeners (including short stiffeners) must not be welded to the tension flange but can be welded to the compression flange. The lower end of the transverse stiffener in the middle of the crane beam should be cut off at 50-100 mm from the lower flange (Figure 6.31a), and not welded to the lower flange, so as not to reduce the fatigue strength. At this time, the connection weld with the web should not be arced or extinguished at the lower end of the stiffener. Sometimes in order to improve the torsional stiffness of the beam, a short Angle can also be added to the lower end of the stiffener to be welded firmly. Still, the short Angle is pressed against the tension flange without welding (Figure 6.31b).

(3) Calculation of supporting stiffener or pressure-bearing stiffener

Supporting stiffeners are set when a large fixed concentrated load is applied on a steel beam and set at the support. They often bear and transmit the concentrated load or the re-

## 6.5 Local Stability of Steel Plate Composite Beam

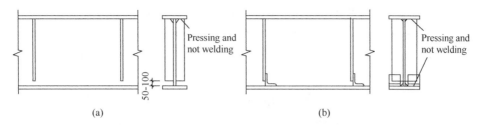

Figure 6.31 Structural requirements of transverse stiffeners of crane beam

action force of the support. Supporting stiffeners should be arranged in pairs on both sides of the web (Figure 6.32), and their cross-section is often larger than that of general transverse stiffeners. The transverse stiffeners at the support of the crane beam should be arranged in pairs on both sides, and they should be flattened and tightened with the upper and lower flanges of the beam. The end supporting stiffeners can be welded with the upper and lower flanges of the beam. There are two main structural forms of support stiffeners, flat support stiffeners (Figures 6.32a, 6.32b) and outstanding flange support stiffeners (Figure 6.32c). The protruding length should not be greater than two times its thickness (Figure 6.32c).

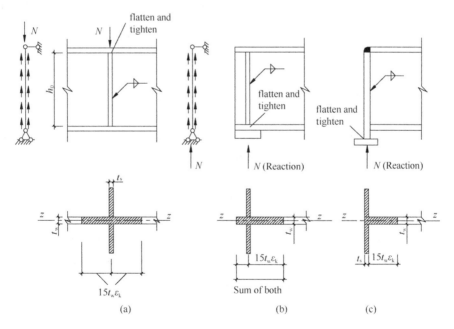

Figure 6.32 Supporting stiffeners

① Stability calculation of supporting stiffener

Supporting stiffeners of beams shall be calculated on the out-of-plane stability of the web as axially loaded compression members under the support reaction or fixed concentrated loads. When the supporting stiffener occurs buckling outside the web plane, the connected web has a certain restraint effect. Therefore, the stiffener's cross-sectional area should add the web area within the width of $15t_w\varepsilon_k$ on both sides of the stiffener when calculating the cross-sectional area of the compression member, as shown in Figure 6.32. Take the actual width (Figure 6.32b) when the actual width of the web on one side is less than $15t_w\varepsilon_k$.

The calculation diagram of the supporting stiffener is shown in Figures 6.32a and 6.32b. Under the action of concentrated load, the reaction force is distributed within the full length of the stiffener, and the effective length can be taken as $h_0$ safely. When checking the stability factor $\varphi$ of the axially loaded compression member, the cross-section shown in Figure 6.32a is a type b cross-section, and the cross-sections shown in Figures 6.32b and 6.32c are type c cross-sections. Calculated as follows:

$$\frac{N}{\varphi A f} \leqslant 1.0 \tag{6-87}$$

where  $N=$ the design value of the concentrated load or bearing reactions;

$A=$ the cross-sectional area of the supporting stiffener. The overall stability is calculated as an axial compression member and adopts according to the shaded area shown in Figure 6.32;

$\varphi=$ the stability factor of the axially loaded compression member is determined by $\lambda$ in Table D.1 or Table D.3 in Appendix D, where $\lambda = \frac{h_0}{i_z} \cdot \frac{1}{\varepsilon_k}$. $i_z = \sqrt{I_z/A}$, $I_z$ is the moment of inertia of the shaded area shown in Figure 6.32 to the $z$-$z$ axis.

② Calculation of bearing strength of end face of supporting stiffener

The end of the beam supporting stiffener is generally planed flat and tightly attached to the flange or column top of the beam, and its end surface compressive strength is calculated as follows:

$$\sigma_{ce} = \frac{N}{A_{ce}} \leqslant f_{ce} \tag{6-88}$$

where  $A_{ce}=$ the pressure-bearing area of the end face, which is the cross-sectional area of the supporting stiffener contacting with the beam flange or column's top plate; the area of the corner loss at the end of the stiffener should be considered to be subtracted;

$f_{ce}=$ the strength design value of steel end bearing pressure (flatten and tighten) can be determined by referring to the Attached Table B.1.

③ Calculation of the connection weld between the stiffener and the web

Assume the weld stress distributes along the full length of the weld evenly under all reaction force or concentrated load, so it is unnecessary to consider the limitation that the effective length $l_w$ of the side fillet weld shall not be greater than $60h_f$. Calculate as follows:

$$\frac{N}{0.7h_\mathrm{f}\sum l_\mathrm{w}} \leqslant f_\mathrm{f}^\mathrm{w} \qquad (6\text{-}89)$$

where $h_\mathrm{f}$ = the leg size of the fillet weld should meet the requirements of construction detail.

Due to the long weld length, the $h_\mathrm{f}$ calculated by Equation (6-89) is very small, which is generally controlled by the minimum weld leg requirements.

## 6.6 Design Considering the Post-Buckling Strength of Beam Web

### 6.6.1 Working performance of beam web after buckling

For welded composite beams subjected to static load and indirect dynamic load, the post-buckling strength of the web should be considered. After the beam web buckles under the action of shear, when the load continues to be applied, although the web produces wave-shaped deformation in the direction of the main compressive stress, it cannot continue to resist the load. The membrane can withstand a large tensile force in the main tensile stress direction due to the membrane's tension, forming a tension field, as shown in Figure 6.33. A web and two flanges and the two stiffeners make the beam like a truss. The upper and lower flanges are equivalent to the upper and lower chords, the tension field of the web is equivalent to the diagonal struts of the truss, and the stiffeners are equivalent to the vertical struts of the truss. In this way, after the web is buckled, it has a greater continuous bearing capacity to use the post-buckling strength. In this regard, the depth-to-thickness ratio of the web in the composite beam can be increased. When the depth-to-thickness ratio reaches 250, it is unnecessary to install longitudinal stiffeners (only transverse stiffeners) to obtain better economic effects.

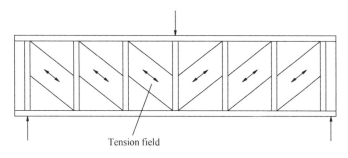

Figure 6.33 Tension field effect of the web

### 6.6.2 Design features after web buckling

(1) Strength calculation

Consider the post-buckling strength of the web, including the shear capacity, the flexural capacity, and their combined effects. The calculation methods stipulated by the design standard GB 50017 will be introduced separately below.

① Calculation of bearing capacity after web buckling

a. Shear capacity

The calculation result of the tension field method is more accurate, but the process is cumbersome. The Chinese code adopts a simplified calculation method. The calculation formula for the design value $V$ of the shear bearing capacity of the beam web considering the buckling strength is as follows:

When $\quad\quad\quad\quad\quad \lambda_{n,s} \leqslant 0.8, V_u = h_w t_w f_v \quad\quad\quad\quad$ (6-90)

When $\quad\quad\quad 0.8 < \lambda_{n,s} \leqslant 1.2, V_u = h_w t_w f_v [1 - 0.5(\lambda_s - 0.8)] \quad\quad$ (6-91)

When $\quad\quad\quad\quad\quad \lambda_{n,s} > 1.2, V_u = \dfrac{h_w t_w f_v}{\lambda_{n,s}^{1.2}} \quad\quad\quad\quad$ (6-92)

where $\lambda_{n,s}$ = the regularized depth-to-thickness ratio for calculating beam webs under pure shear, calculated by Equation (6-64).

From the formula $\tau = V/h_w t_w$, the web shear stress $\tau_u$ considering the strength after buckling can be obtained. Its curve $A'B'C'D'$ is similar to the curve $ABCD$ of the critical shear stress without considering the strength after buckling. Both are composed of three sections, as shown in Figure 6.34. It can be seen from Figure 6.34 that when $\lambda_{n,s} \leqslant 0.8$, $\tau_u = f_v = \tau_{cr}$, there is no post-buckling strength; when $\lambda_{n,s} > 0.8$, the post-buckling strength can be used, and it increases with the increase of $\lambda_{n,s}$.

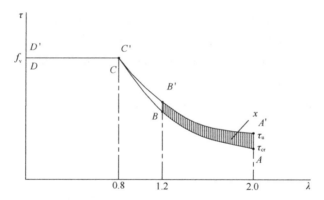

Figure 6.34  Tension field effect of the web

b. Bending capacity

The tension zone is also deducted from this height, as shown in Figure 6.35 so that the neutral axis will not change and the result is safe. Under normal stress, the behavior of the beam web after buckling is different from that under shear. The web buckles under the bending moment; simultaneously, the bending compression zone will undergo concave-convex deformation. As a result, the web in the partial compression zone cannot bear the compressive stress and withdraw from work. In order to calculate the flexural bearing capacity after buckling, the concept of an effective section is adopted. It is assumed that the effective height of the compression zone of the web is $\rho h_c$, which is equally divided at the two ends of $h_c$, and the height of $(1-\rho)h_c$ is deducted in the middle.

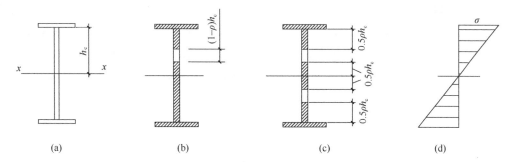

Figure 6.35 Tension field effect of the web

According to the section shown in Figure 6.35(c), the design value of the flexural bearing capacity of the beam is:

$$M_{eu} = \gamma_x \alpha_e W_x f \qquad (6\text{-}93)$$

$$\alpha_e = 1 - \frac{(1-\rho)h_c^3 t_w}{2I_x} \qquad (6\text{-}94)$$

where  $M_{eu}$ = the design value of the flexural bearing capacity of the beam;

$\gamma_x$ = the plastic development coefficient about $x$-axis;

$\alpha_e$ = the reduction factor of the beam section modulus considering the effective depth of the web;

$W_x$ = the modulus of gross cross-section determined by the largest fiber under compression;

$f$ = the steel design strength;

$h_c$ = the depth of the compression zone of the web calculated according to all the beam cross-section;

$t_w$ = the thickness of the web;

$I_x$ = moment of inertia of the gross cross-section about the $x$-axis;

$\rho$ = effective height coefficient of the web compression zone.

Take the $\rho$ value according to the following requirements:

When $\qquad \lambda_{n,b} \leqslant 0.85, \rho = 1.0 \qquad (6\text{-}95)$

When $\qquad 0.85 < \lambda_{n,b} \leqslant 1.25, \rho = 1 - 0.82(\lambda_{n,b} - 0.85) \qquad (6\text{-}96)$

When $\qquad \lambda_{n,b} > 1.25, \rho = \dfrac{1}{\lambda_{n,b}}\left(1 - \dfrac{0.2}{\lambda_{n,b}}\right) \qquad (6\text{-}97)$

where  $\lambda_{n,b}$ = the regularized width-to-thickness ratio for calculating beam webs under pure bending, calculated by Equation (6-59).

② The strength check formula of the web after buckling

Generally, beam webs equipped with transverse stiffeners usually bear bending moment and shear combined action simultaneously. In this case, it is very complicated to accurately calculate the bending and shear resistance of the beam after web buckling. Studies have shown that when the edge normal stress reaches the yield strength, the web of the I-section welded beam can also withstand a shear force of $0.6V_u$. The post-buckling strength under the combined action of bending and shear is similar. When the shear force does not exceed $0.5V_u$, the strength of the web after flexural buckling does not decrease, as shown in Figure 6.36. Therefore, the formula for calculating the strength after considering the buckling of the web given by the design standard GB 50017 is:

$$\left(\frac{V}{0.5V_u} - 1\right)^2 + \frac{M - M_f}{M_{eu} - M_f} \leqslant 1.0 \qquad (6-98)$$

$$M_f = \left(A_{f1}\frac{h_{m1}^2}{h_{m2}} + A_{f2}h_{m2}\right)f \qquad (6-99)$$

where  $M,V =$ the calculated design value of bending moment (N · mm) and design value of shear force (N) of the beam on the same section. When $V \leqslant 0.5V_u$, take $V = 0.5V_u$; when $M \leqslant M_f$, take $M = M_f$;

$M_f =$ the design value of bending moment that the two flanges of the beam can bear (N · mm);

$M_{eu} =$ the design value of the flexural bearing capacity of the beam, calculated by Equation (6-93).

$A_{f1}, h_{m1} =$ the larger flange's cross-sectional area (mm$^2$) and the distance from its centroid to the beam's neutral axis (mm);

$A_{f2}, h_{m2} =$ the smaller flange's cross-sectional area (mm$^2$) and the distance from its centroid to the beam's neutral axis (mm).

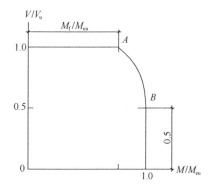

Figure 6.36  The correlation curve of bending moment and shear

(2) Stiffener design

① When only supporting stiffeners cannot meet requirements of Equation (6-98), middle transverse stiffeners shall be arranged on both sides. In addition to meeting the requirements of Equations (6-83) and (6-84), the cross-sectional sizes of the middle transverse stiffener and the middle support stiffener with concentrated pressure on the upper end shall be chacked out-of-plane stability according to axially loaded compression member. The axial pressure should be calculated according to the following equation:

$$N_s = V_u - \tau_{cr}h_w t_w + F \qquad (6-100)$$

where  $V_u =$ calculate according to Equations (6-90)~(6-92) (N);

$h_w =$ web height (mm);

$\tau_{cr} =$ calculate according to Equations (6-65a)~(6-65c) (N/mm$^2$);

$F =$ the concentrated pressure (N) acting on the upper end of the middle support stiffener.

② When the zone near the support and $\lambda_{n,s} > 0.8$, the support stiffener should bear the horizontal component $H$ of the tensile field and the reaction force. Its strength and stability outside the plane of the web should be checked as a compressive-bending member. The cross-section and effective length of the support stiffener shall comply with the provisions of Section 6.5.3. $H$ is at $h_0/4$ from the upper edge of the web's calculated height and shall be calculated according to the following equation:

$$H = (V_u - \tau_{cr} h_w t_w)\sqrt{1 + \left(\frac{a}{h_0}\right)^2} \tag{6-101}$$

where $a$ = the stiffener spacing at the end zone near the support for beams with transverse stiffeners in the middle. For webs without middle stiffeners, take the distance (mm) from the beam support to the point where the shear is zero.

③ When the support stiffener is the same as that shown in Figure 6.37, the calculation can be carried out according to the simplified method. Stiffener one is calculated as an axially loaded compression member bearing the reaction force $R$, and the cross-sectional area of the head stiffener two shall not be less than the value calculated by the following equation:

$$A_c = \frac{3h_0 H}{16ef} \tag{6-102}$$

④ For beams considering the post-buckling strength of the web, the depth-to-thickness ratio of the web should not be greater than 250, and intermediate transverse stiffeners may be provided according to the structural requirements. $a > 2.5h_0$ and for webs without intermediate transverse stiffeners, when the Equation (6-74) is satisfied, the horizontal component force $H = 0$ can be taken.

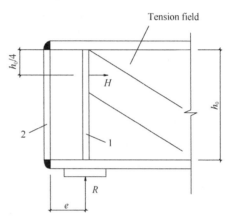

Figure 6.37 Beam end structure with head stiffener
1. supporting stiffener; 2. head stiffener

## 6.7 The Design of the Beam

### 6.7.1 Section selection of beams

(1) Selection of hot-rolled shaped beam

The choice of the hot-rolled shaped beam (universal beam) is relatively simple, only according to the calculation of the maximum bending moment in the beam according to the following equation to find the required net section modulus:

$$W_{nx} = \frac{M_{xmax}}{\gamma_x f} \tag{6-103}$$

where  $M_{xmax}$ = the maximum bending moment generated by load design value;

$\gamma_x$ = the plastic development coefficient;

$f$ = the steel design strength.

Then select the section with a section modulus close to $W_{nx}$ from the section specification table as the initial selecting section. In order to save steel, the design should avoid opening the stud hole on the section acted by the maximum bending moment so as not to weaken the section.

(2) Selection of welded composite beam section

The welded composite beam section selection includes estimating beam depth, web thickness, and flange size.

① Estimation of beam depth

To determine the depth of the beam, the building requirements, the stiffness of the beam, and the beam's economic conditions should be considered. The beam's building height requirement determines the maximum depth $h_{max}$ of the beam, while the building requirement depends on the users' requirement. The stiffness requirement of the beam determines the minimum depth $h_{min}$ of the beam. In order to meet the required modulus of the cross-section, there can be a variety of schemes. The beam can be deep and narrow, or shallow and wide. The former uses less steel for flange and more steel for the web, while the latter, on the contrary, the appropriate plan is to make the total steel minimum. The depth of the beam determined according to this principle is called economic depth $h_e$. With the above three depths, you can choose the beam depth. Appropriate beam depth is between the maximum depth $h_{max}$ and the minimum depth $h_{min}$, as close to the economic depth $h_e$ as possible.

The $h_{min}$ calculation method is illustrated by taking the simply supported beam under uniformly distributed load as the following example. The maximum deflection of the beam shall meet the requirements of design standard (specification):

$$v_{max} = \frac{5q_k l^4}{384EI_x} = \frac{5l^2}{48EI_x} \cdot \frac{q_k l^2}{8} = \frac{5M_{k\,max} l^2}{48EI_x} = \frac{5}{48} \cdot \frac{M_{k\,max} l^2}{EW_x(h/2)} = \frac{5\sigma_{k\,max} l^2}{24Eh} \leqslant [v]$$

$$\tag{6-104}$$

where $v_{max}$ is the maximum deflection of the beam calculated according to the standard load; $l$ is the span of the beam; $q_k$ is the uniformly distributed load standard value; $I_x$ and $W_x$ are the moment of inertia and section modulus of the beam, respectively; $M_{k\,max}$ is the maximum bending moment in the beam span generated by the load standard value; $\sigma_{k\,max}$ is the maximum normal stress of the section at the maximum bending moment in the beam, $\sigma_{k\,max} = M_{k\,max}/W_x$.

It can be seen from Equation (6-104) that the stiffness of the beam is directly related to the depth. In order to give full play to the strength of the beam and ensure the stiffness of the beam, $\sigma_{k\,max} = f/1.3$ (1.3 is the average value of the partial coefficients of permanent load and live load) is taken. From this:

$$h_{min} = \frac{5fl}{31.2E}\left[\frac{l}{v}\right] \tag{6-105}$$

The economic depth can be calculated by the following empirical equation (Unit: cm):

$$h_e = 7\sqrt[3]{W_x} - 30 \tag{6-106}$$

② The web size

After determining the depth of the beam, the height of the web is also determined. The web height is the beam depth minus the thickness of the two flanges. When taking the web's height, the specification and size of the steel plate should be considered. From the economic point of view, the web is thinner than flange, but the thickness of the web should consider the shear strength of the web, the local stability of the web, and the structure requirements. From the view of shear strength, the following equation should be satisfied:

$$\tau_{max} = 1.2\frac{V_{max}}{h_w t_w} \leqslant f_v \tag{6-107}$$

It is assumed that the maximum shear stress of the web is 1.2 times the average shear stress, $V_{max}$ is the maximum shear force of the beam.

The web thickness obtained by Equation (6-107) shall meet the following requirements:

$$t_w \geqslant \frac{1.2V_{max}}{h_w f_v} \tag{6-108}$$

As calculated by the above equation, the $t_w$ value is generally small. In order to meet the requirements of local stability and construction, it is usually estimated according to the following empirical equation:

$$t_w = \frac{\sqrt{h_w}}{3.5} \tag{6-109}$$

The above two Equations (6-108) and (6-109) can determine the thickness of the web. Web thickness should be chosen following the specifications of the steel plate. The choice of the thin web is, of course, to save steel, but in order to ensure local stability, it is necessary to equip the stiffener. The structure is more complex. Simultaneously, too small thickness reduces the bearing capacity easily due to corrosion, and it is also easy to produce large deformation in the manufacturing process. In addition to the economy, too much thickness will make it difficult to manufacture, so when choosing web thickness, the above factors need to be considered comprehensively. Generally speaking, the web thickness should be within the range of 8-22 mm. For some small-span beams, the minimum web thickness can be 6 mm.

The depth-to-thickness ratio of the web must be controlled to avoid setting longitudinal stiffeners and making the structure too complicated.

(3) The flange size

The required net section modulus can be obtained from Equation (6-104), then the moment of inertia required for the whole section is:

$$I_x = W_{nx} \cdot \frac{h}{2} \qquad (6\text{-}110)$$

Since the web has been determined, its moment of inertia is:

$$I_w = \frac{1}{12} t_w h_0^3 \qquad (6\text{-}111)$$

Then the moment of inertia required by the flange is:

$$I_t = I_x - I_w \approx 2bt\,(h_0/2)^2 \qquad (6\text{-}112)$$

By Equation (6-112):

$$bt = \frac{2(I_x - I_w)}{h_0^2} \qquad (6\text{-}113)$$

The entire flange width $b$ or thickness $t$ can be determined by one, then the other can be determined. $b$ usually takes $(1/5\text{-}1/3)h$, and simultaneously, it needs $b/t \leqslant 30\sqrt{f_y/235}$ to ensure local stability. If the section considers the development of partial plasticity, then $b/t \leqslant 26\sqrt{f_y/235}$.

When choosing $b$ and $t$, the size of the steel plate should conform to its specification. $b$ is usually a multiple of 10 mm, $t$ take a multiple of 2 mm and not less than 8 mm.

### 6.7.2  Checking the beam

The checking calculation of the hot-rolled shaped beam includes stiffness, strength, and global stability. For the welded composite beam, the checking calculation needs to add the local stability.

(1) The strength calculation

Strength check includes normal stress, shear stress, local compressive stress, and the equivalent stress check under complex stress.

① Normal stress

The maximum bending moment of the beam section and the place where the maximum normal stress is likely to occur (such as the place where the variable section and the place where the section is greatly weakened) are found out using the course *Material Mechanics*.

② Shear stress

In general, the beam bears both bending moment and shear force. Therefore, when the shear strength of the beam is insufficient (rarely), the most effective method is to increase the web area. However, the height of the web is generally determined by the beam's stiffness conditions and structural requirements, so the design method can increase the beam's shear strength by increasing the thickness of the web.

③ Local compressive stress

Acting on the load on the beam generally takes the form of distributed load and concentrated load. The concentrated load in the actual project also has a certain length, but its distribution range is small. Therefore, when the flange stress is along with the web plane, including concentrated load and reaction force without bearing stiffener or moving concentrated load (such as crane wheel pressure), it should check web local pressure on the high-stress edge.

④ Equivalent stress

The junction of the web and flange of the steel plate composite (or welded) beam is sometimes under a complex stress state. For instance, the large normal stress, shear stress, and local compressive stress at the same time, or large normal stress and shear stress at the same time (such as the central support of the continuous beam or the change of the flange section of the beam, etc.), the converted stress should be checked according to the Equation (6-114).

$$\sqrt{\sigma^2 + \sigma_c^2 - \sigma\sigma_c + 3\tau^2} \leqslant \beta_1 f \qquad (6\text{-}114)$$

(2) Beam stiffness calculation

As all known as, the deflection of the floor beam will give people a sense of discomfort and insecurity, but also make the attachment such as plaster off, affecting the use. Excessive deflection of the crane beam will affect the normal operation of the crane. Therefore, in addition to the bearing capacity meeting the requirements, the beam stiffness should be checked to ensure the beam's normal use. For members with different requirements, the maximum deflection limit values are also different. The design standard GB 50017 gives the allowable deflection values of crane beam, girder, floor beam, roof beam (purlin), working platform beam, and wall frame beam.

There are many methods to calculate the beam's deflection, which can be calculated using material mechanics and structural mechanics methods or the manual of static structural or general mechanical software. Generally, the beam load is uniformly distributed load and concentrated force and can be calculated by Equation (6-115) when the beam load is uniformly distributed at the same section. Unfortunately, under the action of multiple concentrated forces (such as crane beam, floor girder, etc.), it isn't easy to accurately calculate the deflection. However, as the deflection is close to that of the beam under the same maximum bending moment under the action of uniformly distributed load, we can obtain the following simplified calculation equation:

Beam with equal section simply supported,

$$v = \frac{5}{384} \frac{q_k l^4}{EI_x} = \frac{5}{48} \frac{q_k l^4}{8EI_x} \approx \frac{M_{k\max} l^2}{10 EI_x} \qquad (6\text{-}115)$$

Simply supported beams with variable section,

$$v = \frac{M_{k\,max}l^2}{10EI_x}\left(1+\frac{3}{25}\frac{I_x-I_{x1}}{I_x}\right) \qquad (6\text{-}116)$$

where $q_k$ is the standard value of uniform distributed load; $M_{k\,max}$ is the maximum bending moment in the beam span generated by the load standard value; $I_x$ is the moment of inertia of section of beam in the mid-span; $I_{x1}$ is the moment of inertia of the cross-section near the support.

For the composite beam, the beam is taller than $h_w$ when selecting the section, so it is unnecessary to check the stiffness.

(3) Check the overall stability

The beam's overall stability shall be checked when the global stability condition without calculating, as mentioned above, is not satisfied. Even if the maximum bending normal stress of the fiber under compression on the beam section is not more than the critical stress of the overall stability, after considering the resistance component coefficient, it can be obtained:

A beam with unidirectional bending in the main plane of maximum stress:

$$\frac{M_x}{\varphi_b W_x f} \leqslant 1.0 \qquad (6\text{-}117)$$

An H-shaped or I-section beam bent in two main planes:

$$\frac{M_x}{\varphi_b W_x f}+\frac{M_y}{\gamma_y W_y f} \leqslant 1.0 \qquad (6\text{-}118)$$

(4) Check local stability

The hot-rolled shaped beam's local stability has met the requirements, and it is unnecessary to check the calculation. For welded composite beams, the flange can avoid local buckling by limiting the plate's ratio of width-to-thickness. The web is more complex. One method is to ensure that local instability does not occur by setting stiffeners. Another method is to allow local instability of the web and use its post-buckling bearing capacity. The design standard GB 50017 suggests that post-buckling strength should be considered for beams subjected to static and indirect dynamic loads. See the design details of local stability in Section 6.6.

As long as one of the above cross-section check calculations is not met, or all requirements are met but the cross-section is obviously surplus, it is necessary to make corresponding adjustments to the primarily selected cross-section. Then recheck the adjusted section until an appropriate section is obtained.

## 6.8 Working Examples

[**Example**] A simple supporting beam with a span of 6 m bears a uniformly distributed load (acting on the upper flange of the beam). The standard value of the permanent load is

20 kN/m, and the standard value of the variable load is 25 kN/m. Design a welded I-section made of Q235B steel.

**Solution:**

Standard load: $q_k = 20 + 25 = 45$ kN/m

Design load: $q = 1.2 \times 20 + 1.4 \times 25 = 59$ kN/m

Maximum bending moment in beam span: $M_{max} = 59 \times 6^2/8 = 265.5$ kN·m

$[v_T]$ is $l/400$, the minimum depth of the beam is:

$$h_{min} = \frac{5fl}{31.2E}\left(\frac{l}{v_T}\right) = \frac{5 \times 215 \times 6 \times 10^3}{31.2 \times 2.06 \times 10^5}\left(\frac{6 \times 10^3}{6 \times 10^3/400}\right) = 401.4 \text{ mm}$$

The net section modulus is:

$$W_{nx} = \frac{M_{max}}{\gamma_x f} = \frac{265.5 \times 10^6}{1.05 \times 215} = 1.18 \times 10^6 = 1.18 \times 10^3 \text{ cm}^3$$

The economic depth of the beam is:

$$h_e = 7\sqrt[3]{W_x} - 30 = 44 \text{ cm}$$

Therefore, the girder web is 450 mm high.

The maximum shear at the support is:

$$V_{max} = 59 \times 6/2 = 177 \text{ kN}$$

$$t_w \geq \frac{1.2 V_{max}}{h_w f_v} = \frac{1.2 \times 177 \times 10^3}{450 \times 125} = 3.8 \text{ mm}$$

$$t_w = \frac{\sqrt{h_w}}{3.5} = \frac{\sqrt{450}}{3.5} \text{ mm} = 6.1 \text{ mm}$$

The web thickness is: $t_w = 8$ mm, so the web is $-450 \times 8$ steel plate.

Assume that the beam depth is 500 mm and the required net section moment of inertia is:

$$I_{nx} = W_{nx}\frac{h}{2} = 1.18 \times 10^6 \times 500/2 \approx 2.95 \times 10^8 \text{ mm}^4 = 2.95 \times 10^4 \text{ cm}^4$$

The web moment of inertia is:

$$I_w = t_w h_0^3/12 = 0.8 \times 45^3/12 = 6,075 \text{ cm}^4$$

$$bt = \frac{2(I_{nx} - I_w)}{h_0^2} = \frac{2(2.95 \times 10^4 - 6,075)}{45^2} = 23.1 \text{ cm}^2$$

$$b = h_0/3 = 150 \text{ mm}, t = 23.1/15 = 1.54 \text{ cm}, t = 18 \text{ mm}.$$

$$t > b/26 = 5.8 \text{ mm}.$$

The flange is $-150 \times 18$.

The selected section size is shown in Figure 6.38.

The moment of inertia of the section is:

$$I_x = t_w h_0^3/12 + 2bt\left[\frac{1}{2}(h_0 + t)\right]^2$$

$$= 0.8 \times 45^3/12 + 2 \times 15 \times 1.8\left[\frac{1}{2} \times (45 + 1.8)\right]^2$$

$$= 35,643 \text{ cm}^4$$

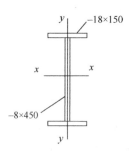

Figure 6.38 Beam cross-section

$$W_x = \frac{I_x}{h/2} = \frac{35,643}{(45+3.6)/2} = 1,467 \text{ cm}^3$$

$$A = 2bt + t_w h_0 = 2 \times 15 \times 1.8 + 0.8 \times 45 = 90 \text{ cm}^2$$

Strength checking calculation:

Beam weight:

$$g = A\gamma = 0.009 \times 7.85 \times 9.8 = 0.69 \text{ kN/m}$$

Design load:

$$q = 1.2 \times (0.69 + 20) + 1.4 \times 25 = 59.83 \text{ kN/m}$$

$$M_{max} = ql^2/8 = 59.83 \times 6^2/8 = 269.2 \text{ kN} \cdot \text{m}$$

$$\sigma = \frac{M_{max}}{\gamma_x W_{nx}} = \frac{269.2 \times 10^6}{1.05 \times 1,467 \times 10^3} = 174.8 \text{ N/mm}^2 < f = 205 \text{ N/mm}^2$$

The shear stress and stiffness need not be checked because the web size and beam depth have been met.

If there are no supporting stiffeners at the support, the local compressive stress should be checked. Generally, the main beam (girder) is provided with supporting stiffeners.

Check the overall stability:

$$I_y = 2 \times 1.8 \times 15^3/12 = 1,012.5 \text{ cm}^4$$

$$i_y = \sqrt{\frac{I_y}{A}} = \sqrt{\frac{1,012.5}{90}} = 3.35 \text{ cm}$$

$$\lambda_y = l_1/i_y = 600/3.35 = 179$$

$$\xi = \frac{l_1 t_1}{b_1 h} = \frac{6 \times 0.018}{0.15 \times 0.486} = 1.481$$

$$\beta_b = 0.69 + 0.13\xi = 0.69 + 0.13 \times 1.481 = 0.883$$

$$\varphi_b = \beta_b \frac{4,320}{\lambda_y^2} \cdot \frac{Ah}{W_x} \left[ \sqrt{1 + \left(\frac{\lambda_y t_1}{4.4h}\right)^2} + \eta_b \right]$$

$$= 0.883 \times \frac{4,320}{179^2} \cdot \frac{90 \times 48.6}{1,467} \sqrt{1 + \left(\frac{179 \times 1.8}{4.4 \times 48.6}\right)^2} = 0.642$$

When $\varphi_b > 0.6$, the following $\varphi_b'$ should be used instead of $\varphi_b$:

$$\varphi_b' = 1.07 - \frac{0.282}{\varphi_b} = 0.631$$

$$\varphi_b f = 0.631 \times 205 = 129.4 \text{ N/mm}^2$$

$$\sigma = \frac{M_{max}}{W_x} = \frac{269.2 \times 10^6}{1,467 \times 10^3}$$

$$= 183.5 \text{ N/mm}^2 > \varphi_b f = 129.4 \text{ N/mm}^2$$

It does not meet the requirement.

If a side bearing point is set in the middle span, a similar method above can be used to obtain:

$$\varphi_b = 2.314$$

$$\varphi'_b = 1.07 - \frac{0.282}{2.314} = 0.948$$

$$\varphi_b f = 0.948 \times 205 = 194.3 \text{ N/mm}^2$$

$\sigma = 183.5 \text{ N/mm}^2 < \varphi_b f = 194.3 \text{ N/mm}^2$, meets the requirements.

Local stability verification:

Flange plate:

$$b/t \approx 8.3 < 26$$

That means the outstanding flange plate width-to-thickness ratio is less than 13 and meets requirements.

Web:

$$h_0/t_w = \frac{450}{8} \approx 56 < 80$$

**O. K.**

## 6.9 Chapter Summary

The component that mainly bears the bending moment is called the flexural component. In steel structures, bending members mainly appear in the form of beams. In engineering, beams are mainly subjected to bending moments and shear forces. This chapter mainly introduces the four aspects of strength, stiffness, overall stability, and local stability in the design of steel beams and expounds on the design methods of steel shaped beams and welded composite beams.

Beginners should focus on the characteristics and concepts of various failure modes of steel beams and the methods to prevent these failures, and the basic calculation equations commonly used in the design and their application scope. In addition, beginners should understand the design of steel beams under various structural requirements.

(1) The steel beams in the project are mostly unidirectional bending beams. In order to make the members have stronger bending resistance, the bending moment should be applied in the plane of maximum stiffness of the beam section. To save steel, the plates that make up the steel beam should be as wide and thin as possible on the premise of meeting local stability requirements. An appropriate I-shaped section should have thicker flange

plates and thinner webs. The section depth of the beam should be within the economic depth range.

(2) The steel beams in the floor structure and working platform structure generally don't have to calculate the overall stability after appropriate structural measures. The design of this type of steel beam is mainly in three aspects: strength, stiffness, and local stability.

(3) Due to the high strength of steel, steel beams tend to appear slender. Overall stability problems often control the design of steel beams that do not guarantee overall stability. The overall instability of steel beams occurs suddenly due to insufficient lateral bending rigidity and torsional rigidity of the beam. There is no warning in advance, and the form of instability is lateral-torsional instability. Taking effective measures in the design can improve the overall stability of the steel beam. The specific methods include: providing lateral support points for the compression flange of the steel beam within the span of the beam, adopting the beam section to strengthen the compression flange, and reducing the position of the load application point.

(4) The steel beam's strength damage and overall instability will directly cause the beam to lose its bearing capacity. The consequences are serious and must be prevented. When the width-to-thickness ratio or the depth-to-thickness ratio of the plates that make up the steel beam is too large, the steel beam may be partially unstable. However, the consequences of local instability of steel beams are not as serious as the consequences of strength failure and overall instability. Generally, the strength of webs after buckling can be used for ordinary steel beams that do not directly bear dynamic loads.

(5) Hot-rolled steel beams generally do not suffer from local instability. The calculation problems are attributed to three aspects: strength, stiffness, and overall stability. When the overall stability of such beams must be calculated, the design is usually controlled by overall stability conditions.

(6) The design of welded composite beams generally has four aspects: strength, stiffness, overall stability, and local stability. However, the strength conditions generally do not have a controlling effect on the design.

(7) The steel beam strength checks calculation object is a certain dangerous point on a certain dangerous section of the beam. The object of the steel beam's overall stability check calculation is the steel beam integral member. Therefore, the strength calculation needs to use the geometric characteristics of the net section, while the stability calculation uses the geometric characteristics of the gross cross-section.

(8) Checking the stiffness belongs to a normal service limit state problem, and the standard combination of loads should be used. In addition, checking the strength, overall stability, and local stability is the ultimate bearing capacity problem, and the basic combination of loads should be used.

(9) Take appropriate structural measures to make the steel beam meet the conditions that it does not need to check the overall stability. These measures include setting enough lateral support points for the compression flange and the rigid plate on the top flange and reliably connected with the flange.

(10) In addition to the necessary calculation requirements, the design of steel beams must also meet various specified structural requirements.

(11) The layout and design of web stiffeners are important to ensure that the composite steel beam does not appear unstable. The types, functions, and structural requirements of various stiffeners should be understood.

# Questions

6.1 What are the contents of the beam strength calculation?

6.2 What is the significance of the section plastic development coefficient? What is the relationship with the section shape coefficient?

6.3 Why is the deformation calculated for the bending member? Why does the axial compression member only need to control the slenderness ratio?

6.4 What is warping normal stress? What effect does it have on the overall stability of the beam?

6.5 What is the difference between the overall instability of the beam and the instability of the axial compression column?

6.6 What is the difference between the strength failure of the beam and the failure of the overall stability? What is the difference between the overall failure and the local failure?

6.7 Is the use of high-strength steel beneficial to improve the stability of the beam?

6.8 What are the main factors that affect the overall stability of the beam? What are the rules?

6.9 To improve the overall stability of the beam, what measures can be adopted in the design? Which of them is the most effective?

6.10 What are the similarities and differences between the local stability of flexural members and axially compressed members?

6.11 What is the general height-to-thickness ratio of panels? What are the advantages of this expression method?

6.12 What are the types of web stiffeners? Which types of instability do stiffeners mainly control?

6.13 How is the web tension field in the I-shaped cross-section composite beam generated? What kind of load-bearing capacity of the beam can be improved by the tension field?

6.14 What are the similarities and differences between beam design and axial compression member design?

# Exercises

6.1 Welded I-section beam simply supported at both ends, as shown in Figure 6.39. Its span is $l=15$ m, biaxially symmetrical section H500×1,200×12×20, the steel is grade Q355B, the cross-section is not weakened. There are two concentrated load design values $P=700$ kN (static load, including the weight of the beam) at the third point of the beam, the supporting length of the concentrated load $a=150$ mm, and the distance between the load acting surface and the top surface of the beam is 90 mm. Support stiffeners are arranged at the beam support and try to check the strength of the beam.

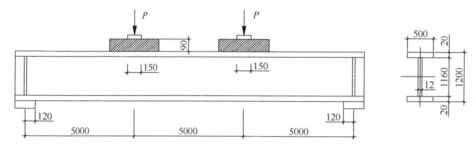

Figure 6.39 Welded I-section beam

6.2 Check the stiffness of the steel beam shown in Figure 6.39. This beam is the main beam in the working platform. The ratio of the permanent load standard value and the variable load standard value to the total load standard value of the beam is 25% and 75%, respectively. When calculating the load design value, the permanent load sub-factor is 1.3, and the variable load sub-factor is 1.5.

6.3 Welded simply supported beam with I-section, the size of the beam section is shown in Figure 6.40. The beam span is $l=12$ m, and there is no lateral support in the middle of the span. The upper flange bears a uniformly distributed load across the entire span: the standard value of the permanent load (including the self-weight of the beam) $q_1=12$ kN/m, the standard value of the variable load $q_2=15$ kN/m, and the beam is made of Q355B steel. Check the overall stability of the beam.

6.4 Reduce the width of the lower flange of the beam in Exercise 6-3 (Figure 6.40) by 100 mm, and increase the width of the upper flange by 100 mm to form a new section that strengthens the upper flange (Figure 6.41). The remaining conditions are the same as in Exercise 6.3. Check the new beam's overall stability and compare the changes before and after the section is changed.

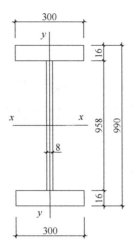
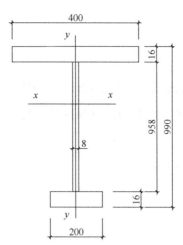

Figure 6.40  Exercise 6.3        Figure 6.41  Exercise 6.4

6.5  Design a simply supported H-shape beam and a simply supported welded H-beam, with a span of 9 m. The upper flange of the beam bears a uniform static load. The standard value of the permanent load is 9 kN/m (not including the weight of the beam), and the standard value of the variable load is 20 kN/m. It is assumed that the compression flange of the beam has a reliable floor, the steel grade is Q355B, the allowable deflection of the beam is $L/250$, and $L$ is the beam span.

6.6  Try to design stiffeners for Exercise 6.3 (Figure 6.40). The steel is grade Q345B, E50 electrode (manual welding). Design according to the method of not considering and considering the post-buckling strength of the web, respectively.

# Chapter 7  TENSION-BENDING MEMBERS AND COMPRESSION-BENDING MEMBERS

**GUIDE TO THIS CHAPTER:**

➢ *Knowledge points*

  *Section form, strength calculation, and stiffness check calculation of tension-bending and compression-bending members; in-plane and out-of-plane elastoplastic overall stability of solid web compression-bending members; equivalent bending moment coefficient in the bending moment action plane and out of the bending moment action plane; the local stability of solid web compression-bending members, including the limit of flange width-to-thickness ratio and web depth-to-thickness ratio; the overall stability of laced compression-bending members with bending moment acting to the virtual axis, including stability in-plane of the bending moment, stability of the split limbs and stability out-plane of the bending moment.*

➢ *Emphasis*

  *The overall stability and local stability of solid web compression-bending members; the overall stability of laced compression-bending members.*

➢ *Nodus*

  *Strength calculation theory of compression-bending members; overall stability theory of solid web compression-bending members.*

## 7.1  Introduction

In the steel structure, a type of member bears the combined action of axial load (parallel to the member's longitudinal axis) and lateral load (perpendicular to the member's longitudinal axis) or end bending moment. When the axial load generates tensile force, it is called a tension-bending member. When the compressive force is generated by axial load, it is called a compression-bending member. Figure 7.1 shows the application of tension-bending and compression-bending members on some actual steel structure buildings and bridges. The deck chord of an arch bridge like Figure 7.1(a) is a typical tension-bending member. Compression-bending members are widely found in the beams and columns of the steel frame and portal frame structure shown in Figures 7.1(b) and 7.1(c). Therefore, its global name is beam-column.

The force diagram of tension-bending members is shown in Figure 7.2(a), including the member subjected to eccentric tension and the tension member with transverse load in the middle. Compression-bending members have many forms of load action. Figure 7.2(b) shows various possible forms of force, including members bearing eccentric compressive force, members bearing lateral loads, and bearing bending moments transmitted by other members. Again, there are multiple forms of superimposed loads in real structures.

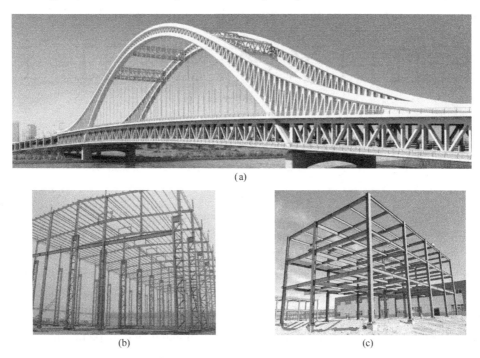

Figure 7.1 The application of tension-bending and compression-bending members
(a) chords of a bridge; (b) laced columns of a workshop; (c) solid web columns of a steel frame

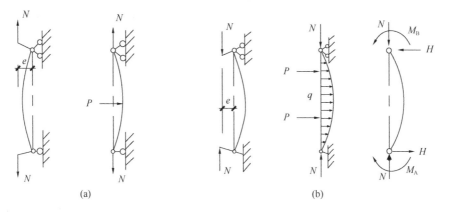

Figure 7.2 Force diagrams of members under combined axial forces and moments
(a) tension-bending members; (b) compression-bending members (beam-columns)

Tension-bending and compression-bending members usually adopt biaxial symmetry, single axisymmetric sections, a solid web type, or laced-type (Figure 7.3). The biaxial symmetric section is usually used for the members with large axial load and small bending moments or positive and negative absolute values of bending moments roughly equal. The common sections are the same as axially loaded compression members, shown in Figure 7.3(a). The single axisymmetric section is often used in the positive and negative bending moment difference of large members. For example, the section of the force of the larger side of the appropriate increase to save steel is shown in Figure 7.3(b).

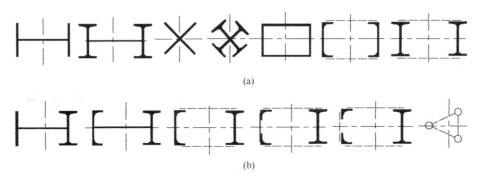

Figure 7.3  Common sectional forms

As with the axially loaded members, the design of tension-bending and compression-bending members should simultaneously meet the requirements of the ultimate bearing capacity and the normal service limit state. Tension-bending members need to check strength and stiffness (limiting slenderness ratio). Compression-bending members need to check strength, overall stability (stability in the bending moment action plane and out of the bending moment action plane), and local stability and stiffness (limiting slenderness ratio).

The allowable slenderness ratio of the tension-bending member is the same as that of the axial tension member. The allowable slenderness ratio of the compression-bending member is the same as that of the axial compression member.

## 7.2  Strength and Stiffness of Tension-Bending and Compression-bending Members

Strength calculations are required for tension-bending members and compression-bending members with holes in the weakened section, or the bending moment at the end of the member is greater than the bending moment between the spans.

A biaxial symmetric section compression-bending (tension-bending) member whose bending moment $M$ acts on only one main plane is taken as an example (Figures 7.4a and 7.4b). The axial force $N$ increases proportionately to the bending moment $M$. The uniform normal stress caused by $N$ is superimposed with the bending normal stress caused by $M$ (dashed lines in Figures 7.4a and 7.4b).

## 7.2 Strength and Stiffness of Tension-Bending and Compression-bending Members

On the cross-section, the outermost fiber on the side with greater compressive (tensile) stress is the maximum compressive (tensile) stress, and the outermost fiber on the other side is the maximum tensile (compressive) stress (when $M$ is larger, see Figure 7.4a ①) or the smallest compressive (tensile) stress (when $M$ is small, see Figure 7.4b①). When the maximum compressive (tensile) stress is less than the steel yield strength $f_y$, the member is in an elastic working state. When $N$ and $M$ continue to increase, and the maximum compressive (tensile) stress reaches $f_y$ (Figures 7.4a and 7.4b②), the strength of the member section reaches the ultimate state in the elastic stage. When $N$ and $M$ increase again, the side of the maximum compressive (tensile) stress develops plastic deformation, and the plastic zone gradually develops inward with the increase of internal force (Figures 7.4a and 7.4b③). The member is in a state of elastoplastic stress. Then, the outermost fiber on the other side of the section reaches the tensile (compressive) yield strength (Figure 7.4a④), and the plastic zone gradually develops inward with the increase of $N$ and $M$ (Figure 7.4a⑤). When the plastic zone on both sides develops to the full section (Figures 7.4a and 7.4b⑥), a plastic hinge is formed. The member reaches the limit state of the plastic force stage, which is the ultimate bearing capacity state of its strength.

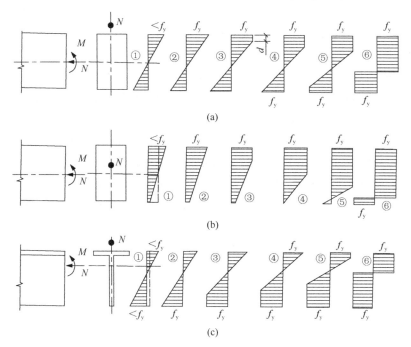

Figure 7.4 The stress distribution of the section of compression-bending members

In structural design, depending on the nature of the load on the member, the shape of the section, and the force's action point, etc., different section stress states are specified as the limit state of the strength calculation. There are usually the following three types.

223

# Chapter 7 TENSION-BENDING MEMBERS AND COMPRESSION-BENDING MEMBERS

## 7.2.1 The ultimate state of bearing capacity calculated based on the edge fiber yield

For example, solid-web tension-bending and compression-bending members directly bear dynamic loads. Because the current research on the members' plastic properties under dynamic loads is still immature, the design codes often stipulate that the edge yield of the section is used as the basis for the members' strength calculation. Another example is the laced tension-bending and compression-bending members. When the bending moment acts around the virtual axis, there are no solid parts in the section's abdomen. Thus, there is little difference between the edge yield and the section's ultimate state in the plastic force stage. For simplicity, it is also stipulated that the limit state of the elastic force stage is used as the basis for strength calculation.

Like the axially loaded member and the flexural member, the strength calculation is based on the net section. Then the member resistance coefficient is considered, and $f$ is substituted for $f_y$. The strength calculation formula of the tension-bending or compression-bending member, in this case, is:

$$N/A_n + M_x/W_{nx} \leqslant f \tag{7-1}$$

where $W_{nx}$ = the net section resistance moment of the maximum tension (compression) fiber of the tension-bending (compression-bending) member;

$A_n$ = the member's net cross-sectional area.

For compression-bending (tension-bending) members with a uniaxial symmetrical section, when the bending moment acts on the main plane of the axis of symmetry and causes the larger flange to be compressed (tension), the largest tension (compression) stress may occur first in the outermost fiber of the smaller flange (Figure 7.4c①). For this kind of compression-bending (tension-bending) member, the elastic limit state strength is controlled by the tensile (compressive) stress of the outermost fiber of the smaller flange (Figure 7.4c②). The net section resistance moment at this point is $W_{nx2}$, and the calculation equation is:

$$|N/A + M_x/W_{nx2}| \leqslant f \tag{7-2}$$

Equations (7-1) and (7-2) can be comprehensively written as:

$$\pm N/A \pm M_x/W_{nx} \leqslant f \tag{7-3}$$

## 7.2.2 Taking the full plastic limit state of the section as the limit state of bearing capacity

In order to give full play to the strength potential of members, the strength calculation can be based on the plastic hinge formed in the section of the solid-web tension-bending and compression-bending members, which are generally subjected to static load or indirect dynamic load, including the laced tension-bending and compression-bending members which are bent around the real axis.

## 7.2 Strength and Stiffness of Tension-Bending and Compression-bending Members

When the most dangerous section of the member is in the plastic working stage, the plastic neutral axis may be in the web or the flange. Then, according to internal and external forces'balance conditions, the relationship between the axial force N and the bending moment can be obtained.

When the axial force is small ($N \leqslant A_w f_y$), the plastic neutral axis is in the web, and the section stress distribution is shown in Figure 7.5. For the sake of simplicity, take $h = h_w$, and $A_f = \alpha A_w$, then:

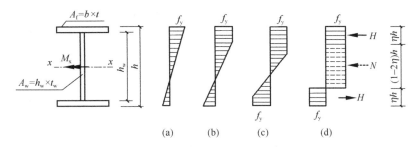

Figure 7.5 Development process of section stress of compression-bending member

The axial force of section yield:
$$N_p = A f_y = (2\alpha + 1) A_w f_y \tag{7-4}$$

Plastic yield bending moment of the section:
$$M_{px} = W_{px} f_y = \alpha A_w f_y h + 0.5 A_w f_y h_w/2 \approx (\alpha + 0.25) A_w h f_y \tag{7-5}$$

In Equation (7-5) is the plastic section modulus. According to the full plastic stress graph, the equilibrium conditions of axial force and bending moment are:
$$N = (1 - 2\eta) h t_w f_y \approx (1 - 2\eta) A_w f_y \tag{7-6}$$
$$M_x = A_f f_f (h - t) + (\eta h - t) t_w f_y (1 - \eta - t) h \approx A_w h f_y (\alpha + \eta - \eta^2) \tag{7-7}$$

Eliminate the $\eta$ in the above two equations, then the related equations of N and $M_x$ are:
$$\frac{(2\alpha + 1)^2}{4\alpha + 1} \cdot \frac{N^2}{N_p^2} + \frac{M_x}{M_{px}} = 1 \tag{7-8a}$$

When the axial force is very large ($N > A_w f_y$), the plastic neutral axis will be located within the flange range. According to the same method as above, we can get:
$$\frac{N}{N_p} + \frac{(4\alpha + 1)}{2(2\alpha + 1)} \cdot \frac{M_x}{M_{px}} = 1 \tag{7-8b}$$

The relational Equations (7-8a) and (7-8b) of $N/N_p$ and $M/M_p$ are convex curves, which are not only related to the cross-sectional shape but also related to $\alpha = A_f/A_w$, the smaller $\alpha$ is, the more convex it is.

Common I-shaped cross-section $\alpha \approx 1.5$, and the curve convex is not much, can be approximated by a straight line. When $N/N_p$ is very small for simplicity of design, $M = M_p$ is calculated; when $N/N_p$ is large, $A_f/A_w = 1.5$ is calculated in Equation (7-8b).

# Chapter 7  TENSION-BENDING MEMBERS AND COMPRESSION-BENDING MEMBERS

## 7.2.3  Taking the plastic development of the section as the strength calculation criterion

Not to cause excessive deformation of members, and considering the second-order bending moment (additional bending moment caused by axial force in flexural members) and shearing force not included in the previous calculation, it is stipulated that when compression-bending and tension-bending members bear static load or indirect dynamic load, their strength shall be calculated according to the following approximate linear equation.

Only when there is a bending moment on a principal plane,

$$N/A_n \pm M_x/(\gamma_x W_{nx}) \leqslant f \tag{7-9a}$$

$$N/A_n \pm M_y/(\gamma_y W_{ny}) \leqslant f \tag{7-9b}$$

The section strength of tension-bending members and compression-bending members (except for round pipe section) whose bending moment acts in the two main planes shall be calculated as follows:

$$\frac{N}{A_n} \pm \frac{M_x}{\gamma_x W_{nx}} \pm \frac{M_y}{\gamma_y W_{ny}} \leqslant f \tag{7-10}$$

The section strength of circular cross-section tension-bending members and compression-bending members whose bending moment acts in the two main planes shall be calculated according to the following provisions:

$$\frac{N}{A_n} + \frac{\sqrt{M_x^2 + M_y^2}}{\gamma_m W_n} \leqslant f \tag{7-11}$$

where  $N$ = design value of axial pressure at the same section (N);

$M_x, M_y$ = the design values of the bending moments on the x-axis and y-axis at the same section (N·m);

$\gamma_x, \gamma_y$ = the cross-section plastic development coefficient is determined according to the pressure plate's internal force distribution. When the cross-section plate width-to-thickness ratio does not meet the requirements of Class S3, it is taken as 1.0. When it meets the requirements of Class S3, it can be determined according to Table 6.1; for tension-bending and compression-bending members requiring fatigue strength verification, 1.0 should be adopted;

$\gamma_m$ = the plastic development coefficient of the circular member section. It is 1.2 for the solid-web round section, it is 1.0 when the sectional width-to-thickness ratio class does not meet the requirements of Class S3, and it is 1.15 when the sectional width-to-thickness ratio class meets the requirements of Class S3. For tension-bending and compression-bending members whose fatigue strength needs to be checked, it is 1.0.

$A_n$ = the net cross-sectional area of the member (mm$^2$);

$W_n$ = the net section modulus of the member (mm$^3$).

When the ratio of the free extension width of the compression flange of the compression bending member to its thickness is greater than $13\varepsilon_k$ but not exceeding $15\varepsilon_k$, it shall be taken $\gamma_x = 1.0$. In addition, tension-bending and compression-bending members that need to check fatigue strength should be taken $\gamma_x = \gamma_y = \gamma_m = 1.0$.

### 7.2.4 Stiffness

As with the axially loaded compressive members, the tension-bending and compression-bending members should have a certain stiffness to meet the normal service limit state's requirements to ensure that the members will not be excessively deformed. The stiffness of tension-bending and compression-bending members is also achieved by limiting the slenderness ratio. When calculating, the allowable slenderness ratio of tension-bending members is the same as that of axially loaded tension members; the allowable slenderness ratio of compression-bending members is the same as that of the axially loaded compressive member.

## 7.3 Global Stability of Solid Web Compression-Bending Members

### 7.3.1 Stability calculation of solid web unidirectional compression-bending members in the plane of bending moment

Global stability usually determines the load-bearing capacity of compression-bending members. Now take the ideal compression-bending member (abbreviated as a one-way compression-bending member) as an example. The bending moment acts in a principal plane to illustrate that the compression-bending member loses global stability (Figure 7.6). Under the simultaneous action of $N$ and $M$, the member deforms the bending moment action plane initially and assumes a bending state. When $N$ and $M$ increase to a certain size at the same time, it reaches the limit. If the limit is exceeded, the bending member may be in the bending moment. As a result, bending instability occurs in the plane of the bending moment (Figure 7.6a), or flexural-torsional instability occurs out the plane of the bending moment (Figure 7.6b). Therefore, the stability of compression-bending members in and out of the bending moment action plane must be calculated separately.

This section first discusses the global stability of the compression-bending member in the plane of the bending moment. When studying the compression-bending member's global stability in the bending moment action plane, it should be noted that the axial pressure $N$ and the bending moment $M$ which has been born may have different loading paths. For example, $N$ and $M$ can be increased in the same proportion (such as eccentric compression members), or $N$ can be added first and then $M$ gradually (such as a high-rise structure), or $M$ can be added first and then $N$ gradually. In the elastic force stage, the bearing capacity of the member has nothing to do with the loading path, only the final load value. In the

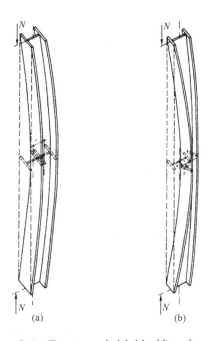

Figure 7.6 Two types of global buckling of compression-bending members (both pin-ends)
(a) buckling in-plane of bending moment (flexural buckling); (b) buckling out-plane of bending moment (flexural-torsional buckling)

stage of elastic-plastic stress, it is related to the loading path. Under normal circumstances, there is little difference in member bearing capacity under different loading methods. The following description is mainly based on the proportional increase of $N$ and $M$.

Now take the hinged bending member shown in Figure 7.7 as an example. In addition to the axial pressure, a bending moment $M$ acts on each end. This compression-bending member's global stability working behavior in the bending moment action plane is the same as the axial compression member with geometric defects such as initial bending and initial eccentricity. Accidental factors cause the initial defects of the bending moment of the axial compression member, and the $M$ value is relatively small. At the same time, the bending moment $M$ in the compression-bending member is the same as the axial force $N$ as the main internal force.

Because of the simultaneous action of the axial force $N$ and the bending moment $M$ of the compression-bending member, the equilibrium branching phenomenon of the ideal axial compression member will not occur when the bending moment is in the plane of action. Besides the inevitable initial bending of the member, the bending deformation occurs at the force's beginning. The compressive load ($N$)-deflection ($Y_m$) curve is shown as Oabc in Figure 7.7(b). The Oa section is an elastic working stage, but it is nonlinear due to the additional bending moment $N_y$. After point a enters the elastoplastic state, the curve ob section rises. The deflection increases with the increase of $N$, and the balance is stable. To maintain the balance of the section, $N$ must be continuously reduced and the deflection continuously increased, which is unstable. Point b is the curve's extreme point from the stable equilibrium state to the unstable equilibrium state. The corresponding $N$ value ($N_u$) is the member's stable ultimate bearing capacity in the bending moment action plane. The corresponding average cross-sectional stress is called the ultimate stress. The position of point b can be obtained according to the $N$-$Y_m$ curve's extreme value problem, $dN/dY_m = 0$.

When a compression-bending member loses stability in the plane of the bending mo-

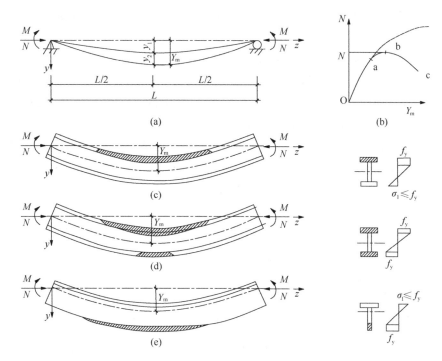

Figure 7.7 The global buckling of the unidirectional compression-bending member in-plane of $M$

ment, it depends on the cross-sectional shape, size, member length, and residual stress distribution. The member's plastic area may only be on the side of the middle part of the length, where the cross-section is the most compressed, or on both sides of the section at the same time, or only on the side of the section under tension (Figures 7.7c, 7.7d, and 7.7e). The last case may occur in a compression-bending member with a uniaxially symmetric section. There are currently three calculation methods for the stability of unidirectional compression-bending members in the bending moment action plane, namely, the method according to the edge fiber yield criterion, the method according to the ultimate load-bearing capacity criterion, and the method of practical calculation formula (single formula or related formula expression). It was introduced separately below.

(1) The method according to the edge yield criterion

As mentioned in Chapter 5, the edge fiber yield criterion is an approximate method that substitutes the stress problem for the stability calculation. It takes the load when the edge fiber with the largest cross-sectional stress of the member begins to yield under the maximum load of the member in the elastic stage, as the stable bearing capacity of the bending member. This criterion is expressed as:

$$N/A + M_{max}/W_{1x} = f_y \quad (7\text{-}12)$$

where  $N$ = the axial compressive load;

$M_{max}$ = the maximum bending moment after considering the effects of $N$ and initial defects;

$A, W_{1x}$ = the member's gross cross-sectional area and the gross cross-sectional resistance moment of the larger compressed edge.

The axially compressive load $N$ could amplify the initial deflection (the deflection curve is a half-sine wave). Therefore, the magnification factor is derived as $1/(1 - N/N_{Ex})$, where $N_{Ex} = \pi^2 EA/\lambda_x^2$ is Euler critical load. This formula can generally approximate the deflection magnification factor of $N$ for compression-bending members under other loads. Now take the compression-bending member subjected to uniform bending moment (Figure 7.7a) as an example (for simplicity, the initial deflection $y_0$ is not considered, then $Y=y$); the solutions of the balance differential equation and deflection curve are as follows:

$$d^2y/dz^2 + k^2 y = -M/EI, k^2 = N/EI, kl = \pi\sqrt{N/N'_{Ex}}$$

Solve and use the boundary conditions $z=0$ and $z=l$ for $y=0$, and we can get:

$$y = \frac{M}{N}\left(\frac{\sin kz + \sin k(l-z)}{\sin kl} - 1\right) = \frac{M}{N}\left(\tan\frac{kl}{2}\sin kz + \cos kz - 1\right) \quad (7\text{-}13)$$

The maximum deflection at the mid-height of the member is:

$$y_{max} = \frac{M}{N}\left(\sec\frac{kl}{2} - 1\right) = \frac{Ml^2}{8EI}\frac{8EI}{Nl^2}\left(\sec\frac{kl}{2} - 1\right) = \delta_0\left[\frac{2\left(\sec\frac{kl}{2} - 1\right)}{(kl/2)^2}\right] \quad (7\text{-}14)$$

where $\delta_0$ = the mid-span deflection of the corresponding simply supported beam (subject to uniform bending moment $M$) without considering $N$, $\delta_0 = Ml^2/8EI$.

The square bracket item is the mid-span deflection magnification factor of the compression-bending member. For example, expands $\sec(kl/2)$ in the above equation into a power series, then:

$$\frac{2\left(\sec\frac{kl}{2} - 1\right)}{(kl/2)^2} = 1 + \frac{5}{12}\left(\frac{kl}{2}\right)^2 + \frac{61}{360}\left(\frac{kl}{2}\right)^4 + \cdots$$

$$= 1 + 1.028 N/N_{Ex} + 1.032 (N/N_{Ex})^2 + \cdots$$

$$\approx 1 + N/N_{Ex} + (N/N_{Ex})^2 + \cdots = 1/(1 - N/N_{Ex}) \quad (7\text{-}15)$$

The deflection magnification factor can also be approximated to $1/(1 - N/N_{Ex})$ for compression-bending members under other loads. In general, $N/N_{Ex} < 0.6$, the error does not exceed 2%. The influence of axial pressure $N$ on the increase of bending moment is analyzed below.

The theoretical solution of the maximum bending moment $M_{max}$ of a compression-bending member under different loads and the increased coefficient compared with the maximum bending moment $M_x$ of the simply supported beam can be obtained from the above differential equation and deflection curve solution method. The bending moment increase factor can also be expanded and written in the form of a power series of $N/N_{Ex}$, which can be expressed as the product of the basic increase term $1/(1 - N/N_{Ex})$ and the correction term $\beta_{mx} = 1 + \eta_1 N/N_{Ex} + \eta_2 (N/N_{Ex})^2 + \cdots \approx 1 + \eta N/N_{Ex}$. When the deflection magnifica-

## 7.3 Global Stability of Solid Web Compression-Bending Members

tion factor is approximate $1/(1-N/N_{Ex})$, the maximum bending moment $M_{max}$ can be expressed as follows:

$$M_{max} \approx M_x + N \cdot \frac{\delta_0}{1-N/N_{Ex}} = \frac{M_x}{1-N/N_{Ex}}\left(1+\eta\frac{N}{N_{Ex}}\right) = \frac{\beta_{mx}M_x}{1-N/N_{Ex}} \quad (7\text{-}16)$$

where $M_x, \delta_0$ = the maximum bending moment and maximum deflection of the simply supported beam (the above equation applies when both are at the mid-height);

$\beta_{mx} = 1 + \eta N/N_{Ex}$ = the bending moment correction coefficient or equivalent bending moment coefficient;

$\beta_{mx}/(1-N/N_{Ex})$ = the bending moment increase coefficient.

The approximate expression of $\eta$ can be obtained from the above equation: $\eta = N_{Ex}\delta_0/M_x - 1$. The value of $\eta$ varies with the load type. For example, when the two simple end-supporting compression-bending members have equal bending moments, $\eta = 0.234$; when the concentrated lateral load at the mid-span, $\eta = -0.178$; when the load is distributed, $\eta = 0.028$; when the load with a half-wave sine curve along the span (i.e., $q_z = \sin\pi z/l$), $\eta = 0$, $\beta_{mx} = 1$.

Using the edge fiber yield criterion to solve the stable bearing capacity of compression-bending members with initial defects is now studied. Various initial defects of members are generally considered comprehensively represented by an equivalent initial bending or equivalent initial eccentricity.

If the equivalent initial bending represents the comprehensive defect, and the value is known, use Equation (7-16) to obtain the following equation.

$$M_{max} = M_x + \frac{N(\delta_0 + v_0^*)}{1-N/N_{Ex}} = \frac{\beta_{mx}M_x + Nv_0^*}{1-N/N_{Ex}} \quad (7\text{-}17)$$

Substitution Equations (7-17) to (7-12), we get:

$$\frac{N}{A} + \frac{\beta_{mx}M_x + Nv_0^*}{W_{1x}(1-N/N_{Ex})} = f_y \quad (7\text{-}18)$$

For $\sigma_0 = N/A$, $\sigma_{Ex} = N_{Ex}/A$, and the relative initial bending (or the relative initial eccentricity of the mid-span section), $\varepsilon_0^* = v_0 A/W_{1x}$ as well as the relative eccentricity of the compression-bending member $\varepsilon = M_x A/NW_{1x}$. Then, the stable load-bearing stress expressed in stress, known as the Perry equation, can be obtained.

$$\sigma_0 = \frac{f_y + (1+\beta_{mx}\varepsilon + \varepsilon_0^*)\sigma_{Ex}}{2} - \sqrt{\left[\frac{f_y + (1+\beta_{mx}\varepsilon + \varepsilon_0^*)\sigma_{Ex}}{2}\right]^2 - f_y\sigma_{Ex}} \quad (7\text{-}19)$$

The various defects in the member are complicated, and their comprehensive influence on the compression-bending member also varies with the section type and load situation. Hence, the equivalent initial bending value $v_0^*$ is often difficult to determine. However, when $M_x = 0$ is usually available, $N$ is equal to the relationship between the stability bearing capacity of the axially loaded compression member after the effect of the equivalent ini-

## Chapter 7 TENSION-BENDING MEMBERS AND COMPRESSION-BENDING MEMBERS

tial bending. Thus, the following equation is obtained for the stability bearing capacity of the flexural member, namely the corresponding equation.

Considering the member resistance coefficient and writing it as a design equation, we get:

$$\frac{N}{\varphi_x A} + \frac{\beta_{mx} M_x}{W_{1x}(1 - \varphi_x N/N_{Ex})} \leqslant \frac{f_y}{\gamma_R} = f \tag{7-20}$$

Equation (7-20) is the relevant equation derived from the compression-bending member's edge fiber yield. The calculation equation for in-plane stability is given in the design standard GB 50017:

$$\frac{N}{\varphi_x A f} + \frac{\beta_{mx} M_x}{\gamma_x W_{1x}\left(1 - 0.8 \dfrac{N}{N'_{Ex}}\right) f} \leqslant 1.0 \tag{7-21}$$

where  $N$ = the design value of axially compressive load within the range of calculated member section;

$M_x$ = the maximum design value of bending moment within the range of calculated member section;

$\varphi_x$ = Global stability factor of axial compression member in the bending moment action plane;

$W_{1x}$ = the gross section modulus of the largest fiber under compression in the bending moment action plane;

$N'_{Ex}$ = a parameter, $N'_{Ex} = \pi^2 EA/(1.1\lambda_x^2)$;

$\beta_{mx}$ = the equivalent bending moment coefficient shall be taken according to the following condition:

① Columns of the frame with sideway inhibited and members braced at both ends:

• When there is no lateral load,

$$\beta_{mx} = 0.6 + 0.4 M_2/M_1 \tag{7-22}$$

where  $M_1, M_2$ = the end bending moments, the same sign is used when the member produces the same curvature (without the point of inflection), and the opposite sign is used when the member has reverse curvature (when there is a point of inflection), $|M_1| \geqslant |M_2|$.

• When there is no end bending moment, but there is lateral load.

Single concentrated load in the middle of the span:

$$\beta_{mx} = 1 - 0.36 N/N_{cr} \tag{7-23}$$

Load uniformly distributed across the span:

$$\beta_{mx} = 1 - 0.18 N/N_{cr} \tag{7-24}$$

$$N_{cr} = \frac{\pi^2 EI}{(\mu l)^2}$$

where  $N_{cr}$ = elastic buckling critical load;

$\mu$ = the effective length factor of the member.

- When the end bending moment and the lateral load act simultaneously, the $\beta_{mx} M_x$ of Equation (7-21) should be calculated as follows:

$$\beta_{mx} M_x = \beta_{mqx} M_{qx} + \beta_{m1x} M_1 \tag{7-25}$$

where  $M_{qx}$ = the maximum bending moment produced by the lateral uniform load;

$M_1$ = the bending moment caused by a single lateral concentrated load in the middle of the span;

$\beta_{m1x}$ = the equivalent bending moment coefficient calculated according to Equation (7-22);

$\beta_{mqx}$ = the equivalent bending moment coefficient is calculated according to Equation (7-23).

② Columns of the frame with sideway uninhibited and cantilever members:

- Frame columns other than those specified below, $\beta_{mx}$ should be calculated as follows:

$$\beta_{mx} = 1 - 0.36 N/N_{cr} \tag{7-26}$$

- The single-story frame column and the bottom-story column of the multi-story frame hinged on the column foot with lateral load, $\beta_{mx} = 1.0$;

- Cantilever column with bending moment acting on the free end:

$$\beta_{mx} = 1 - 0.36(1-m) N/N_{cr} \tag{7-27}$$

where  $m$ = the bending moment's ratio at the free end to the bending moment at the fixed end. The positive sign is taken when the bending moment diagram has no point of inflection, and the negative sign is taken when there is a point of inflection.

When the second-order elastic analysis of the frame's internal forces is adopted, the moment of the column consists of the unshifted moment and the amplified lateral moment, and the two moments can be multiplied by the equivalent moment coefficients of the unshifted column and the lateral column respectively.

(2) The method according to the ultimate bearing capacity criterion

For solid web compression-bending members, the edge fibers can continue to bear loads after yielding (point a in Figure 7.7b) until the vertex b of $N$-$Y_m$ curve real ultimate stability bearing capacity in-plane of bending moment. The ultimate stability bearing capacity $N_u$ is determined by solving the extreme value of the $N$-$Y_m$ curve of the flexural member, which is called the ultimate bearing capacity criterion.

There are many ways to find $N_u$ according to the ultimate bearing capacity criterion. The numerical solution method is currently the most widely used, considering the effects of various members'defects. It is suitable for different boundary conditions and elastic and elastoplastic working stages. In order to find the closed solution of the equilibrium differential equation of the flexural member, a simplified method can be adopted. The most commonly used is the Jezek K. simplified method. The numerical integration method in the numerical method is briefly described below.

The most commonly used method for the ultimate bearing capacity criterion is a numerical solution, which considers the influence of the member's geometric defects and residual stress and solves it with a computer. Section 5.3 briefly introduces a numerical integration method of numerical calculation. It is also applicable to compression-bending members, except that the bending moment is added to the balance equation. This item is slightly revised. Although the numerical integration method is also an approximation method, the accuracy can be improved by increasing member sections and section blocks. In addition, it is convenient to consider the effects of defects such as residual stress. $N_u$ (axial pressure)-$\lambda$(slenderness ratio)-$\varepsilon$ (relative eccentricity) correlation curve cluster can be obtained according to the numerical solution. Figures 7.8(a) and 7.8(b) are examples of the $\sigma_u/f_y$-$\lambda$-$\varepsilon$ curve to the strong axis ($x$-axis) and the weak axis ($y$-axis) of an eccentric compression member with a welded I-shaped cross-section.

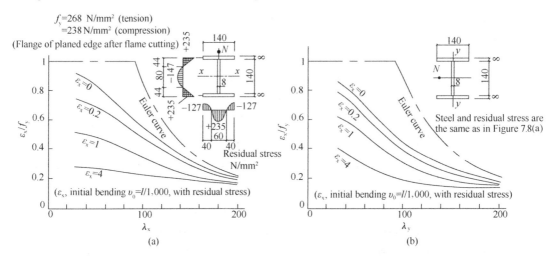

Figure 7.8  $\sigma_u/f_y$-$\lambda$-$\varepsilon$ curve of an eccentric compression member welded I-section

### 7.3.2 Stability calculation of solid web unidirectional compression-bending members out-plane of bending moments

Before losing the global stability in the plane of the bending moment, the solid-web compression-bending member may produce lateral bending deformation, accompanied by torsion around the axis of the torsion center (shear center). That is, the loss of global stability out the plane of moment action or buckling out of the plane of the moment action. Therefore, when we design the compression-bending member, we should ensure that it does not lose the overall stability out-plane of the bending moment.

(1) Elastic buckling of unidirectional ideal compression-bending members out-plane of bending moment

A biaxially symmetrical I-shaped cross-section member hinged at both ends, and both ends bear the axial pressure $N$ and bending moment $M_x = Ne$ (Figure 7.9). The critical

## 7.3 Global Stability of Solid Web Compression-Bending Members

force of this ideal compression-bending member when elastic buckling outside the bending moment plane is solved.

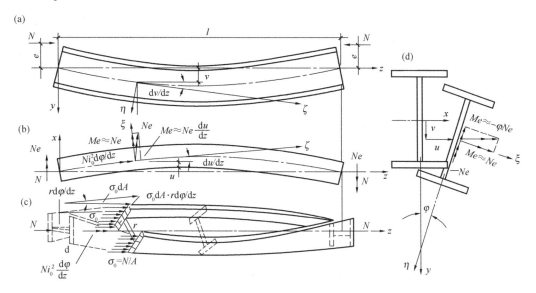

Figure 7.9  Elastic buckling of compression-bending members out-plane of bending moment

The compression-bending members shown in Figures 7.9 are decomposed into two stress situations: pure bending and axial compression. Therefore, when the member undergoes lateral bending and torsion buckling, the corresponding equilibrium differential equations are the bending equation around the y-axis and the torsion equation around the longitudinal torsion axis (the biaxially symmetric section coincides with the centroid axis). Therefore, the second type of force should be added to the differential equation mentioned above for compression-bending members: the additional lateral bending moment and torque term generated by the axial pressure $N$ on the lateral displacement $u$ and torsional deformation $\varphi$.

Because the lateral displacement $u$ will be generated during lateral bending, the axial pressure $N$ will generate a lateral (to the y-axis) bending moment. Therefore, this bending moment should be added to the bending equation to obtain the total lateral bending equation when the compression-bending member buckles.

As a result of torsion, each longitudinal fiber of the member is inclined (Figure 7.9c), and the inclination Angle is $r\,d\varphi/dz$ ($r$ is the distance from the fiber to the shear center). Hence, the transverse member of the longitudinal axial force $dN$ on each longitudinal fiber will generate torque $dN(r\,d\varphi/dz)r$ on the longitudinal torsional shaft.

The total torque produced by all the axial pressure $N$ is $\int_A \sigma_0 (r\,d\varphi/dz) r\, dA = \sigma_0 I_p\, d\varphi/dz = Ni^2\,d\varphi/dz$. This torque should be added to the torsion equation to obtain the total torsion equation for the bending and torsion buckling of the compression-bending member.

The two differential equations of lateral bending and torsion during the flexural-torsional buckling of the compression-bending member include two variables of sideway $u$ and torsion angle $\varphi$. Combining boundary conditions can be solved simultaneously to obtain the critical flexural-torsional buckling force $N_{y\omega}$ (For writing convenience, the following is to replace $N$ for $N_{y\omega}$). The solution of the minimum critical force corresponds to that both $u$ and $\varphi$ are single half-wave Sinusoids. For $u = C_1 \sin\pi z/l$ and $\varphi = C_2 \sin\pi z/l$, the solution can be obtained after substituting into the differential equation:

$$(N_{Ey} - N)(N_\omega - N) - M_x^2/i_0^2 = 0 \qquad (7\text{-}28a)$$

$$(N_{Ey} - N)(N_\omega - N) - N^2 e^2/i_0^2 = 0 \qquad (7\text{-}28b)$$

$$N = \frac{(N_\omega + N_{Ey}) - \sqrt{(N_\omega - N_{Ey})^2 + 4N_\omega N_{Ey} e^2/i_0^2}}{2(1 - e^2/i_0^2)} \qquad (7\text{-}29)$$

where $\quad N_{Ey} = \pi^2 EI_y/l_{0y}^2 =$ the critical buckling force around the y axis under axial compression load;

$N_\omega = (\pi^2 EI_\omega/l_\omega^2 + GI_t)/i_0^2 =$ the critical torsional buckling force under axial compression load;

$I_\omega, I_t =$ the sector moment of inertia and the free torsional moment of inertia of the gross cross-section;

For members connected at both ends, $l_{0y} = l, l_\omega = l$.

It can be seen from the above equation that the critical force $N$ is related to $N_{Ey}$, $N_\omega$, and $e/i_0$. Thus, $N$ is always less than the smaller of the $N_\omega$ and $N_{Ey}$. When the bending moment of a compression-bending member changes along the member's length, whether it is a biaxial or a uniaxial symmetrical section, the differential equation's solution will be more complicated. In general, the solution can only be solved by a numerical method or a simplified approximate solution.

(2) Elastic-plastic buckling of unidirectional ideal compression-bending members out-plane of bending moment

The calculation of an ideal compression-bending member that does not consider any defects enters the elastoplastic working stage. In theory, tangent modulus can be used, $E_t$. However, the stress on the section is not uniform; the deformation modulus of the plastic zone also changes along with the section height. In general, the stress state on the section changes along the member's length, making it difficult to solve it. Therefore, simplified methods are used to calculate the cross-sectional characteristics. For example, the thickness conversion method can reduce the maximum compression flange thickness $t$ whose stress exceeds the proportional limit to $t$ ($E_t/E$) and calculate the relevant section characteristics according to the converted section.

(3) Ultimate bearing capacity of unidirectional compression-bending members out-plane of bending moment

If considering the effects of defects such as initial deflection, initial torsion angle, and residual stress in-plane and out-plane of the bending moment of the compression-bending member, and finding its ultimate bearing capacity out-plane of the bending moment, the problem is no longer out-plane of the bending moment. Balanced branching is the ultimate bearing capacity problem of bidirectional compression-bending members, which numerical solutions can only calculate, such as finite element method, incremental stiffness method, and different method.

Figure 7.10 is an example of the $\sigma_{cr}/f_y$-$\lambda_y$-$\varepsilon_x$ correlation curve of flexural-torsional buckling outside the bending moment action plane (around the y-axis) of a welded I-section cross-section eccentric compression member (bearing $N$ and $M_x = Ne$). The residual stress is considered, and the initial bending and initial torsion angles outside the plane of the bending moment are not considered. Figure 7.10 also shows the influence of bending moment deformation in the plane before buckling on calculation.

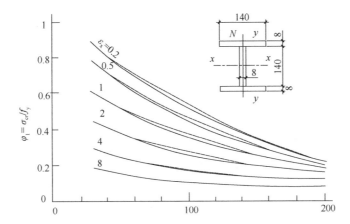

Figure 7.10  $\sigma_u/f_y$-$\lambda$-$\varepsilon$ curve of flexural-torsional buckling of eccentric compression members with welded I-section out-plane of bending moment

Solid line—consider the deformation in the plane of bending moment;

dotted line—do not consider the deformation in the plane of bending moment

(4) The practical calculation formula for global stability out of the plane of bending moment action

The practical formulas for calculating the overall stability of compression-bending members out of the plane of moment action are usually expressed in two ways: single formula and correlation formula. For example, the design code of steel structures stipulates that the correlation formula method is used in China.

① A single team formula for calculation of global stability out-plane of bending moment

The stability calculation of compression-bending member out-plane of bending moment can also be expressed in the form of a single term formula, which is similar to the stability calculation of the axially loaded compression member. The average stress design value of the member calculated from the internal design force does not exceed the critical stress (or ultimate stress) value divided by the partial resistance coefficient:

$$\sigma = \frac{N}{A} = \frac{\sigma_{cr}}{\gamma_R} = \frac{\sigma_{cr}}{f_y} \cdot \frac{f_y}{\gamma_R} = \varphi_1 f \qquad (7\text{-}30)$$

where $\varphi_1 = \sigma_{cr}/f_y$ = the global stability factor of the bending member out-plane of bending moment.

Because many factors affect the value $\varphi_1$, which is more complicated, more tables are required to use the above expression method. Moreover, this expression is not easy to distinguish the magnitude and degree of the influence of various factors. Therefore, to be consistent with the bending moment action plane's stability calculation, many countries use two correlation equations, including $N$ and $M$.

② Relevant formulas for calculation of global stability out-plane of bending moment

In Equation (7-30), the critical bending moment $M_{cr} = i_0 \cdot \sqrt{N_{Ey} N_\omega}$ of the member under pure bending can be obtained as $N = 0$. Substituting $i_0^2 = M_{cr}^2/(N_{Ey} N_\omega)$ into Equation (7-30), we get:

$$\left(1 - \frac{N}{N_{Ey}}\right)\left(1 - \frac{N}{N_\omega}\right) - \left(\frac{M_x}{M_{cr}}\right)^2 = 0$$

or

$$\left(1 - \frac{N}{N_{Ey}}\right)\left(1 - \frac{N}{N_{Ey}} \Big/ \frac{N_\omega}{N_{Ey}}\right) - \left(\frac{M_x}{M_{cr}}\right)^2 = 0 \qquad (7\text{-}31)$$

$N/N_{Ey}$-$N_\omega/N_{Ey}$-$M_x/M_{cr}$ relationship curve is shown in Figure 7.11. Generally, $N_\omega/N_{Ey} \geqslant 1$; if approximate $N_\omega/N_{Ey} = 1$, then get a simple straight line equation:

$$N/N_{Ey} + M_x/M_{cr} = 1 \qquad (7\text{-}32)$$

The above equation is approximately derived from the elastic buckling equation of bi-axially symmetrical I-shaped cross-section compression-bending parts. Therefore, after analysis, it is also approximately applicable to elastic-plastic buckling and uniaxially symmetrical cross-section members. Generally, $N_\omega > N_{Ey}$, which partially safe (Figure 7.11).

Replace $N_{Ey}$ and $M_{cr}$ with $\varphi_y f_y A$ and $\varphi_b f_y W_x$, respectively. If the bending moment graph is not a rectangular distribution under other loads, it will be multiplied by the equivalent bending moment coefficient $\beta_{tx}$. And use $f$ instead of $f_y$. By introducing the cross-section influence coefficient $\eta$ and the

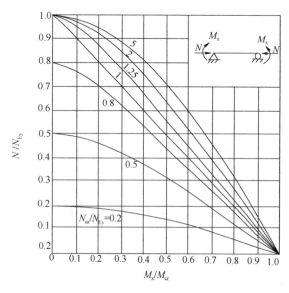

Figure 7.11 The relationship curve of flexural-torsional buckling out-plane of bending moment

## 7.3 Global Stability of Solid Web Compression-Bending Members

plastic development coefficient $\gamma_x$, Equation (7-33) can be written:

$$\frac{N}{\varphi_y A} + \eta \frac{\beta_{tx} M_x}{\gamma_x \varphi_b W_x} \leqslant f \tag{7-33}$$

The equation used in the code to calculate the stability of the unidirectional compression-bending member out-plane of bending moment:

$$\frac{N}{\varphi_y A f} + \eta \frac{\beta_{tx} M_x}{\varphi_b W_{1x} f} \leqslant 1.0 \tag{7-34}$$

$$\left| \frac{N}{A f} - \frac{\beta_{mx} M_x}{\gamma_x W_{2x}(1 - 1.25 N/N'_{Ex}) f} \right| \leqslant 1.0 \tag{7-35}$$

where  $N$ = the design value of axial pressure within the range of calculated members (N);

$N'_{Ex}$ = a parameter, $N'_{Ex} = \pi^2 EA/(1.1\lambda_x^2)$;

$M_x$ = the design value of the maximum bending moment within the range of the calculated member section (N · mm);

$W_{1x}$ = the gross section modulus of the most compressed fiber in the bending moment action plane (mm³);

$\varphi_y$ = global stability factor of the axial compression member outside the bending moment action plane;

$\varphi_b$ = the global stability factor of uniformly bent flexural members shall be calculated following Appendix C. Among them, I-section and T-shaped non-cantilevered members may be determined by the provisions of Appendix C; For closed cross-section, $\varphi_b = 1.0$;

$\eta$ = section influence factor, for closed section, $\eta = 0.7$; for the other sections, $\eta = 1.0$;

$\beta_{tx}$ = equivalent bending moment coefficient, the member section braced at both ends shall be the ratio of the maximum bending moment within 1/3 of the center to the maximum bending moment of the whole section, but not less than 0.5; the cantilever section is taken $\beta_{tx} = 1$;

$W_{2x}$ = the gross section modulus of the flangeless end (mm³).

The equivalent bending moment coefficient shall be adopted according to the following regulations $\beta_{tx}$:

a. The members with support out of the bending moment action plane should be determined according to the load and internal force in the member section between two adjacent supports:

When there is no transverse load,

$$\beta_{tx} = 0.65 + 0.35 \frac{M_2}{M_1}$$

When the end bending moment and transverse load act simultaneously:

Make the member produce the same curvature: $\beta_{tx} = 1.0$;

Make the member produce reverse curvature: $\beta_{tx} = 0.85$;

When the endless bending moment has a transverse load, $\beta_{tx} = 1.0$.

b. Out-plane of the bending moment is a cantilever member, $\beta_{tx} = 1.0$.

### 7.3.3 Stability calculation of solid web bidirectional compression-bending members

Bidirectional compression-bending members refer to compression-bending members whose bending moment acts on the section's two main planes. The instability of two-way compression-bending members is a form of spatial instability, and the theoretical calculation is relatively complicated, especially when the steel enters the elastoplastic stage. Therefore, numerical methods are commonly used for analysis and solution at present. Because the member has bending and torsion deformation in both directions before buckling, it is an extreme problem from buckling properties. In order to facilitate the design and application and to connect with the calculation of unidirectional bending members, the expression form of the related formula is used to calculate. The relative formulas are approximated by formulas that include three simple superpositions. The design standard GB 50017 stipulates that the stability of biaxially symmetrical solid-web I-setion and box-shaped compression-bending members with bending moments acting in two main planes.

The stability of biaxially symmetric solid web-shaped (including H-shaped) and box-shaped (closed) compression-bending members with bending moment acting in the two main planes shall be calculated according to the following equation:

$$\frac{N}{\varphi_x Af} + \frac{\beta_{mx} M_x}{\gamma_x W_x \left(1 - 0.8 \frac{N}{N'_{Ex}}\right)f} + \eta \frac{M_y}{\varphi_{by} \gamma_y W_y f} \leqslant 1 \qquad (7-36)$$

$$\frac{N}{\varphi_y Af} + \eta \frac{M_x}{\varphi_{bx} \gamma_x W_x f} + \frac{\beta_{my} M_y}{\gamma_y W_y \left(1 - 0.8 \frac{N}{N'_{Ey}}\right)f} \leqslant 1 \qquad (7-37)$$

$$N'_{Ey} = \pi^2 EA / (1.1\lambda_y^2) \qquad (7-38)$$

where  $\varphi_x, \varphi_y$ = the global stability factor of the axially loaded compression member for the strong axis $x$-$x$ and the weak axis $y$-$y$;

$\varphi_{bx}, \varphi_{by}$ = the global stability factor of the flexural member considering the influence of the bending moment change and the load position shall be taken as specified in the design standard GB 50017 ($M_{cr}$ is calculated according to the simply supported beam). The non-cantilever member of the I-section (including H-shapes) section can also be following simplified formula in Appendix C is determined, $\varphi_{by}$ can be taken as 1.0; for closed section, take $\varphi_{bx} = \varphi_{by} = 1.0$.

$M_x, M_y$ = the design value of the maximum bending moment for the strong axis $x$-$x$ and the weak axis $y$-$y$ within the calculated member section;

$W_x, W_y$ = the gross section modulus of the strong axis $x$-$x$ and the weak axis $y$-$y$;

$\beta_{mx}, \beta_{my}$ = the relevant regulations for stable calculation shall adopt the equivalent bending moment coefficient in the standard bending moment action plane.

## 7.4 Local Stability of Solid Web Compression-Bending Members

The plates that make up the cross-section of solid-web compression-bending members are similar to axial compression plates and bending members. Under the action of uniform compressive stress (such as compression flange, uniform or nearly uniform compression), or uneven compressive stress and shear stress (such as web), when the stress reaches a certain level, it may deviate from its plane position and cause wavy convex buckling, that is, the buckling of the plate, is called local buckling or loss of local stability for the member.

The local stability of the flexural member is often guaranteed by limiting the width (depth)-to-thickness ratio of the plate.

### 7.4.1 The width-to-thickness ratio of the compression-bending members

(1) Limiting the width-to-thickness ratio of the flange of the compression-bending member

The maximum compression flanges of I-section and box-shaped section compression-bending members are mainly subjected to normal stress, and the shear stress is very small and can be ignored. The maximum compressive stress may be lower than $f_y$ when the slenderness ratio is relatively large, and $N$ is the main bearing condition. When the slenderness ratio is relatively small, or $M$ is dominant, its value may be large and often reach or even enter the plastic zone. When considering the plastic development of the section, all the compression flanges form a plastic zone. It can be seen that the stress state of the flange of the compression-bending member is the same as the compression flange of the axial compression or bending member, and the loss of stability of the flange under uniform compressive stress is the same as that of this member. The limitation of the width-to-thickness ratio of the compressive flange of the solid web compression-bending member is listed in Table 3.1.

(2) Limiting the width-to-thickness ratio of the web of the compression-bending member

In addition to the uneven compressive stress, the web of the compression-bending member also has shear stress. The uneven compressive stress may be in an elastic state or an elastoplastic state, so its stability calculation is quite complicated. The limitation of the depth-to-thickness ratio of solid web compression-bending members is also shown in Table 3.1.

## Chapter 7 TENSION-BENDING MEMBERS AND COMPRESSION-BENDING MEMBERS

**7.4.2 The post-buckling strength of the compression-bending members**

When the web depth-to-thickness ratio of compression-bending members with I-section and box-shaped cross-sections exceeds the S4 class specified in Table 3.1, the effective section should be used instead of the actual section to calculate the bearing capacity.

(1) The effective width of the web

The effective width of the compression zone of the web shall be taken as:

$$h_e = \rho h_c \quad (7\text{-}39)$$

When $\quad \lambda_{n,p} \leqslant 0.75, \rho = 1.0 \quad (7\text{-}40a)$

When $\quad \lambda_{n,p} > 0.75, \rho = \dfrac{1}{\lambda_p^{re}}\left(1 - \dfrac{0.19}{\lambda_p^{re}}\right) \quad (7\text{-}40b)$

$$\lambda_p^{re} = \dfrac{h_w/t_w}{28.1\sqrt{k_\sigma}} \cdot \dfrac{1}{\varepsilon_k} \quad (7\text{-}41)$$

$$k_\sigma = \dfrac{16}{2 - \alpha_0 + \sqrt{(2-\alpha_0)^2 + 0.112\alpha_0^2}} \quad (7\text{-}42)$$

where $h_c, h_e$ = the width and effective width of the web compression zone. When the web is fully compressed, $h_c = h_w$ (mm);

$\rho$ = the effective width coefficient is calculated according to Equation (7-40);

$\alpha_0$ = a parameter, $\alpha_0 = (\sigma_{max} - \sigma_{min})/\sigma_{max}$.

(2) Distribution of effective width of I-section cross-section web

The effective width $h_e$ of the web shall be calculated according to the following equation:

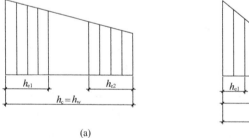
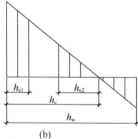

Figure 7.12 Distribution of effective width

When the section is fully compressed, $\alpha_0 \leqslant 1$ (Figure 7.12a).

$$h_{e1} = 2h_e/(4+\alpha_0) \quad (7\text{-}43)$$

$$h_{e2} = h_e - h_{e1} \quad (7\text{-}44)$$

When the section is partially compressed, $\alpha_0 > 1$ (Figure 7.12b).

$$h_{e1} = 0.4h_e \quad (7\text{-}45)$$

$$h_{e2} = 0.6h_e \tag{7-46}$$

When the slenderness ratio of the flange of the box-shaped section compression-bending member exceeds the limit, the effective width shall also be calculated according to Equation (7-39), $k_\sigma = 4$ shall be taken. Thus, the effective width is evenly distributed on both sides.

## 7.5 The Effective Length of the Compression-Bending Members

### 7.5.1 The effective length of the single bending members

As mentioned earlier, the slenderness ratio is used in the stability calculation of compression-bending members. Therefore, it is necessary to know the effective length of the member to calculate the slenderness ratio of the member. The effective length is derived from the ideal axial compression member to convert the axial compression member's length under different support conditions into the length of the equivalent hinged support, which simplifies the calculation.

For a single compression-bending member, when determining effective length $l_0$, the influence of the bending moment can be approximately ignored. Therefore, the calculation can be done by determining the effective length of the axial compression member $l_0 = \mu l$. Where $l$ is the member's geometrical length, $\mu$ is the effective length factor. In order to calculate the effective length factor, different values are used depending on the supporting conditions at both ends of the member, and the same value is taken approximately according to the axial compression member.

### 7.5.2 The effective length of the frame column

Most compression-bending members are not isolated single members but are part of the frame, and their ends are subject to various constraints of other members connected to them. When the frame column buckles, it will inevitably lead to the deformation of some members. In this way, studying the frame columns' buckling must take the entire frame or a part of the frame for analysis.

There are currently two methods for the stable design of frame columns. One is to adopt the first-order theory, that is, without considering the second-order influence of frame deformation, calculate the internal force generated by various load design values of the frame, and then design the frame column as a separate bending member. Then, when the column design is controlled by stability, the effective length is used instead of the actual length to consider the constraint effect of the members connected to the column. This method is simple and has many applications, referred to as the effective length method.

Another method is to analyze the framework as a whole according to the second-order theory. When calculating the frame column section based on stability, the actual geometric length calculates the slenderness ratio. According to the second-order theory, the analysis is more troublesome and inconvenient to apply, so an approximation method is used to consider the effect. First, a virtual horizontal load is introduced in the internal force analysis to consider the frame's lateral displacement. Then, a first-order analysis is carried out with the actual horizontal load and a vertical load of the frame to solve the internal force's design value of the frame column. Finally, when designing the frame column section based on stability, the member's actual geometric length calculates the slenderness ratio. This method is called the $P$-$\Delta$ design method.

This section mainly introduces the calculating method of the effective length of frame columns.

### 7.5.3 The effective length of single-story frame with equal section column in the frame plane

Figure. 7.13 shows a symmetrical single-story and single-bay uniform cross-section column frame. When determining the critical load and effective length of the frame column, which is usually assumed:

(1) The frame only bears the vertical load acting on the beam-to-column joints. As shown in Figure 7.13, the load acting on the frame beam can be moved to the top of the column, ignoring the influence of the beam's main bending moment. After this simplification, for the frame column's stability (without considering defects), the entire frame balance branch's critical load is used instead of the actual ultimate load. But when calculating the column's strength and stability, in addition to the axial compressive force, the internal force of the column must still be included in the bending moment caused by the load, that is, the calculation of the compression-bending member.

(2) The load increases proportionally at the same time, and each column loses its stability at the same time.

(3) The deformation is small and in the elastic stage.

(4) The members are free of defects.

The bracings, shear walls, elevator shafts, or other structures with greater lateral stiffness (5 times the lateral stiffness of the frame itself) can be provided for the frame structure to prevent the frame from being stressed or lost. When the frame joint moves sideways but is stable, it is called an inhibited sideway frame or a braced frame. Without bracings, shear walls, or others, the frame may cause the frame joints'sideway during stress or instability (which cannot be ignored in the calculation), called an uninhibited sideway frame or unbraced frame. When the frame is buckled (instability) without lateral movement, the frame joints have no lateral movement but have rotation angles, and the deformation is roughly symmetrical. Thus, it is often called symmetrical buckling (Fig-

7.5 The Effective Length of the Compression-Bending Members

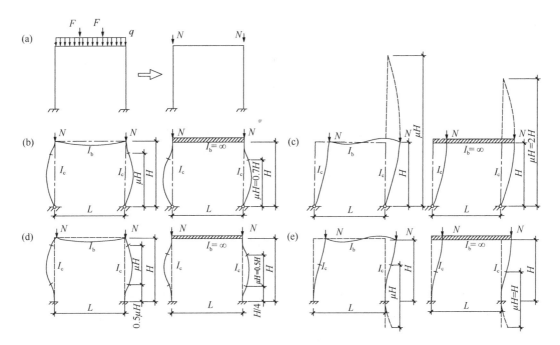

Figure 7.13 Effective length of a single-story and single-bay symmetrical frame with equal section columns

ures 7.13b and 7.13d). When there is lateral buckling of the frame, the joints on the same story of the frame will have equal lateral displacements and rotation angles in the same direction. Thus, the deformation is roughly left-right anti-symmetric buckling, often called anti-symmetric buckling (Figures 7.13c and 7.13e). Theoretically, symmetric buckling may occur in frames with lateral buckling, but its critical load is significantly greater than anti-symmetric buckling and is not controlled.

There are many methods for calculating frame stability, and the displacement method is more common. The characteristic of using the displacement method to calculate the frame stability is to introduce the axial load $N$'s influence into the angular displacement equation.

Figures 7.13(b) and 7.13(c) show the rigid connection between the column and beam, as well as the hinged connection between the column and foundation.

Such as $k^2 = N/EI_c$, $u = kH = H\sqrt{N/EI_c}$, $i_c = EI_c/H$ (column rotational stiffness), $i_b = EI_b/L$ (beam rotational stiffness), $K_1 = i_b/i_c$ (stiffness ratio of the beam to column) in the two cases of symmetric buckling (Figure 7.13b) and anti-symmetric buckling (Figure 7.13c). The following stable equations can be obtained separately:

$$\frac{u^2 \tan u}{\tan u - u} = -2K_1 \tag{7-47}$$

$$u \tan u = 6K_1 \tag{7-48}$$

If the value $K_1$ is known, the value $u$ can be obtained from Equations (7-47) or (7-

48). Hence, the critical load ($u = H\sqrt{N/EI_c}$) can be calculated by:

$$N_{cr} = \frac{u^2 EI_c}{H^2} = \frac{\pi^2 EI_c}{(\pi H/u)^2} = \frac{u^2 EI_c}{(\mu H)^2} \tag{7-49}$$

From this equation, the effective length factor $\mu$ is:

$$\mu = \pi/u \tag{7-50}$$

Figures 7.13(d) and 7.13(e) show the rigid connection between the top of the column and the beam and the rigid connection between the bottom of the column and the foundation in the two cases of symmetric buckling (Figure 7.13d) and anti-symmetric buckling (Figure 7.13e). The stability equations are:

$$\frac{u(\tan u - u)}{\tan u \left(2\tan \dfrac{u}{2} - u\right)} = -2K_1 \tag{7-51}$$

$$u/\tan u = -6K_1 \tag{7-52}$$

For the beam and column are fix connected, the column foot is hinged, and there is no lateral displacement and buckling, $\mu = 0.7 \sim 1.0$; when there is lateral buckling, $\mu = 2 \sim \infty$. For the beam and column, the column foot is fixed, and there is no lateral displacement and buckling, $\mu = 0.5 \sim 0.7$; when there is lateral buckling, $\mu = 1.0 \sim 2.0$. The geometric significance of the effective length in the four cases (the distance between the inflection points) is shown in Figures 7.13(b)~7.13(e).

### 7.5.4 The effective length of multi-story and multi-bay frame with equal section column in the frame plane

The instability forms of multi-story and multi-bay frames are also divided into two cases, no lateral instability (Figure 7.14a) and lateral instability (Figure 7.14b). The basic assumptions for calculation are the same as those of single-story frames. The pure frame structure without supporting structure (support frame, shear wall, anti-shearing simplified) belongs to anti-symmetric instability with lateral displacement. Braced frames can be divided into strong support frames and weak support frames according to the size of anti-side movement stiffness.

Regardless of the type of multi-story frame instability, each column must be affected by the column end members and the distal members. Because of the large number of unknown joint displacements in the multi-story and multi-bay frame, it is necessary to expand the high-order determinant. And to solve complex transcendental equations, the computational workload is large and difficult. Therefore, in practical engineering design, simplifying the constraint conditions of the column end is introduced. The frame is simplified into the calculation unit as shown in Figures 7.14(c) and 7.14(d), and the constraint effect of the members directly connected with the column end is considered.

In order to determine the effective length of the column, it is assumed that when the

## 7.5 The Effective Length of the Compression-Bending Members

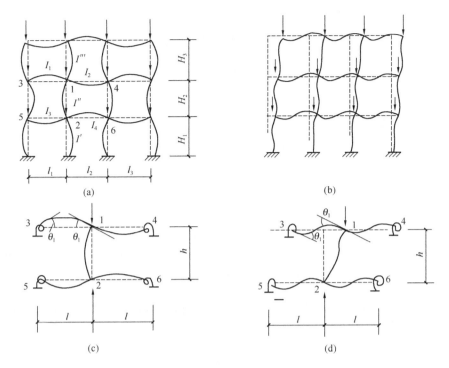

Figure 7.14 Instability form of a multi-story frame

column starts to be unstable, the constrained bending moment provided by the beam intersecting with the joints at the upper and lower ends of the column is allocated to the column according to its ratio to the sum of the rotational stiffness of the joints at the upper and lower ends of the column. Where, $K_1$ is the ratio of the sum of beam's rotational stiffness at the joint intersecting at the upper end of the column to the sum of column rotational stiffness; $K_2$ is the ratio of the sum of beam's rotational stiffness at the joint intersecting at the lower end of the column to the sum of column rotational stiffness.

(1) Unbraced frame

The effective length factor $\mu$ of the frame column is determined according to the effective length factor of the sideway uninhibited frame column in Appendix E.1, or it can be calculated according to the following simplified formula:

$$\mu = \sqrt{\frac{7.5K_1K_2 + 4(K_1+K_2)+1.52}{7.5K_1K_2+K_1+K_2}} \tag{7-53}$$

where $K_1, K_2$ = the ratio of the sum of the beam's rotational stiffness intersecting at the upper end and the lower end of the column to the sum of the column rotational stiffness, respectively.

(2) Braced frame

When the support system meets the Equation (7-54), it is a strong support frame. The frame column's effective length factor is determined according to the frame column's effective length factor $\mu$ with sideway inhibited in Appendix E.2, or it can be calculated according to Equation (7-55).

$$S_b \geq 4.4\left[\left(1+\frac{100}{f_y}\right)\sum N_{bi} - \sum N_{0i}\right] \qquad (7\text{-}54)$$

$$\mu = \sqrt{\frac{(1+0.41K_1)(1+0.41K_2)}{(1+0.82K_1)(1+0.82K_2)}} \qquad (7\text{-}55)$$

where $\sum N_{bi}, \sum N_{0i}$ = the sum of the stability bearing capacity of the axial compression column calculated by the effective length factor of the frame column without side-way and the frame column with side-way, respectively;

$S_b$ = the story lateral stiffness of the support system (N);

$K_1, K_2$ = the ratio of the sum of the beam's rotational stiffness intersecting at the upper end and the lower end of the column to the sum of the column's rotational stiffness, respectively.

### 7.5.5 The effective length of single-story frame stepped column in the frame plane

There are often larger bridge cranes in single-story factories. To support the crane beams, stepped (single-stage or double-step) column frames are often used, in which the lower section of the frame column is larger than the upper section of the column (Figure 7.15).

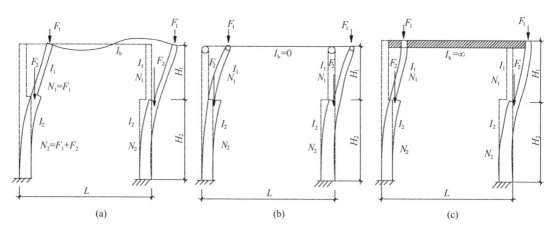

Figure 7.15  Effective length of a stepped column of the single-story frame

The frame of a single-story factory building is generally unbraced frame. The buckling is always anti-symmetric, and the column and the foundation are generally fixed. The beam and column can be rigidly connected (Figure 7.15a) or hinged (Figure 7.15b). Generally, the cross beams are mostly trapezoidal trusses, which have greater stiffness. Compared with the upper column, it can often be considered $I_b = \infty$ (Figure 7.15c).

For the fixed stepped column at the lower end of the single-story factory building frame, the effective length in the frame plane shall be determined according to the follow-

## 7.5 The Effective Length of the Compression-Bending Members

ing regulations:

(1) Single-step column

The effective length factor $\mu_2$ of the lower column: when the upper end of the column is hinged to the beam, the reduction factor in Table 7.1 shall be multiplied by the value in Table E.3 of Appendix E. When the upper end of the column is connected with the truss, the reduction factor in Table 7.1 shall be multiplied by the value in Table E.4 of Appendix E.

When the upper end of the column is fixed with the beam, the effective length factor of the lower column shall be multiplied by the reduction coefficient in Table 7.1 by the factor $\mu_2^1$ calculated in the following equation. The coefficient $\mu_2^1$ shall not be greater than the value $\mu_2$ obtained when the upper end of the column is hinged with the beam. At the same time, the factor $\mu_2^1$ not less than the value $\mu_2$ obtained when the upper end of the column is fixed with the truss-type beam.

$$K_b = \frac{I_b H_1}{l_b I_1} \tag{7-56}$$

$$K_1 = \frac{I_1 H_2}{H_1 I_2} \tag{7-57}$$

$$\mu_2^1 = \frac{\eta_1^2}{2(\eta_1+1)} \cdot \sqrt[3]{\frac{\eta_1 - K_b}{K_b}} + (\eta_1 - 0.5)K_1 + 2 \tag{7-58}$$

$$\eta_1 = \frac{H_1}{H_2}\sqrt{\frac{N_1}{N_2} \cdot \frac{I_2}{I_1}} \tag{7-59}$$

where  $I_b, l_b$ = the moment of inertia and span of solid-web steel beam;

$I_1, H_1$ = the moment of inertia and height of the upper column of the stepped column;

$I_2, H_2$ = the moment of inertia and column height of the lower column of the stepped column;

$K_b$ = the ratio of the beam rotational stiffness to the rotational stiffness of the upper column;

$K_1$ = the ratio of upper section column stiffness to lower section column stiffness.

The actual structure is inconsistent with the above assumption of stability of the calculation framework. The stepped columns of a single-story factory building mainly bear the crane load. When one column reaches the maximum load, the other columns of the same frame generally do not simultaneously reach the maximum load. When a column with a large load is to be buckled, it is bound to be supported by a column with a small load, which increases the critical load and reduces the effective length. Simultaneously, the factory building is often equipped with longitudinal horizontal bracings along the longitudinal direction, and some have large roof slab structures. These members can play a spatial role, allowing the adjacent columns with less force to limit and reduce the column's sideway, thereby improving its stability.

**Reduction factor of the effective length of a stepped column of single-story workshop**

Table 7.1

| Workshop type | | | | Reduction factor |
|---|---|---|---|---|
| Single span or multiple spans | The number of columns in a column in the longitudinal temperature section | Roof situation | Whether there are longitudinal and horizontal roof bracings on both sides of the plant | |
| Single span | Less than or equal to 6 | — | — | 0.9 |
| | More than 6 | Roof with non-large concrete roof slabs | No longitudinal horizontal bracing | |
| | | | Longitudinal horizontal bracing | 0.8 |
| | | Roof with large concrete roof slabs | — | |
| Multiple spans | — | Roof with non-large concrete roof slabs | No longitudinal horizontal bracing | |
| | | | Longitudinal horizontal bracing | 0.7 |
| | | Roof with large concrete roof slabs | — | |

The effective length factor of the upper column $\mu_1$ shall be calculated as follows:

$$\mu_1 = \frac{\mu_2}{\eta_1} \tag{7-60}$$

(2) Double-step column

The effective length factor $\mu_3$ of the lower column: when the upper end of the column is hinged to the beam, the value in Table E.5 of Appendix E (the upper end of the column is a free double-step column) multiplied by the reduction factor in Table 7.1. When the beam is fixed, it is equal to the value in Table E.6 of Appendix E (double-step column with a movable but non-rotating upper end of the column) multiplied by the reduction factor in Table 7.1.

The effective length factors of the upper column $\mu_1$ and the middle column $\mu_2$ shall be calculated according to the following equations:

$$\mu_1 = \frac{\mu_3}{\eta_1} \tag{7-61a}$$

$$\mu_2 = \frac{\mu_3}{\eta_2} \tag{7-61b}$$

### 7.5.6 The effective length of the frame column out of the frame plane

The space frame is subjected to bending moments in both directions, and the effective length in both directions is determined by the same method.

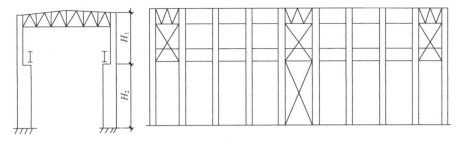

Figure 7.16  Bracings between frame columns

The arrangement of the supporting members generally determines the effective length of the frame column out of the frame plane. The bracing system provides the out-plane support points of the column, and the effective length of the column out the plane depends on the distance between the support points. These support points should prevent the column from moving laterally along the longitudinal direction of the factory building. For example in Figure 7.16, for a single-story factory building frame column, the support point of the lower section of the column is usually the surface of the foundation and the lower flange of the crane beam. The support point of the upper section of the column is the brake beam on the upper flange of the crane beam and is longitudinally horizontally bracing of the lower chord of the roof truss or is the chord of the bracket.

## 7.6  Design of Solid Web Compression-Bending Members

### 7.6.1  Design requirements

The cross-section design of solid-web compression-bending members should meet the requirements of strength, stiffness, global stability, and local stability. First, according to the force, the use of requirements, and structural requirements, choosing the appropriate type of section. When the bending moment is small, or positive and negative bending moments, the absolute value difference is small, and the asymmetrical section can be used. When the absolute value difference between positive and negative moments is large, the asymmetric flange section should be adopted, even if one of the flanges of the section is enlarged. For example, the side column supporting the crane beam of a single-story factory building often uses a larger flange. When satisfying local stability, using requirements, and structural requirements, the section should be made as large as possible in outline size and thinner in a plate. The distribution of section area is as far away as possible from the axis of the section. In this way, the same cross-sectional area can get a larger moment of inertia and radius of gyration and give full play to steel's effectiveness, thereby saving steel. According to the magnitude of the bending moment, the section depth should be appropriately larger than the section width to reduce the bending stress; and try to make the global stability in and out of the plane of the bending moment closer.

Because the calculation of compression-bending members is more complicated, it is generally assumed that the appropriate section is first before checking. When the check calculation is inappropriate, make necessary modifications and repeat the check calculation until satisfied. When assuming the cross-section, refer to the existing similar design and make the necessary estimation. The designed cross-section should also make the structure simple, easy to construct, and easy to connect with other members. In addition, the steel materials and specifications used should be easily available.

The frame structure is usually statically indeterminate. Therefore, when calculating the internal force, it is necessary to assume the moment of inertia between the beam and the column and the upper column and the lower column of the stepped column. When the moment of inertia ratio of each member's selected section does not differ from the assumed ratio by more than 30%, it is generally unnecessary to recalculate.

### 7.6.2 Construction requirements

The structural requirements of solid-web compression-bending members are similar to those of solid-web axial compression members. For example, in the case of $h_0/t_w > 80$, to prevent the web from deforming during construction and transportation and prevent the web from buckling when the shear force is large. Moreover, the transverse stiffeners should be provided for reinforcement spacing should not be greater. When longitudinal stiffeners are provided on the web, transverse stiffeners should support longitudinal stiffeners regardless of the size. Diaphragms should be set at both ends of each conveying unit and place under greater transverse force for large solid-web compression-bending members. When the member is long, the middle partition should be set, and the distance should not be greater than nine times of the larger width of the member section and 8 m. Its function is to keep the cross-sectional shape unchanged, improve the torsional rigidity of members, and prevent deformation during construction and transportation.

When setting the lateral support point of the member, the member with a small section height can only be supported in the central part of the web (ribbed or separated). The members with larger section height or greater stress should be supported in the two flange surfaces simultaneously. For example, the support between the frame columns should be equipped with a limb in the two flange planes, and the straps should be connected.

### 7.6.3 The selection of cross-section of solid web compression-bending members

The specific steps for section selection of solid-web compression-bending members are:

(1) Determining the design value of the internal force that the member bears

## 7.6 Design of Solid Web Compression-Bending Members

By the method of structural mechanics, the design value of the bending moment $M$, the design value of the axially compressive force $N$, and the design value of the shear $V$ are solved.

(2) Choosing section

When the bending moment is small for compression-bending members, the cross-section form is the same as general axial compression members (Figure 5.3). When the bending moment is large, it is advisable to adopt a double-section with a larger section height in-plane bending moment. Thus, the axisymmetric section or uniaxial symmetric section (Figure 7.3) is adopted.

(3) Determining the steel and strength design value

Compression-bending members should be made of steel with higher yield strength, which can achieve better economic efficiency. Determine the strength design value of each part of the material according to the thickness of the steel plate.

(4) Determining the effective length of the compressive-bending member

The effective length of the compression-bending member should be determined according to the followings: the type of frame structure where the member is located (single-story or multi-story, single-bay or multi-bay, side-way inhibited or side-way uninhibited, equal section column or stepped column), column foundation form (a hinged base or a fixed base), the form of the connection between the beam and the column (hinged, rigid or semi-rigid) and other factors.

(5) Primary selection of cross-section size based on experience or available data

(6) Checking calculations for the primary selected section

① Strength check

The strength of the compression-bending member bears unidirectional bending moment is calculated by Equation (7-9) or Equation (7-11).

Non-round pipe section

$$N/A_n \pm M_x/(\gamma_x W_{nx}) \leqslant f$$
$$N/A_n \pm M_y/(\gamma_y W_{ny}) \leqslant f$$

Round pipe section

$$\frac{N}{A_n} + \frac{\sqrt{M_x^2 + M_y^2}}{\gamma_m W_n} \leqslant f$$

When the cross-section has no weakening and the axial force values and the bending moment are the same as the values of the overall stability check, and the equivalent bending moment coefficient is 1.0, the strength check is not necessary.

② Stiffness check

The allowable slenderness ratio of compression-bending members is the same as that of the axially loaded compressive member. Therefore, the slenderness ratio of the bending member should not exceed the allowable slenderness ratio limit value specified in Table 5.2.

③ Global stability checks in-plane of the bending moment

The overall stability calculation in-plane of the bending moment of the solid-web compression-bending member functions is calculated using Equation (7-21).

$$\frac{N}{\varphi_x Af} + \frac{\beta_{mx} M_x}{\gamma_x W_{1x}\left(1-0.8\frac{N}{N'_{Ex}}\right)f} \leqslant 1.0$$

For single-axis symmetrical cross-section compression-bending members (such as T-shaped cross-sections, including hot-rolled T-shapes and double-angle T-shaped cross-sections, etc.), the calculation should also be carried out according to Equation (7-35).

$$\left|\frac{N}{Af} - \frac{\beta_{mx} M_x}{\gamma_x W_{2x}(1-1.25N/N'_{Ex})f}\right| \leqslant 1.0$$

④ Global stability checks out-plane of the bending moment

The overall stability calculation out-plane of the bending moment of the solid-web compression-bending member functions is calculated using Equation (7-34).

$$\frac{N}{\varphi_y Af} + \eta\frac{\beta_{tx} M_x}{\varphi_b W_{1x} f} \leqslant 1.0$$

For the meaning of symbols in each equation, see Section 7.2 and Section 7.3 of this chapter.

⑤ Local stability check

The width-to-thickness ratio of the flange and web of the combined cross-section compressive-bending member shall meet the requirements of Section 7.4.

If the check calculation does not meet the requirements, the trial section shall be modified and recalculated until satisfactory.

## 7.7 Design of Laced Compression-Bending Members

When the bending moment is relatively small or the positive and negative bending moments are likely to appear, and their absolute value difference is small, the symmetrical section can be used. When the bending moment is large or the positive and negative bending moments are likely to appear, their absolute values differ greatly from the single axisymmetric cross-section. Thus, the cross-section of one of the two limbs is larger than the other. Each limb itself can be made of rolled steel or welded by steel plates to form a combined section. When the distance between the two limbs is relatively large, the shapes should be used as the lacings. Under the premise of meeting the functional requirements, the design of the member section should be simple in structure, less in steel, and convenient in construction.

The steps for selecting the cross-section of a laced compression-bending member are similar to those of solid-web compression-bending members, except for considering the characteristics of laced compression-bending members in the calculation and split limb calculation stability and the design of lacings should be added. Because the real axis and the virtual axis are different, the laced compression-bending member has unique characteristics of strength, stiffness, global stability, and local stability.

When the bending moment acts around the virtual axis of laced compression-bending members and tension-bending members, the strength calculation basis is based on the section's edge fiber yield.

When calculating the stiffness and evaluating the global stability, it must adopt the converted slenderness ratio for the slenderness ratio around the laced member's virtual axis. The calculation of the converted slenderness ratio is the same as that of the laced axial compression member. When the deflection of the laced members is calculated, or the linear stiffness is needed to be calculated in the process of internal force analysis or stability analysis of statically indeterminate structures, the moment of inertia of the laced section with respect to the virtual axis should be multiplied by an appropriate reduction factor according to the calculated value. The equivalent section laced members can be generally used 0.9.

The calculation of the global stability of the flexural laced member around the virtual axis is slightly different from that of the solid web flexural member. In the local stability calculation of flexural laced members, the width (depth)-to-thickness ratio of the members is the same as that of the full-belly flexural members. There is also the problem of calculating the split limbs.

### 7.7.1 Stability calculation of laced compression-bending members

The stability calculation of double-limbed laced compression-bending members in different bending moments includes global stability and split-limb stability calculation.

(1) Stability calculation when bending moment acts around the virtual axis

When the bending moment acts around the virtual axis ($x$-axis) of the laced bending member (Figure 7.17), the global stability in the plane of bending moment and the stability of the limb in its own two principle axes should be calculated.

① Calculation of global stability in-plane of the bending moment

For laced compression-bending members with bending moment acting around the virtual axis ($x$-axis), the global stability in-plane of the bending moment should be calculated as follows:

$$\frac{N}{\varphi_x A f} + \frac{\beta_{mx} M_x}{W_{1x}\left(1 - \dfrac{N}{N'_{Ex}}\right) f} \leqslant 1 \qquad (7\text{-}62)$$

$$W_{1x} = I_x/y_0 \qquad (7\text{-}63)$$

where  $I_x$ = the moment of inertia of the gross cross-section about the $x$-axis;

$y_0$ = the axial distance from the $x$-axis to the larger pressure split limb or the distance to the outer edge of the web of the larger pressure split limb, whichever is greater;

$\varphi_x, N'_{Ex}$ = the global stability factor and parameters of the axially loaded compression member in-plane of the bending moment, determined by the conversion slenderness ratio $\lambda_{0x}$ about the virtual axis ($x$-axis).

The global stability out-plane of the bending moment does not have to calculate. Still, the split limb's global stability should be calculated, and the axial force of the split limb should be calculated according to the truss chord. For the batten plate column's split limbs, the local bending moment caused by the shear should be considered.

② Global stability out-plane of the bending moment

The global stability out-plane of the bending moment may not be calculated, but the stability of the split limbs should be calculated. Due to the two split limbs of the laced compression-bending member are only connected by the lacings, and the batten plates are only connected

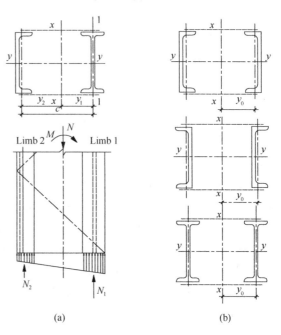

Figure 7.17  Stability calculation when the bending moment acts around the virtual axis of the laced compression-bending member

to the two split limbs in the plane of the lacings. In other words, when one split limb tends to occur bending displacement in-plane of the lacings, the other split limb will contain and support through the lacings. However, the lateral stiffness of the lacing out its plane is very weak. Therefore, when one split limb tends to bend or flexion to move out-plane of the split limb, the other split limb can only give a very weak restraint through the lacings. Therefore, when the bending moment acts around the virtual axis of the laced compression-bending member, it is necessary to ensure the member's out-plane overall stability of the bending moment (that is, perpendicular to the lacing plane). Therefore, it is mainly required that the two limbs' out-plane stability is guaranteed. Therefore, the check of limb stability replaces checking the whole member's out-plane overall stability of bending moment.

③ The stability of the split limbs

## 7.7 Design of Laced Compression-Bending Members

Each limb of the laced compression-bending member is a separate axial compression (tension) or compression-bending (tension-bending) member. Therefore, the stability of each divided limb in and out of the plane of the bending moment shall be ensured. For double-limbs laced compression bending members with bending moment acting around the virtual axis, the split limbs can be regarded as the chord of the truss to calculate the axial force of each split limb (Figure 7.17a, ignoring the additional bending moment):

Limb 1: $\quad\quad\quad\quad N_1 = Ny_2/c + M_x/c \quad\quad\quad\quad$ (7-64a)

Limb 2: $\quad\quad\quad\quad N_2 = Ny_1/c - M_x/c = N - N_1 \quad\quad\quad\quad$ (7-64b)

where $y_1, y_2$ = the distances from the centroid axis to the virtual axis ($x$-axis) of the limbs 1 and 2, respectively;

$c$ = the distance between the centroid axis of the two limbs.

For the split limbs of the compression-bending member with lacing bars, its stability shall be calculated according to the axially loaded compression member bearing the axially compressive force $N_1$ or $N_2$. For the split limbs of the compression-bending member with batten plates, the local bending moment $M_{x1}$ of the split limbs caused by shear should be considered. And the split limbs themselves become compression-bending members. When calculating the local bending moment, the shear of the member shall be the actual shear or the shear calculated according to the formula $V = \dfrac{AF}{85}\sqrt{\dfrac{f_y}{235}}$, whichever is greater. The multi-story frame calculation diagram is adopted, and the point of inflection is assumed at the midpoint of limbs and batten plates in each story.

To calculate the length of the limbs, take the internode length of the lacing bar system in the plane of the lacings (axis 1-1 in Figure 7.17) or the net distance between the two batten plates; outside the plane of the lacings, take the distance between the two lateral support points of the entire structure.

(2) Stability calculation when bending moment acts around the real axis

When the bending moment acts around the real axis ($y$-axis) of the laced compression-bending member (Figure 7.18), the global stability in and out of the plane of the bending moment and the stability of the split limbs in the direction of its two principal axes should be calculated.

For laced compression-bending members with bending moment acting around the real axis, the stability calculations in-plane and out-plane of the bending moment are the same as those of solid web members. How-

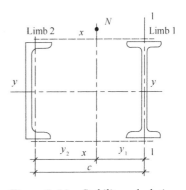

Figure 7.18 Stability calculation when the bending moment acts around the real axis of the laced compression-bending member

## Chapter 7 TENSION-BENDING MEMBERS AND COMPRESSION-BENDING MEMBERS

ever, when calculating the global stability out-plane of the bending moment, the slenderness ratio should be the converted slenderness ratio, and $\varphi_b$ should equal 1.0.

Stability calculation of split limbs: the distribution of axial load between the two limbs is determined by the principle that the distance from the split limb's axis to the $x$-axis is inversely proportional. The distribution of bending moment $M_y$ between the two split limbs is proportional to the moment of inertia of the limb about the $y$-axis and inversely proportional to the distance from the limb's axis about the $x$-axis to maintain balance and deformation coordination:

The axial force of limb 1: $N_1 = N y_2 / c$ \hfill (7-65a)

The axial force of limb 2: $N_2 = N y_1 / c = N - N_1$ \hfill (7-65b)

The bending moment of limb 1: $M_{y1} = M_y \dfrac{I_1/y_1}{I_1/y_1 + I_2/y_2}$ \hfill (7-66a)

The bending moment of limb 2: $M_{y2} = M_y \dfrac{I_2/y_2}{I_1/y_1 + I_2/y_2} = M_y - M_{y1}$ \hfill (7-66b)

where $I_1$, $I_2$ is the moment of inertia of the limbs 1, 2 about the $y$-axis.

(3) Stability calculation when bending moment acts in both main planes

The stability of double-limbs laced compression-bending members with bending moment acting in the two main planes shall be calculated according to the following regulations:

① Global stability calculation

$$\frac{N}{\varphi_x A f} + \frac{\beta_{tm} M_x}{W_{1x}\left(1 - \dfrac{N}{N'_{Ex}}\right)f} + \frac{M_y}{W_{1y} f} \leqslant 1.0 \qquad (7\text{-}67)$$

where $W_{1y}$ = the modulus of the gross cross-section of the larger compressed fiber under the action of $M_y$;

$\varphi_x$ = the global stability factor and parameters of the axially loaded compression member in-plane of the bending moment, determined by the conversion slenderness ratio $\lambda_{0x}$ on the virtual axis ($x$-axis).

② Stability calculation of the limb

Under the action of $N$ and $M_x$, the split limbs are used as truss chords to calculate the axial force, and $M_y$ is assigned to the two split limbs according to Equations (7-68a) and (7-68b) (Figure 7-19), and then the stability of the split limb is calculated.

Limb 1: $\qquad M_{y1} = \dfrac{I_1/y_1}{I_1/y_1 + I_2/y_2} \cdot M_y$ \hfill (7-68a)

Limb 2: $\qquad M_{y2} = \dfrac{I_2/y_2}{I_1/y_1 + I_2/y_2} \cdot M_y$ \hfill (7-68b)

where $I_1, I_2$ = the moment of inertia of limb 1, limb 2 about the $y$-axis;

$y_1, y_2$ = the distance from the acting principal axis plane about the centroid axis of limb 1, limb 2.

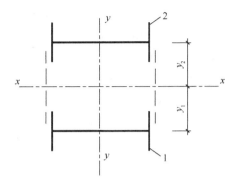

Figure 7.19 Cross-section of the laced member

### 7.7.2 Lacing calculation and construction requirements

The larger value between the actual shear of the member and the shear calculated according to the equation $V = Af/85\varepsilon_k$ should be taken when calculating the lacings of laced compression-bending members.

Laced compression-bending members are the same as laced axially loaded compression members. Transverse diaphragms should be provided at the place subject to greater horizontal force and the end of the transportation unit to ensure that the cross-sectional shape remains unchanged and the member's torsional rigidity is improved. When the member is long, an intermediate partition shall be provided, and the spacing shall not be greater than nine times of the larger width of the member's section or 8 m. The diaphragm can be made of steel plate or cross angles.

## 7.8 Beam-to-Column Connections of the Frame

The connection between the beam and the steel frame column plays the role of transferring bending moment and shear force (the conversion of shear force and axial force) between the two types of members. Therefore, it is the main component of the frame and should be taken seriously.

The classification method of frame beam and column connection is not uniform. In recent years, it has been more inclined to use the relative rotation restraint at the connection between the beam end and the column joint—the relationship between the bending moment $M$ and the relative rotation angle between the beam and column. Beam-to-column connections are divided into flexible connection, semi-rigid connection, and rigid connection (Figures 7.20a and 7.20b).

Its meaning is: ①flexible connection refers to the connection under the action of external force, the change of the angle between the beam and the column axis will reach more than 80% of the rotation angle of the ideal articulation (referring to the connection that can rotate freely), such as, only the beam web is connected to the column through the angle with high strength bolts; ②rigid connection refers to the connection under the action of external force, the rotation constraints can reach the ideal rigid connection (the angle between the beam and the column axis remains the same connection) of more than 90%, such as beam flange and web are welded with the column; ③semi-rigid connection means that its rotation restraint performance is between the above two. For example, beam ends are welded to end plates and connected to column flanges with high-strength bolts.

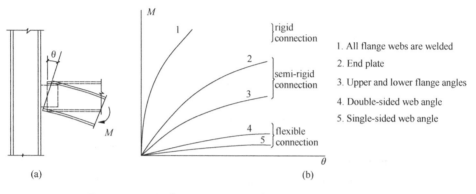

Figure 7.20　Relative rotation angle of beam-to-column and typical bending moment-relative angle curve

### 7.8.1　Flexible connection

The flexible connection of frame beams and columns can only bear a small bending moment and is generally regarded as a hinged connection during design. As a result, ideal articulation rarely exists in actual engineering. In a single-story frame, beams may connect to the column from the top or column side. The multi-story frame generally adopts the column through type, so the beam is connected with the column from the side. Figure 7.21 (a) shows the flexible connection between the beam web and the column flange through the vertical connecting plate (or angle) with welds and high-strength bolts, mainly to transmit shear forces. This connection is easy to install, and a flexible connection can be used in a supporting frame (relying on the support to resist lateral force).

### 7.8.2　Rigid connection

In unbraced frames, the connection joints of beams and columns must have strong bending stiffness to resist lateral forces, and rigid connections are used. In practice, the bracing frame's beam and column connections are also rigidly connected to enhance its overall performance.

Figure 7.21(c) shows one of the rigid connections between single-story frame columns and beams. The beam end bending moment is transmitted to the column through the cover plate and support using high-strength bolts (welds can also be used). The shear is transmitted through the high-strength bolts or welds of the connecting angle on the beam web. In order to reduce the welding workload on-site and simplify the joint structure, it can be designed as a column element with a cantilever beam section. The beam and column connection joints are all-welded in the factory, and the beams are spliced with high-strength bolts on-site (Figure 7.21d). High-strength bolts can connect the light single-story frame beam and column connection with the inclined end plate shown in Figure 7.21 (e).

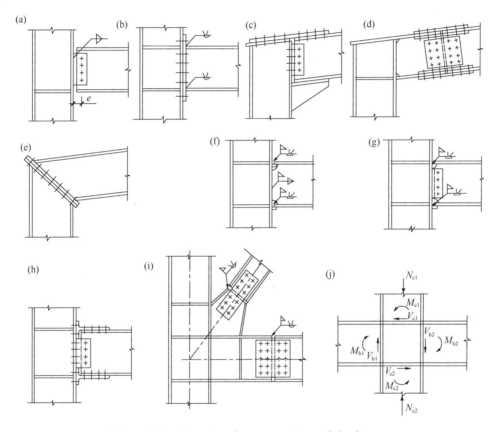

Figure 7.21  Beam-to-column connections of the frame.

Figure 7.21(f) shows the full welded rigid connection of the I-section beam and the I-section column of the multi-story frame. The beam flange and column flange are welded with grooves, and arc pilot plates are provided. In order to set the welding backing plate and facilitate welding, arc notches ($R=35$ mm) are made at the corners of the upper and lower ends of the beam web. The weld bears the tension generated by the bending moment and axial force. Fillet welds connect the beam webs and column flanges to transmit shear forces. High-strength bolts can also be used to connect the beam web and the column, as shown in Figure 7.21(g). This kind of bolt-weld hybrid connection is more convenient to

*261*

install. Figure 7.21(h) is one of full the high-strength bolt connection methods. The upper and lower flanges of the beam are connected to the column flange by two T-shaped steel sections with high-strength bolts. The T-shaped steel can be cut from H-shaped steel or cast. The web connection is the same as in Figure 7.21(c). This connection can simplify manufacturing and installation but is greatly affected by the performance of T-shaped steel and high-strength bolts and sometimes fails to meet the requirements of rigid connection, which can be regarded as a semi-rigid connection.

When the beam-to-column is rigidly connected, transverse stiffeners should be provided at the column web corresponding to the beam flange, and the panel zone should meet the following requirements:

(1) When the thickness of the transverse stiffener is not less than the thickness of the beam flange, the shear regularized slenderness ratio $\lambda_s^{re}$ of the panel zone shall not be greater than 0.8; for single-story and low-rise light buildings, it shall not be greater than 1.2. The shear regularized slenderness ratio $\lambda_s^{re}$ of the panel zone shall be calculated as follows:

When $h_c/h_b \geqslant 10$,

$$\lambda_s^{re} = \frac{h_b/t_w}{37\sqrt{5.34 + 4\,(h_b/h_c)^2}} \cdot \frac{1}{\varepsilon_k} \tag{7-69a}$$

When $h_c/h_b < 10$,

$$\lambda_s^{re} = \frac{h_b/t_w}{37\sqrt{4 + 5.34\,(h_b/h_c)^2}} \cdot \frac{1}{\varepsilon_k} \tag{7-69b}$$

where $h_b, h_c$ = the width and height of the panel zone web, respectively.

(2) The bearing capacity of the panel zone shall meet the requirements of the following equation:

$$\frac{M_{b1} + M_{b2}}{V_p} \leqslant f_{ps} \tag{7-70a}$$

H-shaped column:

$$V_p = h_{b1} h_{c1} t_w \tag{7-70b}$$

Box-shape column:

$$V_p = 1.8 h_{b1} h_{c1} t_w \tag{7-70c}$$

Round column:

$$V_p = (\pi/2) h_{b1} d_c t_c \tag{7-70d}$$

where $M_{b1}, M_{b2}$ = the design values of bending moments at both sides of the beam ends in the panel zone, respectively;

$V_p$ = the volume of the panel zone;

$h_{c1}$ = the width between the center lines of column flanges and the height of beam webs;

$h_{bl}$ = the height between the center lines of beam flanges;

$t_w$ = the thickness of the column web in the panel zone;

$d_c$ = the distance between the centerline of the pipe wall on the diameter line of the steel pipe;

$t_c$ = wall thickness of steel pipe in the panel zone;

$f_{ps}$ = the shear strength of the panel zone.

(3) The shear capacity of the panel zone shall be determined according to the regularized slenderness ratio $\lambda_s^{re}$ of the panel zone under shear according to the following regulations:

When $\lambda_s^{re} \leqslant 0.6$, $$f_{ps} = \frac{4}{3} f_v \tag{7-71a}$$

When $0.6 < \lambda_s^{re} \leqslant 0.8$, $$f_{ps} = \frac{1}{3}(7 - 5\lambda_s^{re}) f_v \tag{7-71b}$$

When $0.8 < \lambda_s^{re} \leqslant 1.2$, $$f_{ps} = [1 - 0.75(\lambda_s^{re} - 0.8)] f_v \tag{7-71c}$$

When the axial compression ratio is $\frac{N}{Af} > 0.4$, the shear capacity $f_{ps}$ should be multiplied by the correction factor. When $\lambda_s^{re} \leqslant 0.8$, the correction factor can be taken as $\sqrt{1 - \left(\frac{N}{Af}\right)^2}$.

(4) When the thickness of the panel zone does not meet the requirements of Equation (7-70), the following reinforcement measures can be adopted for the panel zone of the H-shaped column:

① Thicken the column web of the panel zone. The thickness of the web shall extend beyond the upper and lower flanges of the beam by no less than 150 mm.

② Welding and attaching reinforcing plates at the panel zone to strengthen. Fillet welds can connect the reinforcing plate and the column stiffener and flange, and the column web is connected by plug welding as a whole. The distance between the plug welding points should not be greater than $21\varepsilon_k$ times the thickness of the thinner weldment.

③ Set up diagonal stiffeners to strengthen the panel zone.

(5) In the beam-to-column rigid joint, the H-shaped beam's flange is connected with the flange of the H-shaped column by a T-shaped groove weld. The column web is not provided with horizontal stiffeners; the column flange and web thickness shall conform to the following provisions.

① At the compression flange of the beam, the thickness of the column web $t_w$ should satisfy:

$$t_w \geqslant \frac{A_{fb} f_b}{b_e f_c} \tag{7-72a}$$

Chapter 7  TENSION-BENDING MEMBERS AND COMPRESSION-BENDING MEMBERS

$$t_w \geqslant \frac{h_c}{30} \cdot \frac{1}{\varepsilon_{k,c}} \qquad (7\text{-}72b)$$

$$b_e = t_f + 5h_y \qquad (7\text{-}72c)$$

② At the tension flange of the beam, the thickness $t_c$ of the column flange should meet the requirements of the following formula:

$$t_c \geqslant 0.4\sqrt{A_{ft}f_b/f_c} \qquad (7\text{-}72d)$$

where  $A_{fb}$ = cross-sectional area of beam compression flange (mm²);
  $f_b, f_c$ = the design values of tensile and compressive strength of beam and column steel, respectively (N/mm²);
  $b_e$ = the assumed distribution length of the compressive stress at the edge of the column web calculated height (mm), under the action of the concentrated pressure perpendicular to the column flange;
  $h_y$ = the distance from the column's top surface to the upper edge of the calculated height of the web. The distance from the column flange edge to the starting point of the inner arc for the rolled section, and the thickness of the column flange (mm) for the welded section;
  $t_f$ = the thickness of beam compression flange (mm);
  $h_c$ = depth of column web (mm);
  $\varepsilon_{k,c}$ = the steel grade correction coefficient of the column;
  $A_{ft}$ = cross-sectional area of beam tension flange (mm²).

### 7.8.3 Semi-rigid connection

There are many types of semi-rigid connection beams and columns. Figure 7.21(b) shows the beam end welding where the endplates are connected to the column flanges with high-strength bolts, often called end plate connections, generally a semi-rigid connection. This kind of connection has strict requirements on the length of the beam. Therefore, the semi-rigid connection frame calculation is more complicated because need to know the $M\text{-}\theta$ relationship curve of the connection.

## 7.9 Working Examples

[Example 7-1]

Design Figure 7.22(a) shows the welded I-shaped section size of the member. The flange plate is flame cut, the section is not weakened, and the steel is Q235B. The design value of the component bearing axial load is $N=1,000$ kN (standard value $N_k=770$ kN), there is a lateral fulcrum at the midpoint of the component length and a lateral load, the design value is $F=250$ kN (standard value $F_k=190$ kN), both are static load.

## 7.9 Working Examples

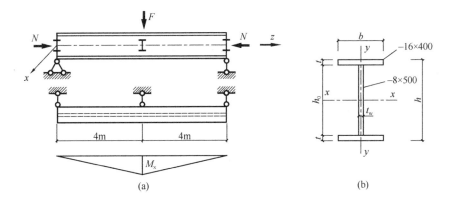

Figure 7.22 Example 7-1

**Solution:**

(1) Trial selection section

① Find the equivalent axial compressive force

$l_{0x} = 8$ m ; $l_{0y} = 4$ m ; $N = 1,000$ kN

Maximum bending moment design value: $M_x = \frac{1}{4}Fl = \frac{1}{4} \times 250 \times 8 = 500$ kN·m

The member has no lateral load between two adjacent lateral fulcrums, so:

$$\beta_{tx} = 0.65 + 0.35 \frac{M_2}{M_1} = 0.65 + 0.35 \times \frac{0}{500} = 0.65$$

In the vertical plane, there is a concentrated load in the middle of the component span, so:

$$\beta_{mx} = 1.0$$

$$N_{x,eq} = N + \frac{\beta_{mx} M_x}{25} = 1,000 + \frac{1 \times 500 \times 10^2}{25} = 3,000 \text{ kN}$$

$$N_{y,eq} = N + \frac{\beta_{tx} M_x}{18} = 1,000 + \frac{0.65 \times 500 \times 10^2}{18} = 2,806 \text{ kN}$$

② Trial selection section

$\lambda_x = 40$, checked by the type b cross-section, $\varphi_x = 0.899$

$\lambda_y = 50$, checked by the type b cross-section, $\varphi_y = 0.856$

Grade Q235: $f = 315$ N/mm² (assume $t \leqslant 16$ mm)

Required cross-sectional area:

$$A_x \geqslant \frac{N_{x,eq}}{\varphi_x f} = \frac{3,000 \times 10^3}{0.899 \times 215} \times 10^{-2} = 155.2 \text{ cm}^2$$

$$A_y \geqslant \frac{N_{y,eq}}{\varphi_y f} = \frac{2,806 \times 10^3}{0.856 \times 215} \times 10^{-2} = 152.5 \text{ cm}^2$$

Find the outer dimensions of the section from the approximate radius of gyration:

$$h \geqslant \frac{i_x}{0.43} = \frac{800}{40 \times 0.43} = 46.5 \text{ cm}$$

## Chapter 7 TENSION-BENDING MEMBERS AND COMPRESSION-BENDING MEMBERS

$$b \geqslant \frac{i_y}{0.24} = \frac{400}{50 \times 0.24} = 33.3 \text{ cm}$$

According to the above analysis, the cross-section of the test member is (Figure 7.22b):

Flange: $2 - 16 \times 400$
Web: $1 - 8 \times 500$

$$h = 532 \text{ mm} > 465 \text{ mm}, b = 400 \text{ mm} > 333 \text{ mm};$$
$$A = 2 \times 1.6 \times 40 + 1 \times 0.8 \times 50 = 168 \text{ cm}^2 > 155.2 \text{ cm}^2.$$

(2) Sectional geometric characteristics and related parameters

$$I_x = \frac{1}{12}(40 \times 53.2^3 + 39.2 \times 50^3) = 93,563 \text{ cm}^4$$

$$I_y = 2 \times \frac{1}{12} \times 1.6 \times 40^3 = 17,067 \text{ cm}^4$$

$$W_x = W_{1x} = \frac{I_x}{h/2} = \frac{93,563}{26.6} = 3,517 \text{ cm}^3$$

$$i_x = \sqrt{\frac{I_x}{A}} = \sqrt{\frac{93,563}{168}} = 23.60 \text{ cm}$$

$$i_y = \sqrt{\frac{I_y}{A}} = \sqrt{\frac{17,067}{168}} = 10.08 \text{ cm}$$

$$\lambda_x = \frac{l_{ox}}{i_x} = \frac{800}{23.60} = 33.9 < [\lambda] = 150$$

$$\lambda_y = \frac{l_{oy}}{i_y} = \frac{400}{10.08} = 39.7 < [\lambda] = 150$$

The stiffness meets the requirements.

Global stability factor of axially loaded compression member:

$$\varphi_x = 0.922, \varphi_y = 0.900.$$

Global stability factor of bending member:

$$\varphi_b = 1.07 - \frac{\lambda_y^2}{44,000} = 1.07 - \frac{39.7^2}{44,000} = 1.034 > 1.0, \text{ take } \varphi_b = 1.0.$$

$$N'_{Ex} = \frac{\pi^2 EA}{1.1\lambda_x^2} = \frac{\pi^2 \times 206 \times 10^3 \times 168 \times 10^2}{1.1 \times 3.9^2} \times 10^{-3} = 27,020 \text{ kN}$$

$$\frac{N}{N'_{Ex}} = \frac{1,000}{27,020} = 0.037$$

$$\beta_{tx} = 0.65, \beta_{mx} = 1.0$$

(3) Strength and stability check

① Sectional strength

$$\frac{N}{A_n} + \frac{M_x}{\gamma_x M_{nx}} = \frac{1,000 \times 10^3}{168 \times 10^2} + \frac{500 \times 10^6}{1.05 \times 3,517 \times 10^3}$$
$$= 59.5 + 135.4 = 194.9 \text{ N/mm}^2 < f = 215 \text{ N/mm}^2$$

The strength meets the requirements.

② Stability in-plane of the bending moment

$$\frac{N}{\varphi_x A} + \frac{\beta_{mx} M_x}{\gamma_x W_{nx}\left(1 - 0.8 \dfrac{N}{N'_{Ex}}\right)} = \frac{1,000 \times 10^3}{0.922 \times 168 \times 10^2} + \frac{500 \times 10^6}{1.05 \times 3,517 \times 10^3 (1 - 0.8 \times 0.037)}$$

$$= 64.6 + 139.5 = 204.1 \text{ N/mm}^2 < f = 215 \text{ N/mm}^2$$

③ Stability out-plane of the bending moment

$$\frac{N}{\varphi_y A_n} + \eta \frac{\beta_{tx} M_x}{\varphi_b M_{1x}} = \frac{1,000 \times 10^3}{0.900 \times 168 \times 10^2} + \frac{0.65 \times 500 \times 10^6}{1.0 \times 3,517 \times 10^3}$$

$$= 66.1 + 92.4 = 158.5 \text{ N/mm}^2 \leqslant f = 215 \text{ N/mm}^2$$

④ Local stability checking

Flange:

$$\frac{b_1}{t} = \frac{(400-8)/2}{16} = 12.3 < 13\sqrt{\frac{235}{f_y}} = 13 \text{ , meets the requirement.}$$

Web:

$$\sigma_{\max} = \frac{N}{A} + \frac{M_x}{I_x} y = \frac{1,000 \times 10^3}{168 \times 10^2} + \frac{500 \times 10^6}{93,563 \times 10^4} \times 250$$

$$= 59.5 + 133.6 = 193.1 \text{ N/mm}^2$$

$$\sigma_{\min} = \frac{N}{A} - \frac{M_x}{I_x} y = 59.5 - 133.6 = -74.1 \text{ N/mm}^2$$

$$\alpha_0 = \frac{\sigma_{\max} - \sigma_{\min}}{\sigma_{\max}} = \frac{193.1 - (-74.1)}{193.1} = 1.38 < 1.6$$

$$(16\alpha_0 + 0.5\lambda + 25)\sqrt{\frac{235}{f_y}} = (16 \times 1.38 + 0.5 \times 33.9 + 25) \times 1 = 64.0$$

$\dfrac{h_0}{t_w} = 62.5 < 64.0$ , meets the requirement.

⑤ Because $\dfrac{h_0}{t_w} = 62.5 < 80$ , the transverse stiffeners of the webs are not required.

⑥ Deflection

$$\frac{v}{l} = \frac{F_k l^2}{480 E I_x} \cdot \frac{1}{1 - \dfrac{N_k}{N_{Ex}}} = \frac{190 \times 10^3 \times 8,000^2}{48 \times 206 \times 10^3 \times 93,563 \times 10^4} \cdot \frac{1}{1 - \dfrac{770}{29,722}}$$

$$= \frac{1}{761} \cdot \frac{1}{0.974} = \frac{1}{741} < \left[\frac{1}{400}\right]$$

Deflection meets the requirement.

**Conclusion:**

The above checking calculations all meet the requirements, and the selected section size is appropriate.

[**Example 7-2**]

Design a laced unidirectional compression-bending double-limb column with a column height of 6 m, hinged at both ends. There is lateral support along the virtual axis ($x$-axis) direction at the midpoint of the column height, and the section is not weakened. The de-

# Chapter 7 TENSION-BENDING MEMBERS AND COMPRESSION-BENDING MEMBERS

sign value of the column top load is the axial load $N=600$ kN, the bending moment 150kN·m, and the column bottom has no bending moment. Refer to Figure 7.23(a), and the load is static. The steel is Q235B.

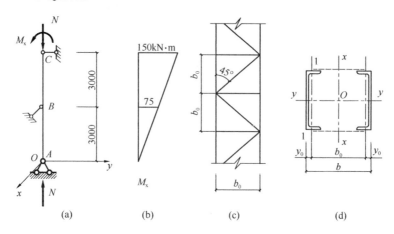

Figure 7.23  Example 7-2

**Solution:**

(1) Width $b$ of selected column section

According to the construction detail and rigidity requirements:

$$b = \left(\frac{1}{15} - \frac{1}{22}\right)H = \left(\frac{1}{15} - \frac{1}{22}\right) \times 6000 = 400 - 273 \text{ mm}$$

Use $b=400$ mm.

(2) Determine the split-limb section

Because the columns bear positive and negative bending moments and have the same value, a biaxially symmetrical section is adopted, the split-limb section adopts a hot-rolled channel. Assuming the distance between the axis of the channel and its back, the distance between the axis of the two limbs is (Figure 7.23d):

$$b_0 = b - 2y_0 = 400 - 2 \times 20 = 360 \text{ mm}$$

The maximum axial pressure in the split limb is:

$$N_1 = \frac{N}{2} + \frac{M_x}{b_0} = \frac{600}{2} + \frac{150}{0.36} = 716.7 \text{ kN}$$

The effective length of the split-limb to the $y$-axis:

$$l_{0y} = \frac{H}{2} = \frac{600}{2} = 300 \text{ cm}$$

Suppose the angle between the oblique strip and the axis of the split limb is 45°, refer to Figure 7.23(c), the effective length of the split limb to the 1-1 axis:

$$l_{01} = b_0 = 36 \text{ cm}$$

Set split limb $\lambda_y = 35$, find $\varphi = 0.918$ according to type b cross-section.

Need to split limbs cross-sectional area:

268

$$A_1 = \frac{N_1}{\varphi f} = \frac{716.7 \times 10^3}{0.918 \times 215} \times 10^{-2} = 36.3 \text{ cm}^2$$

Need radius of gyration:

$$i_y = \frac{l_{0y}}{\lambda_y} = \frac{300}{35} = 8.57 \text{ cm}$$

The required $A_1$ and $i_y$ can be found from the Table F.4 [25b can meet both requirements at the same time, and the cross-sectional characteristics, according to the coordinates shown in Figure 7.23(d), are:

$$A_1 = 39.917 \approx 39.92 \text{ cm}^2, I_y = 3,530 \text{ cm}^4$$

$$i_y = 9.41 \text{ cm}, I_1 = 196 \text{ cm}^4, i_1 = 2.22 \text{ cm}, y_0 = 1.98 \text{ cm}$$

(3) Design lacing bars

Shear in the column:

$$V_{max} = \frac{M_x}{H} = \frac{150}{6} = 25 \text{ kN}$$

$$V = \frac{Af}{85}\sqrt{\frac{f_y}{235}} = \frac{(2 \times 39.92 \times 10^2) \times 215}{85} \times 1 \times 10^{-3} = 20.2 \text{ kN}$$

Take $V_{max} = 25$ kN.

The internal force in a diagonal strip:

$$N_d = \frac{V_{max}/2}{\sin 45°} = \frac{25}{2 \times 0.707} = 17.7 \text{ kN}$$

The length of adiagonal strip:

$$l_d = \frac{b_0}{\cos 45°} = \frac{400 - 2 \times 19.8}{0.707} = 510 \text{ mm}$$

The cross-section of the diagonal lacing bar is one∠45 × 4 (minimum angle), providing:

$A_d = 3.486 \approx 3.49 \text{ cm}^2$, $i_{min} = i_v = 0.89$ cm.

Slenderness ratio:

$$\lambda_d = \frac{l_d}{i_{min}} = \frac{51.0}{0.89} = 57.3 < [\lambda] = 150$$

**O. K.**

According to the type b cross-section: $\varphi = 0.822$

When the single-sided equilateral single angle is checked and calculated according to the axial compression, the design strength reduction factor is as follows:

$$\eta_f = 0.6 + 0.0015\lambda = 0.6 + 0.0015 \times 57.3 = 0.686$$

Stability of the diagonal lacing bar:

$$\frac{N_d}{\varphi A_d} = \frac{17.7 \times 10^3}{0.822 \times 3.49 \times 10^2} = 61.7 \text{ N/mm}^2 < \eta_f f = 147.5 \text{ N/mm}^2$$

The calculation of the fillet weld connection between the strip and the column limbs is omitted here.

(4) Checking calculations of laced columns

① Geometric characteristics of the entire column section (Figure 7.23d)
Cross-sectional area:
$$A = 2A_1 = 2 \times 39.92 = 79.84 \text{ cm}^2$$
Moment of inertia:
$$I_x = 2[196 + 39.92 \times (20 - 1.98)^2] = 26,318 \text{ cm}^4$$
The radius of gyration:
$$i_x = \sqrt{\frac{I_x}{A}} = \sqrt{\frac{26,318}{79.84}} = 18.16 \text{ cm}$$
Sectional modulus of elasticity:
$$W_{1x} = W_{nx} = \frac{I_x}{b/2} = \frac{26,318}{20} = 1,316 \text{ cm}^3$$

② Stability in the plane of bending moment
$$\lambda_x = \frac{l_{0x}}{i_x} = \frac{600}{18.16} = 33.0$$
The cross-sectional area of the two diagonal lacing bars:
$$A_{dx} = 2 \times 3.49 = 6.98 \text{ cm}^2$$
Converted slenderness ratio:
$$\lambda_{0x} = \sqrt{\lambda_x^2 + 27\frac{A}{A_{dx}}} = \sqrt{33.0^2 + 27 \times \frac{79.84}{6.98}} = 37.4$$
According to the type b cross-section and $\lambda_{0x} = 37.4$: $\varphi_x = 0.908$
$$N'_{Ex} = \frac{\pi^2 \times 206 \times 10^3 \times 79.84 \times 10^2}{1.1 \times 37.4^2} \times 10^{-3} = 10,550 \text{ kN}$$
Refer to the bending moment diagram in Figure 7.23(b):
$$M_1 = 150 \text{ kN} \cdot \text{m}, \quad M_2 = 0 \text{ kN} \cdot \text{m}$$
$$\beta_{mx} = 0.60 + 0.4\frac{M_2}{M_1} = 0.60$$

$$\frac{N}{\varphi_x A} + \frac{\beta_{mx} M_x}{W_{1x}(1 - \varphi_x \frac{N}{N'_{Ex}})} = \frac{600 \times 10^3}{0.908 \times 79.84 \times 10^2} + \frac{0.60 \times 150 \times 10^6}{1,316 \times 10^3 (1 - 0.908 \times \frac{600}{10,550})}$$
$$= 82.8 + 72.1 = 154.9 \text{ N/mm}^2 < f = 215 \text{ N/mm}^2$$
**O.K.**

③ The stability out-plane of the bending moment is ensured by the stability calculation of the split limbs, and calculation is unnecessary here.

④ The stability of the split limbs
Limb internal force:
$$N_1 = \frac{N}{2} + \frac{M_x}{b_0} = \frac{600}{2} + \frac{150 \times 10^3}{400 - 2 \times 19.8} = 716.2 \text{ kN}$$
The slenderness ratio of the limb to the 1-1 axis:
$$\lambda_1 = \frac{b_0}{i_1} = \frac{400 - 2 \times 19.8}{22.2} = 16.2$$

The slenderness ratio of the limb to the $y$-axis:

$$\lambda_y = \frac{l_{0y}}{i_1} = \frac{300}{9.41} = 31.9$$

$$\lambda_1 = 16.2 < \lambda_y = 31.9 < [\lambda] = 150$$

According to the type b cross-section and $\lambda_y = 31.9$: $\varphi_1 = 0.929$

$$\frac{N_1}{\varphi_1 A_1} = \frac{716.2 \times 10^3}{0.929 \times 39.92 \times 10^2} = 193.1 \text{ N/mm}^2 < f = 215 \text{ N/mm}^2$$

⑤ Strength of full section

$$\frac{N}{A_n} + \frac{M_x}{\gamma_x W_{nx}} = \frac{600 \times 10^3}{79.84 \times 10^2} + \frac{150 \times 10^6}{1.0 \times 1316 \times 10^3} = 75.2 + 114.0$$

$$= 189.2 \text{ N/mm}^2 < f = 215 \text{ N/mm}^2$$

**Conclusion:**

The above checking calculations all meet the requirements, and the selected section is appropriate.

(5) Diaphragm setting

A 10 mm thick steel plate is used as the diaphragm. The spacing should not be greater than nine times the larger width of the column section and 8 m, not greater than 3.6 m ($9 \times 0.4 = 3.6$ m). The height of the current column is 6 m. In addition to the upper and lower ends, a transverse partition is set at the middle point. The spacing of the transverse partition is 3 m which meets the requirements.

**[Example 7-3]**

The geometric dimensions of two single-story single-bay frames shown in Figure 7.24 are the same. Still, the column feet are hinged and fixed, respectively. Calculate the effective length factor of the frame column.

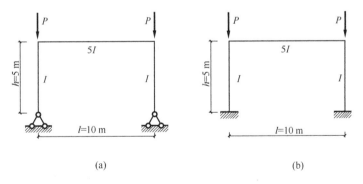

Figure 7.24  Example 7-3

**Solution:**

(1) The symmetrical frame is shown in Figure 7.24(a)

This frame is an uninhibited sideway frame.

Rotational stiffness of beam:

$$i_b = \frac{5I}{10} = 0.5I$$

Rotational stiffness of column:
$$i_c = \frac{I}{5} = 0.2I$$
$$K_1 = \frac{0.5I}{0.2I} = 2.5, \; K_2 = 0$$

Find out from Appendix E:
$$\mu = \frac{2.17 + 2.11}{2} = 2.14$$

(2) The symmetrical frame is shown in Figure 7.24(b)
This frame is an uninhibited sideway frame.
Rotational stiffness of beam:
$$i_b = \frac{5I}{10} = 0.5I$$

Rotational stiffness of column:
$$i_c = \frac{I}{5} = 0.2I$$
$$K_1 = \frac{0.5I}{0.2I} = 2.5, \; K_2 = 10$$

Find out from Appendix E:
$$\mu = 1.08$$

[**Example 7-4**]

Analyze the effective length factor of the frame column $AB$ shown in Figure 7.25.

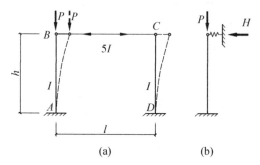

Figure 7.25  Example 7-4

**Solution:**

The two columns of this frame are subjected to different forces, the $AB$ column is subjected to the pressure $P$, and the $CD$ column is not. If this situation is not considered, simply use the Table E.1 it the Appendix E to find $\mu$ according to the table with the uninhibited sideway frame, and the result is:

$K_1 = 0$ (hinged to the beam)
$K_2 = 10$ (fixed at the bottom)

Check the Table E.1 to get $\mu = 2.03$, equivalent to the $AB$ column as an axially loaded compression member with one end fixed and one end free.

**Discussion:**

In the actual situation, when the $AB$ column is unstable, it must be shown by the dashed line in Figure 7.25(a), and the $CD$ column has the same top displacement as the $AB$ column through the beam $BC$. In this way, the $CD$ column is like a cantilever member with a horizontal force at the top. The bending rigidity of the $CD$ column will make the $AB$ column form an axially loaded compression member with a fixed lower end and elastic support at the upper end, as shown in Figure 7.25(b). The effective length factor is not $\mu = 2.03$, but $\mu < 2.03$ (see structural stability theory books for the specific method of $\mu$ value, which will be omitted here). If $\mu = 2.03$, it is on the safe side for this example. In some cases, calculating the $\mu$ value according to the look-up table may lead to unsafe results. Therefore, when the stiffness parameters of the columns are quite different, the effective length factor $\mu$ of the frame columns must be analyzed separately.

**[Example 7-5]**

Figure 7.26 shows the rigid connection joint of an H-shaped beam and H-shaped composite column. The column cross-sections are $2-16 \times 300$ and $1-10 \times 300$, and the flange weld $h_f = 8$ mm. The cross-sections of the two beams are both $2-14 \times 200$ and $1-8 \times 400$. The steel is Q235B. E43 type welding electrode is used for connection, manual welding. The required design values of internal forces in beams and columns are shown in Figure 7.26(a), the unit of bending moment is kN·m, and the unit of shear is kN. Check this joint and connection (without considering seismic design).

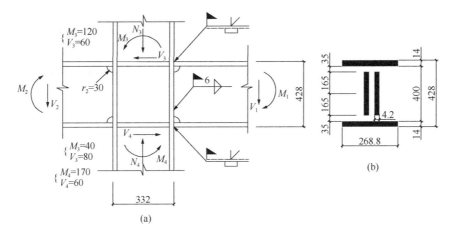

Figure 7.26 Example 7-5

**Solution:**

(1) Checking calculation of beam end section

Moment of inertia of two flanges to the cross-sectional centroid axis:

$$I_f = 2 \times 1.4 \times 20 \times \left(\frac{40 + 1.4}{2}\right)^2 = 23,995 \text{ cm}^4$$

# Chapter 7 TENSION-BENDING MEMBERS AND COMPRESSION-BENDING MEMBERS

Moment of inertia of the web to the cross-sectional centroid axis:
$$I_w = \frac{1}{12} \times 0.8 \times (40 - 2 \times 3)^3 = 2,620 \text{ cm}^4$$

Section modulus at the outermost fiber of the beam section:
$$W_x = \frac{I_x}{h/2} = \frac{26,615}{21.4} = 1,244 \text{ cm}^3$$

Section modulus at the top of the web:
$$W_{x1} = \frac{26,615}{(40 - 2 \times 3)/2} = \frac{26,615}{17} = 1,566 \text{ cm}^3$$

Bending normal stress at the outermost fiber of the right beam flange:
$$\sigma = \frac{M_1}{W_x} = \frac{250 \times 10^6}{1,244 \times 10^3} = 201.0 \text{ N/mm}^2 < f = 215 \text{ N/mm}^2$$

**O. K.**

Bending normal stress at both ends of the web:
$$\sigma_1 = \frac{M_1}{W_{x1}} = \frac{250 \times 10^6}{1,566 \times 10^3} = 159.6 \text{ N/mm}^2$$

The average shear stress of the web section of the right beam:
$$\tau_1 = \frac{V_1}{A_w} = \frac{48 \times 10^3}{0.8(40 - 2 \times 3) \times 10^2} = 17.6 \text{ N/mm}^2 < f = 125 \text{ N/mm}^2$$

**O. K.**

Converted stress:
$$\sqrt{\sigma_1^2 + 3\tau_1^2} = \sqrt{159.6^2 + 3 \times 17.6^2} = 162.5 \text{ N/mm}^2 < 1.1f = 1.1 \times 215 = 236.5 \text{ N/mm}^2$$

**O. K.**

(2) Strength check calculation of beam end weld

When the weld quality level is second, the design value of the bending strength of the groove weld is $f_t^w = 215 \text{ N/mm}^2$.

Tensile, compressive, and shear strength design values of fillet welds are $f_f^w = 160 \text{ N/mm}^2$.

The length of the butt weld of the flange plate $b = 200$ mm (weld with the pilot arc plate) is converted into the length equivalent to the fillet weld according to the strength design value:
$$b' = 200 \times \frac{215}{160} = 268.8 \text{ mm}$$

Fillet weld at the web:
Effective thickness, $0.7h_f = 0.7 \times 6 = 4.2$ mm.
Effective length $l_w = 400 - 2 \times 30 - 2 \times 6 = 328$ mm $\approx 330$ mm.

The effective cross-sections of all equivalent fillet welds for rigid connection of beams and columns are shown in Figure 7.26(b).

Moment of inertia of the effective section of the weld:
$$I_x^w = 2 \times 1.4 \times 26.88 \left(\frac{41.4}{2}\right)^2 + 2 \times \frac{1}{12} \times 0.42 \times 33^3 = 34,766 \text{ cm}^4$$

Section modulus at the outer edge of the flange:
$$W_x^w = \frac{34,766}{21.4} = 1,625 \text{ cm}^3$$

Section modulus at both ends of the web:
$$W_{x1}^w = \frac{34,766}{16.5} = 2,107 \text{ cm}^3$$

The effective area of fillet welds on both sides of the web:
$$A_f^w = 2 \times 0.42 \times 33 = 27.72 \text{ cm}^2$$

Bending stress in the weld on the outer edge of the flange:
$$\sigma_f^M = \frac{M_1}{W_x^w} = \frac{250 \times 10^6}{1,625 \times 10^3} = 153.8 \text{ N/mm}^2 < f_f^w = 160 \text{ N/mm}^2$$

**O. K.**

The weld stress at both ends of the web is:
$$\sigma_f^M = \frac{M_1}{W_{x1}^w} = \frac{250 \times 10^6}{2,107 \times 10^3} = 118.7 \text{ N/mm}^2$$

$$\tau_f^v = \frac{V_1}{A_f^w} = \frac{48 \times 10^3}{27.72 \times 10^2} = 17.3 \text{ N/mm}^2$$

$$\sqrt{\left(\frac{\sigma_f^M}{\beta_f}\right)^2 + (\tau_f^v)^2} = \sqrt{\left(\frac{118.7}{1.22}\right)^2 + 17.3^2} = 98.8 \text{ N/mm}^2 < f_f^w = 160 \text{ N/mm}^2$$

**O. K.**

(3) The local compressive strength and local stability of the column web at the connection point of the beam compression flange

Assuming that the bending moment of the beam flange plate is $M_f = M_1 \cdot \frac{I_f}{I_x}$, the axial compressive force in the beam flange is:

$$N_{fb} = \frac{M_f}{d_b} = 250 \times \frac{23,995}{26,615} \times \frac{1}{0.414} = 544.4 \text{ kN}$$

The assumed distribution length of the local compressive stress in the column web (take $\beta = 2.5$):
$$l_z = t_{bf} + 2\beta(t_{cf} + k) = 14 + 2 \times 2.5(16 + 8) = 134 \text{ mm}$$

Local compressive stress in the column web:
$$\sigma_c = \frac{N_{fb}}{l_x t_{cw}} = \frac{544.4 \times 10^3}{134 \times 10} = 406.3 \text{ N/mm}^2 > f = 215 \text{ N/mm}^2$$

It does not meet the requirements.

The thickness requirements of the column web are:
$$t_w \geqslant \frac{A_{fb} f_b}{b_e f_c} = \frac{14 \times 200}{14 + 5 \times 16} \cdot 1.0 = 29.79 \text{ mm} \approx 30 \text{ mm}$$

and
$$t_w \geqslant \frac{h_c}{30}\sqrt{f_{yc}/235} = \frac{300}{30} = 10 \text{ mm}$$

Because $t_w = 10 \text{ mm} < 30 \text{ mm}$, the column web shall be provided with transverse stiffeners at the horizontal position of the compression flange.

The cross-sectional area of the required stiffener is:
$$A_s \geq \frac{N_{fb} - t_{cw} l_z f_c}{f_s} = \frac{544.4 \times 10^3 - 10 \times 134 \times 215}{215} \times 10^{-2} = 11.92 \text{ cm}^2$$

The width of each stiffener:
$$b_s \geq \frac{b_{bf}}{3} - \frac{t_{cw}}{2} = \frac{200}{3} - \frac{10}{2} = 61.7 \text{ mm}$$

Take $b_s = 90 \text{ mm} < \dfrac{300 - 10}{2} = 145$ mm (300mm is the width of the column flange, 10mm is the thickness of the column web).

The corner of the stiffener needs to be cut off at the connection with the column flange, $\dfrac{b_s}{3} = 30$ mm.

Therefore, the required stiffener thickness is:
$$t_s \geq \frac{11.92 \times 10^2}{2(90 - 30)} = 9.93 \text{ mm}$$

Take
$$t_s = 12 \text{ mm} > \frac{t_{bf}}{2} = 7 \text{ mm}$$

$\dfrac{b_s}{t_s} = \dfrac{90}{12} = 7.5 < 15$, meets the requirements.

Adopt horizontal stiffener $2-12 \times 90$.

The end of the stiffener is connected with the column flange plate with a groove weld, and the effective cross-sectional area of the supply weld is:
$$2 \times 1.2(9.0 - 3.0 - 2t_s) = 8.64 \text{ cm}^2 < 11.92 \text{ cm}^2$$

It does not meet the requirements.

Change the stiffener from $2-12 \times 90$ to $2-10 \times 120$.

The effective cross-sectional area of the weld becomes 14 cm², which can meet the requirements.

The connection of the four fillet welds between the two stiffeners and the column web should be able to resist the algebraic sum of the axial forces of the beam flange plates caused by the bending moments at the two beam ends, which can be obtained:
$$N_{fb1} + N_{fb2} \leq 4 \times 0.7 h_f l_w f_f^w$$

get
$$h_f \geq \frac{544.4\left(1 + \dfrac{40}{250}\right) \times 10^3}{4 \times 0.7(300 - 2 \times 30 - 20) \times 160} = 6.41 \text{ mm}$$

According to construction requirements:
$$h_f \geq 1.5\sqrt{t} = 1.5\sqrt{10} = 4.74 \text{ mm}$$

Take $h_f = 8$ mm.

(4) The thickness requirement of column flange at the connection point of the beam

compression flange

In order to prevent the column flange connected to the beam tension flange plate from lateral bending due to tension, the minimum thickness of the column flange is:

$$t_c \geq 0.4\sqrt{A_{ft} f_b/f_c} = 0.4\sqrt{14 \times 200 \times 1.0} = 21.2 \text{ mm}$$

However $t_{cf} = 16$ mm $< 21.2$ mm, transverse stiffeners should also be provided on the column webs at the horizontal position of the tension flange, the size of which is the same as that used for the beam's compression flange mentioned above, that is, $2-10 \times 120$.

(5) Checking the shear strength of the panel zone at the connection of beam-to-column

$$h_c/h_b = 300/400 = 0.75 < 10$$

$$\lambda_{n,s} = \frac{h_b/t_w}{37\sqrt{4+5.34(h_b/h_c)^2}} \frac{1}{\varepsilon_k} = \frac{400/10}{37\sqrt{4+5.34(400/300)^2}} = 0.2943 < 0.6$$

$$f_{ps} = \frac{4}{3} f_v = \frac{4}{3} \times 125 = 166.67 \text{ N/mm}^2$$

I-shaped column:

$$V_p = h_{b1} h_{c1} t_w = 414 \times 312 \times 10 = 1.308 \times 10^6 \text{ mm}^3$$

The pannel zone must meet the requirement:

$$\frac{M_{b1} + M_{b2}}{V_p} = \frac{(250+40) \times 10^3}{1.308 \times 10^6} = 221.67 \text{ N/mm}^2 > f_{ps} = 166.67 \text{ N/mm}^2$$

It does not meet the requirement.

## 7.10 Chapter Summary

Both tension-bending member and compression-bending member must bear the combined action of axial force and bending moment. This chapter mainly discusses the strength and stability calculation of members when two loads are acting simultaneously. When analyzing the strength design of tension-bending and compression-bending members, this chapter first describes the members' elastic design and plastic design respectively, and finally describes the elastoplastic design. The difficulty of understanding lies in the correlation between axial force and bending moment and simplifying theoretical formulas to design formulas. When studying this chapter, you should pay attention to the content of the first two chapters.

(1) When the bending moment is small, and the axial force is large, the tension-bending and compression-bending members often adopt the same biaxially symmetrical section as the axially loaded member. When the bending moment is large, in order to save materials, a larger section is used on the bending and compression side, and a smaller section is used on the bending and tension side, forming a uniaxially symmetric section. When the length of the member is relatively large, and there is a bending moment, a larger moment

of inertia of the section is required, and laced (built-up) members can be used.

(2) Tension and bending members are generally used when the axial tension is large and the bending moment is small. In the frame structure, beams are usually designed according to the bending member, and the column is usually designed according to the compression-bending member. Therefore, the normal service limit state check calculations of tension-bending and compression-bending members generally only check the slenderness ratio. If the bending moment of the member is large, or the frame beam is calculated according to the compression-bending member. In that case, the bending and compression-bending members also need to check the deflection according to the bending member.

(3) For cold-formed thin-walled members and members that require fatigue checking, the ultimate state of bearing capacity can be used when the material strength reaches the yield strength, and elastic design methods can be adopted. For other members, it is necessary to design according to the plastic development of the cross-section.

(4) A compression-bending member subjected to bending moments in a plane has in-plane flexural instability and out-plane flexural-torsional instability. The actual member has defects such as initial bending and longitudinal residual stress. Instability is an extreme point instability problem that needs to be solved using a numerical integration method. The design formulas for in-plane instability and out-plane instability of compression-bending members are very similar, and attention should be paid to distinguish them. It can be understood from the superposition of axial stress and bending stress.

(5) The flange width-to-thickness ratio of the compression-bending member is the same as that of the bending member, described in Chapter 6. The web stability of compression-bending members requires the concept of stress gradient. The limit of the depth-to-thickness ratio is related to the slenderness ratio of the member and the stress gradient on the section. The box-shaped cross-section members have high out-plane torsion rigidity and bending stiffness and are suitable for compression-bending members that are unsupported out-of-plane when the force is large.

(6) Laced compression-bending members with bending moment acting around the virtual axis are often used in heavy-duty industrial plants. The stability design includes the stability design in-plane of the bending moment and the split limb stability design. The stability design out-plane of the bending moment is the same as the split limb design. The biggest difference between the in-plane stability design of the laced compression-bending member and the in-plane stability design of the solid web compression-bending member is that the laced member's in-plane stability design cannot be considered partial plastic development of the section. Nevertheless, the design formulas of the two are very similar, and their subtle differences should be noted.

## Questions

7.1  Are there any similarities between the strength calculation formulas of tension-bending and compression-bending members and the strength calculation formulas of axially load Compressive members and bending members?

7.2  In the design of tension-bending and compression-bending members, when should elastoplastic design be adopted? Conversely, when should elastic design be adopted?

7.3  How to check the stiffness of tension-bending and compression-bending members?

7.4  What are the instability forms of solid web compression-bending members?

7.5  Under what circumstances will the in-plane instability of a solid-web flexural member subjected to unidirectional bending occur? Under what circumstances will out-plane instability occur?

7.6  What is the equivalent in-plane bending moment coefficient of a compression-bending member? What is its function?

7.7  How many design methods are there for the stability design in-plane of the bending moment? Describe the applicable members of each method.

7.8  How to prevent out-plane instability of bending moment of compression-bending members?

7.9  What items need to be calculated for the design of solid-web compression-bending members? What about laced compression-bending members?

7.10  What is the meaning of each symbol in the calculation of strength and stability of compression-bending members?

7.11  What is the stress gradient? When the stress gradient equals 0 and 2, which force state of the compression-bending member is represented respectively?

## Exercises

7.1  A tension-bending member which is made of Q345B HN400 × 200 × 8 × 13 is shown in Figure 7.27. There is no weakening of openings on the section, and the two ends of the member are pin-ends. Regardless of the stiffness, please calculate the maximum value of tension force $N$ that can be carried.

7.2  Figure 7.28 shows the biaxially sym-

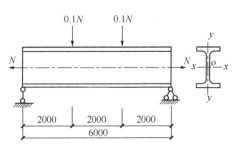

Figure 7.27  Exercise 7.1

metric welded H-shaped cross-section column. The cross-section is H900×400×14×20, the column is made of Q355B, the flange has flame-cut edges, and the cross-section of the column is not weakened. An axial compressive force is $N=1,500$ kN and the horizontal force is $H=150$ kN on the top of the column. The horizontal force is in the strong-axis plane. There is effective support at 1/2 the height of the column in the weak-axis plane, and the support point can be regarded as an out-plane fixed point. The lower end of the column is fixed, and the upper end can move freely. Try to calculate whether the welded H-shaped section column meets the requirements.

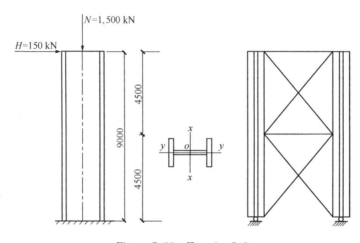

Figure 7.28　Exercise 7.2

7.3　An eccentric compression column withtwo pin-ends, the length is 3 m and is made of Q345B HN400×200×8×13 (without weakening the section). The design value of the compressive force is 750 kN, and the eccentricity at both ends is the same as 15 cm. Check whether its bearing capacity meets the design requirements.

7.4　The compression-bending component shown in Figure 7.29 adopts a welded H-shaped section H800×400×14×20, the flange is a rolled edge, and the cross-section is

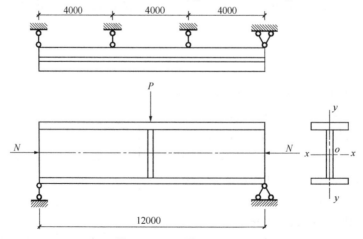

Figure 7.29　Exercise 7.4

not weakened. Made of Q355B steel, 12 m length, hinged at both ends, axial compressive force $N=3,000$ kN, lateral concentrated load $P=300$ kN acts at mid-span, lateral support point distribution out-plane of the bending moment is shown in Figure 7.29. Check whether the section meets the requirements.

7.5 The cross-section of a laced column is shown in Figure 7.30. The columns on both sides use HN $650 \times 300 \times 11 \times 17$, and both the horizontal and diagonal columns use ∟ $125 \times 8$. The upper end of the column is elastic support with sideway, and the lower end is fixed. The effective length of the column $l_{0x} = 30$ m and $l_{0y} = 12$ m, the steel is Q355B, the maximum design internal force is $N=3,000$ kN, the bending moment is $M_x = \pm 2,200$ kN·m acting around the $x$-axis, check the bearing capacity and stiffness of this column.

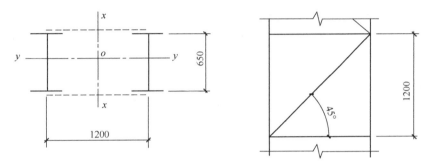

Figure 7.30 Exercise 7.5

7.6 In Figure 7.31, the column is a solid-web column with a constant cross-section in the rigid-connected frame, and the beam is a truss-type. Try to determine the effective length of the column.

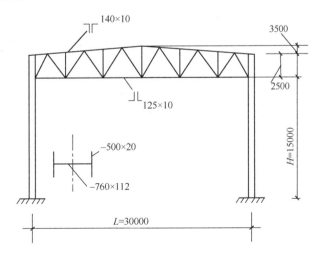

Figure 7.31 Exercise 7.6

# APPENDIX A  CHEMICAL COMPOSITION OF STEEL

The chemical composition of carbon structural steel (According to GB/T 700—2006)    Table A.1

| Grade | Unified digital code | Quality level | Thickness (or Diameter) (mm) | Deoxygenation method | Chemical composition (mass fraction) (%) | | | | |
|---|---|---|---|---|---|---|---|---|---|
| | | | | | C | Si | Mn | P | S |
| Q195 | U11952 | — | — | F and Z | ≤0.12 | ≤0.30 | ≤0.50 | ≤0.035 | ≤0.040 |
| Q215 | U12152 | A | — | F and Z | ≤0.15 | ≤0.35 | ≤1.20 | ≤0.045 | ≤0.050 |
| | U12155 | B | | | | | | | ≤0.045 |
| Q235 | U12352 | A | — | F and Z | ≤0.22 | ≤0.35 | ≤1.40 | ≤0.045 | ≤0.050 |
| | U12355 | B | | | ≤0.20[b] | | | | ≤0.045 |
| | U12358 | C | | Z | ≤0.17 | | | ≤0.040 | ≤0.040 |
| | U12359 | D | | TZ | | | | ≤0.035 | ≤0.035 |
| Q275 | U12752 | A | — | F and Z | ≤0.24 | ≤0.35 | ≤1.50 | ≤0.045 | ≤0.050 |
| | U12755 | B | ≤40 | Z | ≤0.21 | | | ≤0.045 | ≤0.045 |
| | | | >40 | | ≤0.22 | | | | |
| | U12758 | C | — | Z | ≤0.20 | | | ≤0.040 | ≤0.040 |
| | U125759 | D | | TZ | | | | ≤0.035 | ≤0.035 |

Note: 1. The unified numbers are grades of killed steels and special killed steels.

The unified digital codes of boiling steel grade are: Q195F——U11950; Q215AF——U12150, Q215BF——U12153; Q235AF——U12350, Q235BF——U12353; Q275AF——U12750;

2. With the purchaser's consent, the carbon content of Q235B may not exceed 0.22%.

The chemical composition of low alloy high-strength steel (according to GB/T 1591—2018)

Table A.2

| Grade | Quality level | Chemical composition (mass fraction) (%) | | | | | | | | | | | | | | |
|---|---|---|---|---|---|---|---|---|---|---|---|---|---|---|---|---|
| | | C | Si | Mn | P | S | Nb | V | Ti | Cr | Ni | Cu | N | Mo | B | Als |
| Q345 | A | ≤0.20 | ≤0.50 | ≤1.70 | ≤0.035 | ≤0.035 | ≤0.07 | ≤0.15 | ≤0.20 | ≤0.30 | ≤0.50 | ≤0.30 | ≤0.012 | ≤0.10 | — | — |
| | B | | | | ≤0.035 | ≤0.035 | | | | | | | | | | |
| | C | | | | ≤0.030 | ≤0.030 | | | | | | | | | | |
| | D | ≤0.18 | | | ≤0.030 | ≤0.025 | | | | | | | | | | ≥0.015 |
| | E | | | | ≤0.025 | ≤0.020 | | | | | | | | | | |
| Q390 | A | ≤0.20 | ≤0.50 | ≤1.70 | ≤0.035 | ≤0.035 | ≤0.07 | ≤0.15 | ≤0.20 | ≤0.30 | ≤0.50 | ≤0.30 | ≤0.012 | ≤0.10 | — | — |
| | B | | | | ≤0.035 | ≤0.035 | | | | | | | | | | |
| | C | | | | ≤0.030 | ≤0.030 | | | | | | | | | | |
| | D | | | | ≤0.030 | ≤0.025 | | | | | | | | | | ≥0.015 |
| | E | | | | ≤0.025 | ≤0.020 | | | | | | | | | | |

## APPENDIX A  CHEMICAL COMPOSITION OF STEEL

continued

| Grade | Quality level | Chemical composition (mass fraction) (%) | | | | | | | | | | | | | | |
|---|---|---|---|---|---|---|---|---|---|---|---|---|---|---|---|---|
| | | C | Si | Mn | P | S | Nb | V | Ti | Cr | Ni | Cu | N | Mo | B | Als |
| Q420 | A | ≤0.20 | ≤0.50 | ≤1.70 | ≤0.035 | ≤0.035 | ≤0.07 | ≤0.20 | ≤0.20 | ≤0.30 | ≤0.80 | ≤0.30 | ≤0.015 | ≤0.20 | — | — |
| | B | | | | ≤0.035 | ≤0.035 | | | | | | | | | | |
| | C | | | | ≤0.030 | ≤0.030 | | | | | | | | | | |
| | D | | | | ≤0.030 | ≤0.025 | | | | | | | | | | ≥0.015 |
| | E | | | | ≤0.025 | ≤0.020 | | | | | | | | | | |
| Q460 | C | ≤0.20 | ≤0.60 | ≤1.80 | ≤0.030 | ≤0.030 | ≤0.11 | ≤0.20 | ≤0.20 | ≤0.30 | ≤0.80 | ≤0.55 | ≤0.015 | ≤0.20 | ≤0.004 | ≥0.015 |
| | D | | | | ≤0.030 | ≤0.025 | | | | | | | | | | |
| | E | | | | ≤0.025 | ≤0.020 | | | | | | | | | | |
| Q500 | C | ≤0.18 | ≤0.60 | ≤1.80 | ≤0.030 | ≤0.030 | ≤0.11 | ≤0.20 | ≤0.20 | ≤0.30 | ≤0.80 | ≤0.55 | ≤0.015 | ≤0.20 | ≤0.004 | ≥0.015 |
| | D | | | | ≤0.030 | ≤0.025 | | | | | | | | | | |
| | E | | | | ≤0.025 | ≤0.020 | | | | | | | | | | |
| Q550 | C | ≤0.18 | ≤0.60 | ≤1.80 | ≤0.030 | ≤0.030 | ≤0.11 | ≤0.12 | ≤0.20 | ≤0.80 | ≤0.80 | ≤0.80 | ≤0.015 | ≤0.30 | ≤0.004 | ≥0.015 |
| | D | | | | ≤0.030 | ≤0.025 | | | | | | | | | | |
| | E | | | | ≤0.025 | ≤0.020 | | | | | | | | | | |
| Q620 | C | ≤0.18 | ≤0.60 | ≤2.00 | ≤0.030 | ≤0.030 | ≤0.11 | ≤0.12 | ≤0.20 | ≤1.00 | ≤0.80 | ≤0.80 | ≤0.015 | ≤0.30 | ≤0.004 | ≥0.015 |
| | D | | | | ≤0.030 | ≤0.025 | | | | | | | | | | |
| | E | | | | ≤0.025 | ≤0.020 | | | | | | | | | | |
| Q690 | C | ≤0.18 | ≤0.60 | ≤2.00 | ≤0.030 | ≤0.030 | ≤0.11 | ≤0.12 | ≤0.20 | ≤1.00 | ≤0.80 | ≤0.80 | ≤0.015 | ≤0.30 | ≤0.004 | ≥0.015 |
| | D | | | | ≤0.030 | ≤0.025 | | | | | | | | | | |
| | E | | | | ≤0.025 | ≤0.020 | | | | | | | | | | |

Note: 1. The P and S content of shapes and bars can be increased by 0.005%, of which the upper limit of grade A steel can be 0.045%;

2. When adding a combination of grain refinement elements, 20 (Nb+V+Ti) ≤0.22%, 20 (Mo+Cr) ≤0.30%.

**The chemical composition of Q345GJ (according to GB/T 19879—2015)    Table A.3**

| Grade | Quality level | Thickness (or Diameter) (mm) | Chemical composition (mass fraction) (%) | | | | | | | | | | | |
|---|---|---|---|---|---|---|---|---|---|---|---|---|---|---|
| | | | C | Si | Mn | P | S | V | Nb | Ti | Als | Cr | Cu | Ni |
| Q345 GJ | B | 6-100 | ≤0.20 | ≤0.55 | ≤1.60 | ≤0.025 | ≤0.015 | 0.020 - 0.150 | 0.015 - 0.060 | 0.010 - 0.030 | ≥0.015 | ≤0.30 | ≤0.30 | ≤0.30 |
| | C | | | | | | | | | | | | | |
| | D | | ≤0.18 | | | ≤0.020 | | | | | | | | |
| | E | | | | | | | | | | | | | |

# APPENDIX B  DESIGN STRENGTH OF STEEL

Design strength of steel (Unit: N/mm$^2$)   Table B.1

| Grade | | Thickness (or Diameter) (mm) | Design strength | | | Yield strength $f_y$ | Tensile strength $f_u$ |
|---|---|---|---|---|---|---|---|
| | | | Tension Compression Bending $f$ | Shear $f_v$ | End pressure $f_{cu}$ | | |
| Carbon structural steel | Q235 | ≤16 | 215 | 125 | 320 | 235 | 370 |
| | | >16, ≤40 | 205 | 120 | | 225 | |
| | | >40, ≤100 | 200 | 115 | | 215 | |
| Low alloy structural steel | Q355 | ≤16 | 305 | 175 | 400 | 355 | 470 |
| | | >16, ≤40 | 295 | 170 | | 345 | |
| | | >40, ≤63 | 290 | 165 | | 335 | |
| | | >63, ≤80 | 280 | 160 | | 325 | |
| | | >80, ≤100 | 270 | 155 | | 315 | |
| | Q390 | ≤16 | 345 | 200 | 415 | 390 | 490 |
| | | >16, ≤40 | 330 | 190 | | 370 | |
| | | >40, ≤63 | 310 | 180 | | 350 | |
| | | >63, ≤100 | 295 | 170 | | 330 | |
| | Q420 | ≤16 | 375 | 215 | 440 | 420 | 520 |
| | | >16, ≤40 | 355 | 205 | | 400 | |
| | | >40, ≤63 | 320 | 185 | | 380 | |
| | | >63, ≤100 | 305 | 175 | | 360 | |
| | Q460 | ≤16 | 410 | 235 | 470 | 460 | 550 |
| | | >16, ≤40 | 390 | 225 | | 440 | |
| | | >40, ≤63 | 355 | 205 | | 420 | |
| | | >63, ≤100 | 340 | 195 | | 400 | |
| Steel for building structure | Q345GJ | >16, ≤50 | 325 | 190 | 415 | 345 | 490 |
| | | >50, ≤100 | 300 | 175 | | 335 | |

Note: The diameter in the table refers to the diameter of the solid bar, the thickness refers to the thickness of the steel or steel pipe wall at the calculation point, and the thickness of the thicker plate in the section for axial tension and compression members.

# APPENDIX B  DESIGN STRENGTH OF STEEL

**Design strength of welds (Unit: N/mm²)**  Table B.2

| Welding method and electrode model | Grade of steel | | Compressive $f_c^w$ | The design strength of groove welds — When the weld quality is in the following grades, tensile strength | | Shear $f_v^w$ | Tensile, compressive, and shear $f_f^w$ | Tensile strength of groove weld | Tensile, compressive, and shear strength of fillet welds |
|---|---|---|---|---|---|---|---|---|---|
| | Grade | Thickness or diameter (mm) | | Level 1 / Level 2 | Level 3 | | | | |
| Automatic welding, semi-automatic welding, and manual welding with E43 electrodes | Q235 | ≤16 | 215 | 215 | 185 | 125 | 160 | 415 | 240 |
| | | >16, ≤40 | 205 | 205 | 175 | 120 | | | |
| | | >40, ≤100 | 200 | 200 | 170 | 115 | | | |
| Automatic welding, semi-automatic welding and manual welding with E50 and E55 electrodes | Q355 | ≤16 | 305 | 305 | 260 | 175 | 200 | 480 (E50) / 540 (E55) | 280 (E50) / 315 (E55) |
| | | >16, ≤40 | 295 | 295 | 250 | 170 | | | |
| | | >40, ≤63 | 290 | 290 | 245 | 165 | | | |
| | | >63, ≤80 | 280 | 280 | 240 | 160 | | | |
| | | >80, ≤100 | 270 | 270 | 230 | 155 | | | |
| | Q390 | ≤16 | 345 | 345 | 295 | 200 | 200 (E50) / 220 (E55) | | |
| | | >16, ≤40 | 330 | 330 | 280 | 190 | | | |
| | | >40, ≤63 | 310 | 310 | 265 | 180 | | | |
| | | >63, ≤100 | 295 | 295 | 250 | 170 | | | |
| Automatic welding, semi-automatic welding and manual welding with E55 and E60 electrodes | Q420 | ≤16 | 375 | 375 | 320 | 215 | 220 (E55) / 240 (E60) | 540 (E55) / 590 (E60) | 315 (E55) / 340 (E60) |
| | | >16, ≤40 | 355 | 355 | 300 | 205 | | | |
| | | >40, ≤63 | 320 | 320 | 270 | 185 | | | |
| | | >63, ≤100 | 305 | 305 | 260 | 175 | | | |
| Automatic welding, semi-automatic welding and manual welding with E55 and E60 electrodes | Q460 | ≤16 | 410 | 410 | 350 | 235 | 220 (E55) / 240 (E60) | 540 (E55) / 590 (E60) | 315 (E55) / 340 (E60) |
| | | >16, ≤40 | 390 | 390 | 330 | 225 | | | |
| | | >40, ≤63 | 355 | 355 | 300 | 205 | | | |
| | | >63, ≤100 | 340 | 340 | 290 | 195 | | | |
| Automatic welding, semi-automatic welding and manual welding with E50 and E55 electrodes | Q345GJ | >16, ≤35 | 310 | 310 | 265 | 180 | 200 | 480 (E55) / 540 (E60) | 280 (E50) / 315 (E55) |
| | | >35, ≤50 | 290 | 290 | 245 | 170 | | | |
| | | >50, ≤100 | 285 | 285 | 240 | 165 | | | |

Note: The thickness in the table refers to the thickness of the steel at the calculation point. Axial tension and axial compression members refer to the thickness of the thicker plate in the section.

# APPENDIX C   GLOBAL STABILITY FACTORS OF BEAM

## C.1   Welded I-section and rolled H-shapes simply supported beams with the constant cross-section

The global stability factor $\varphi_b$ of welded I-section and rolled H-shapes simply supported beams of the constant cross-section (Figure C.1) shall be calculated according to the following formula:

$$\varphi_b = \beta_b \frac{4,320}{\lambda_y^2} \cdot \frac{Ah}{W_x} \left[ \sqrt{1 + \left(\frac{\lambda_y t_1}{4.4h}\right)^2} + \eta_b \right] \varepsilon_k^2 \tag{C-1}$$

$$\lambda_y = \frac{l_1}{i_y} \tag{C-2}$$

The cross-sectional asymmetry influence factor $\eta_b$ should be calculated according to the following formula:

For the biaxially symmetrical section (Figures C.1a and C.1d):

$$\eta_b = 0 \tag{C-3}$$

For the uniaxial symmetrical I-shaped section (Figures C.1b and C.1c):
Reinforced the compression flange:

$$\eta_b = 0.8(2\alpha_b - 1) \tag{C-4}$$

Reinforced the tension flange:

$$\eta_b = 2\alpha_b - 1 \tag{C-5}$$

$$\alpha_b = \frac{I_1}{I_1 + I_2} \tag{C-6}$$

When $\varphi_b$ calculated according to Equation (C-1) is greater than 0.6, $\varphi'_b$ calculated by Equation (C-7) should be used instead of $\varphi_b$:

$$\varphi'_b = 1.07 - \frac{0.282}{\varphi_b} \leqslant 1.0 \tag{C-7}$$

where   $\beta_b$ = the equivalent bending moment factor for the global stability of the beam shall be adopted following Table C.1;

$\lambda_y$ = the slenderness ratio of the beam to the weak axis (y-axis) of the section be-

APPENDIX C  GLOBAL STABILITY FACTORS OF BEAM

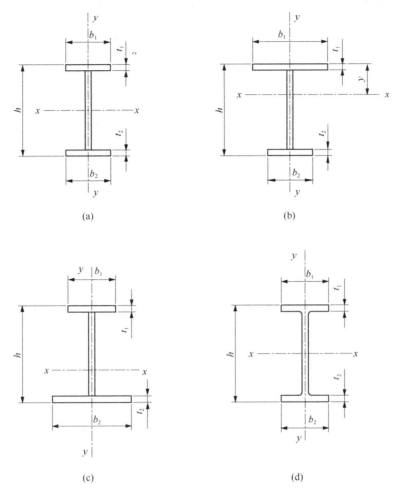

Figure C.1  Welded I-section and rolled H-shapes
(a) biaxially symmetric welded I-shape;
(b) uniaxially symmetric welded I-shape of reinforced compression flange;
(c) uniaxially symmetric welded I-shape of reinforced tension flange; (d) rolled H-shape

tween the lateral support points;

$A$ = the gross cross-sectional area of the beam;

$h, t_1$ = the full height of the beam section and the thickness of the compression flange, the thickness of the compression flange $t_1$ of the simple-supported beam with constant section riveted (or high-strength bolt connection) includes the thickness of the flange angle;

$l_1$ = the distance between the lateral support points of the beam compression flange;

$i_y$ = the radius of gyration of the beam section to the y-axis;

$I_1, I_2$ = the moment of inertia of the compression flange and the tension flange to the y-axis.

287

# APPENDIX C  GLOBAL STABILITY FACTORS OF BEAM

**Factor $\beta_b$ of H-shapes and I-section simply supported beams with equal cross-section**

Table C. 1

| Items | Lateral support | Load | | $\xi \leqslant 2.0$ | $\xi > 2.0$ | Scope of application |
|---|---|---|---|---|---|---|
| 1 | No lateral support in the middle span | Uniformly distributed load | Top flange | $0.69 + 0.13\xi$ | 0.95 | Cross-sections of Figures C. 1a, C. 1b, and C. 1d |
| 2 | | | Bottom flange | $1.73 - 0.20\xi$ | 1.33 | |
| 3 | | Concentrated load | Top flange | $0.73 + 0.18\xi$ | 1.09 | |
| 4 | | | Bottom flange | $2.23 - 0.28\xi$ | 1.67 | |
| 5 | A lateral support point in the middle of the span | Uniformly distributed load | Top flange | 1.15 | | All sections in Figure C. 1 |
| 6 | | | Bottom flange | 1.40 | | |
| 7 | | The concentrated load is applied at any position of the height of the section | | 1.75 | | |
| 8 | Not less than two equidistant lateral support points in the middle span | Any load acting on | Top flange | 1.20 | | |
| 9 | | | Bottom flange | 1.40 | | |
| 10 | Bending moment at the end of the beam, but there is no load in the middle span | $1.75 - 1.05\left(\dfrac{M_2}{M_1}\right) + 0.3\left(\dfrac{M_2}{M_1}\right)^2 \leqslant 2.3$ | | | | |

Note: 1. $\xi$ is a parameter, $\xi = \dfrac{l_1 t_1}{b_1 h}$, where $b_1$ is the width of the compressed flange;

2. $M_1$ and $M_2$ are bending moments of the beam end. When the beam has a curvature in the same direction, $M_1$ and $M_2$ take the same sign, and when the curvature is reversed, take the different sign, $|M_1| \geqslant |M_2|$;

3. Items 3, 4, and 7 in Table C. 1 refer to the situation where one or a few concentrated loads are located near the span center. The concentrated loads should follow items 1, 2, 5, and 6 for other situations;

4. The $\beta_b$ of items 8 and 9 in the table, when concentrated load acts on the lateral support point, take $\beta_b = 1.2$;

5. Load acting on the upper flange means that the load acting point is on the flange surface, and the direction points to the section centroid. Load acting on the lower flange means that the load acting point is on the flange surface, and the direction is away from the section centroid;

6. For the reinforced compression flange I-section with $\alpha_b \geqslant 0.8$, the $\beta_b$ value in the following cases should be multiplied by the corresponding factor:
   Item 1: When $\xi \leqslant 1$, multiply by 0.95;
   Item 3: When $\xi \leqslant 0.5$, multiply by 0.90; when $0.5 \leqslant \xi \leqslant 1$, multiply by 0.95.

## C. 2  Rolled ordinary I-shape simply supported beams

The global stability factor $\varphi_b$ of rolled ordinary I-shape simply supported beams shall be adopted following Table C. 2. When the obtained value of $\varphi_b$ is greater than 0.6, the substitute value calculated by Equation (C-7) shall be used.

APPENDIX C  GLOBAL STABILITY FACTORS OF BEAM

**Rolled ordinary I-shape simply supported beam**   Table C. 2

| Items | Load conditions | | | I-shape model | Free length $l_1$ (m) | | | | | | | | |
|---|---|---|---|---|---|---|---|---|---|---|---|---|---|
| | | | | | 2 | 3 | 4 | 5 | 6 | 7 | 8 | 9 | 10 |
| 1 | A beam without lateral support in the middle span | Concentrated load | Top flange | 10-20 | 2.00 | 1.30 | 0.99 | 0.80 | 0.68 | 0.58 | 0.53 | 0.48 | 0.43 |
| | | | | 22-32 | 2.40 | 1.48 | 1.09 | 0.86 | 0.72 | 0.62 | 0.54 | 0.49 | 0.45 |
| | | | | 36-63 | 2.80 | 1.60 | 1.07 | 0.83 | 0.68 | 0.56 | 0.50 | 0.45 | 0.40 |
| 2 | | | Bottom flange | 10-20 | 3.10 | 1.95 | 1.34 | 1.01 | 0.82 | 0.69 | 0.63 | 0.57 | 0.52 |
| | | | | 22-40 | 5.50 | 2.80 | 1.84 | 1.37 | 1.07 | 0.86 | 0.73 | 0.64 | 0.56 |
| | | | | 45-63 | 7.30 | 3.60 | 2.30 | 1.62 | 1.20 | 0.96 | 0.80 | 0.69 | 0.60 |
| 3 | | Uniform load | Top flange | 10-20 | 1.70 | 1.12 | 0.84 | 0.68 | 0.57 | 0.50 | 0.45 | 0.41 | 0.37 |
| | | | | 22-40 | 2.10 | 1.30 | 0.93 | 0.73 | 0.60 | 0.51 | 0.45 | 0.40 | 0.36 |
| | | | | 45-63 | 2.60 | 1.45 | 0.97 | 0.73 | 0.59 | 0.50 | 0.44 | 0.38 | 0.35 |
| 4 | | | Bottom flange | 10-20 | 2.50 | 1.55 | 1.08 | 0.83 | 0.68 | 0.56 | 0.52 | 0.47 | 0.42 |
| | | | | 22-40 | 4.00 | 2.20 | 1.45 | 1.10 | 0.85 | 0.70 | 0.60 | 0.52 | 0.46 |
| | | | | 45-63 | 5.60 | 2.80 | 1.80 | 1.25 | 0.95 | 0.78 | 0.65 | 0.55 | 0.49 |
| 5 | A beam with lateral support in the mid-span (regardless of the load position at the section depth) | | | 10-20 | 2.20 | 1.39 | 1.01 | 0.79 | 0.66 | 0.57 | 0.52 | 0.47 | 0.42 |
| | | | | 22-40 | 3.00 | 1.80 | 1.24 | 0.96 | 0.76 | 0.65 | 0.56 | 0.49 | 0.43 |
| | | | | 45-63 | 4.00 | 2.20 | 1.38 | 1.01 | 0.80 | 0.66 | 0.56 | 0.49 | 0.43 |

Note: 1. Same as Note 3 and Note 5 of Table C. 1;
    2. The $\varphi_b$ applies to Q235 steel. For other steel grades, the value should be multiplied by $\varepsilon_k^2$.

## C. 3  Rolled ordinary channel simply supported beams

The global stability factor $\varphi_b$ of the rolled channel simply supported beam, regardless of the load form and the load position on the section depth, can be calculated as follows:

$$\varphi_b = \frac{570bt}{l_1 h} \cdot \varepsilon_k^2 \quad \text{(C-8)}$$

where $h, b, t$ = the depth of the channel, the width of the flange, and the average thickness.

When the value $\varphi_b$ calculated according to Equation (C-8) is greater than 0.6, the corresponding $\varphi'_b$ value should be calculated according to Equation (C-7) instead of value $\varphi_b$.

## C. 4  Two-axis symmetrical I-section cantilever beam with constant cross-section

The global stability factor of a biaxially symmetrical I-section cantilever beam with constant cross-section can be calculated according to Equation (C-1). Still, the coefficient $\beta_b$ in the equation should be checked according to Table C. 4. The slenderness ratio $\lambda_y$, $l_1$ is

APPENDIX C  GLOBAL STABILITY FACTORS OF BEAM

the overhang length of the cantilever beam. When the obtained value $\varphi_b$ is greater than 0.6, the value $\varphi_b$ calculated according to Equation (C-7) shall be used instead of the value $\varphi_b$.

**Coefficient $\beta_b$ of biaxially symmetric I-shaped cantilever beam**　　　Table C.4

| Item | Load form | | $0.60 \leqslant \xi \leqslant 1.24$ | $1.24 \leqslant \xi \leqslant 1.96$ | $1.96 \leqslant \xi \leqslant 3.10$ |
|---|---|---|---|---|---|
| 1 | A concentrated load at the free end acts on | Upper flange | $0.21+0.67\xi$ | $0.72+0.26\xi$ | $1.17+0.03\xi$ |
| 2 | | Bottom flange | $2.94-0.65\xi$ | $2.64-0.40\xi$ | $2.15-0.15\xi$ |
| 3 | Uniform load acting on the upper flange | | $0.62+0.82\xi$ | $1.25+0.31\xi$ | $1.66+0.10\xi$ |

Note: 1. This table is determined according to the condition that the supporting end is fixed. When used for the overhanging beam extending from the adjacent span, measures should be taken to strengthen the supporting place's torsion resistance in the structure;
2. See Table C.1 Note 1.

## C.5 Approximate Calculation of global stability factors of bending members

For uniformly bent members, when $\lambda_y \leqslant 120\varepsilon_k$, the global stability factor $\varphi_b$ can be calculated according to the approximate formula.

(1) I-section section

Biaxial symmetry:

$$\varphi_b = 1.07 - \frac{\lambda_y^2}{44,000\varepsilon_k^2} \tag{C-9}$$

Uniaxial symmetry:

$$\varphi_b = 1.07 - \frac{W_x}{(2\alpha_b+0.1)Ah} \cdot \frac{\lambda_y^2}{14,000\varepsilon_k^2} \tag{C-10}$$

(2) T-shaped section (the bending moment acts on the plane of the symmetrical axis, around the $x$-axis)

① When the bending moment compresses the flange:

Double-angle T-shaped section:

$$\varphi_b = 1 - 0.0017\lambda_y/\varepsilon_k \tag{C-11}$$

Structural Tee and two-plate combined T-shaped Section:

$$\varphi_b = 1 - 0.0022\lambda_y/\varepsilon_k \tag{C-12}$$

② When the bending moment causes the flange to be tensiled, and the web's depth-to-thickness ratio is not greater than $18\varepsilon_k$:

$$\varphi_b = 1 - 0.0005\lambda_y/\varepsilon_k \tag{C-13}$$

When the value $\varphi_b$ calculated according to Equation (C-9) and Equation (C-10) is greater than 1.0, take $\varphi_b = 1.0$.

# APPENDIX D  GLOBAL STABILITY FACTORS OF AXIALLY LOADED COMPRESSION MEMBERS

## D.1  Global stability factors of a-section of the axially loaded compression member

Stability factors $\varphi$ of a-section of the axially loaded compression member    Table D.1

| $\lambda/\varepsilon_k$ | 0 | 1 | 2 | 3 | 4 | 5 | 6 | 7 | 8 | 9 |
|---|---|---|---|---|---|---|---|---|---|---|
| 0 | 1.000 | 1.000 | 1.000 | 1.000 | 0.999 | 0.999 | 0.998 | 0.998 | 0.997 | 0.996 |
| 10 | 0.995 | 0.994 | 0.993 | 0.992 | 0.991 | 0.989 | 0.988 | 0.986 | 0.985 | 0.983 |
| 20 | 0.981 | 0.979 | 0.977 | 0.976 | 0.974 | 0.972 | 0.970 | 0.968 | 0.966 | 0.964 |
| 30 | 0.963 | 0.961 | 0.959 | 0.957 | 0.954 | 0.952 | 0.950 | 0.948 | 0.946 | 0.944 |
| 40 | 0.941 | 0.939 | 0.937 | 0.934 | 0.932 | 0.929 | 0.927 | 0.924 | 0.921 | 0.918 |
| 50 | 0.916 | 0.913 | 0.910 | 0.907 | 0.903 | 0.900 | 0.897 | 0.893 | 0.890 | 0.886 |
| 60 | 0.883 | 0.879 | 0.875 | 0.871 | 0.867 | 0.862 | 0.858 | 0.854 | 0.849 | 0.844 |
| 70 | 0.839 | 0.834 | 0.829 | 0.824 | 0.818 | 0.813 | 0.807 | 0.801 | 0.795 | 0.789 |
| 80 | 0.783 | 0.776 | 0.770 | 0.763 | 0.756 | 0.749 | 0.742 | 0.735 | 0.728 | 0.721 |
| 90 | 0.713 | 9.706 | 0.698 | 0.691 | 0.683 | 0.676 | 0.668 | 0.660 | 0.653 | 0.645 |
| 100 | 0.637 | 0.630 | 0.622 | 0.614 | 0.607 | 0.599 | 0.592 | 0.584 | 0.577 | 0.569 |
| 110 | 0.562 | 0.555 | 0.548 | 0.541 | 0.534 | 0.527 | 0.520 | 0.513 | 0.507 | 0.500 |
| 120 | 0.494 | 0.487 | 0.481 | 0.475 | 0.469 | 0.463 | 0.457 | 0.451 | 0.445 | 0.439 |
| 130 | 0.434 | 0.428 | 0.423 | 0.417 | 0.412 | 0.407 | 0.402 | 0.397 | 0.392 | 0.387 |
| 140 | 0.382 | 0.378 | 0.373 | 0.368 | 0.364 | 0.360 | 0.355 | 0.351 | 0.347 | 0.343 |
| 150 | 0.339 | 0.335 | 0.331 | 0.327 | 0.323 | 0.319 | 0.316 | 0.312 | 0.308 | 0.305 |
| 160 | 0.302 | 0.298 | 0.295 | 0.292 | 0.288 | 0.285 | 0.282 | 0.279 | 0.276 | 0.273 |
| 170 | 0.270 | 0.267 | 0.264 | 0.261 | 0.259 | 0.256 | 0.253 | 0.250 | 0.248 | 0.245 |
| 180 | 0.243 | 0.240 | 0.238 | 0.235 | 0.233 | 0.231 | 0.228 | 0.226 | 0.224 | 0.222 |
| 190 | 0.219 | 0.217 | 0.215 | 0.213 | 0.211 | 0.209 | 0.207 | 0.205 | 0.203 | 0.201 |
| 200 | 0.199 | 0.197 | 0.196 | 0.194 | 0.192 | 0.190 | 0.188 | 0.187 | 0.185 | 0.183 |
| 210 | 0.182 | 0.180 | 0.178 | 0.177 | 0.175 | 0.174 | 0.172 | 0.171 | 0.169 | 0.168 |
| 220 | 0.166 | 0.165 | 0.163 | 0.162 | 0.161 | 0.159 | 0.158 | 0.157 | 0.155 | 0.154 |
| 230 | 0.153 | 0.151 | 0.150 | 0.149 | 0.148 | 0.147 | 0.145 | 0.144 | 0.143 | 0.142 |
| 240 | 0.141 | 0.140 | 0.139 | 0.137 | 0.136 | 0.135 | 0.134 | 0.133 | 0.132 | 0.131 |
| 250 | 0.130 | — | — | — | — | — | — | — | — | — |

Note: The values are calculated according to the formula in Appendix D.5.

# APPENDIX D  GLOBAL STABILITY FACTORS OF AXIALLY LOADED COMPRESSION MEMBERS

## D. 2  Global stability factors of b-section of the axially loaded compression member

**Stability factors $\varphi$ of b-section of the axially loaded compression member**    Table D. 2

| $\lambda/\varepsilon_k$ | 0 | 1 | 2 | 3 | 4 | 5 | 6 | 7 | 8 | 9 |
|---|---|---|---|---|---|---|---|---|---|---|
| 0 | 1.000 | 1.000 | 1.000 | 0.999 | 0.999 | 0.998 | 0.997 | 0.996 | 0.995 | 0.994 |
| 10 | 0.992 | 0.991 | 0.989 | 0.987 | 0.985 | 0.983 | 0.981 | 0.978 | 0.976 | 0.973 |
| 20 | 0.970 | 0.967 | 0.963 | 0.960 | 0.957 | 0.953 | 0.950 | 0.946 | 0.943 | 0.939 |
| 30 | 0.936 | 0.932 | 0.929 | 0.925 | 0.921 | 0.918 | 0.914 | 0.910 | 0.906 | 0.903 |
| 40 | 0.899 | 0.895 | 0.891 | 0.886 | 0.882 | 0.878 | 0.874 | 0.870 | 0.865 | 0.861 |
| 50 | 0.856 | 0.852 | 0.847 | 0.842 | 0.837 | 0.833 | 0.828 | 0.823 | 0.818 | 0.812 |
| 60 | 0.807 | 0.802 | 0.796 | 0.791 | 0.785 | 0.780 | 0.774 | 0.768 | 0.762 | 0.757 |
| 70 | 0.751 | 0.745 | 0.738 | 0.732 | 0.726 | 0.720 | 0.713 | 0.707 | 0.701 | 0.694 |
| 80 | 0.687 | 0.681 | 0.674 | 0.668 | 0.661 | 0.654 | 0.648 | 0.641 | 0.634 | 0.628 |
| 90 | 0.621 | 0.614 | 0.607 | 0.601 | 0.594 | 0.587 | 0.581 | 0.574 | 0.568 | 0.561 |
| 100 | 0.555 | 0.548 | 0.542 | 0.535 | 0.529 | 0.523 | 0.517 | 0.511 | 0.504 | 0.498 |
| 110 | 0.492 | 0.487 | 0.481 | 0.475 | 0.469 | 0.465 | 0.458 | 0.453 | 0.447 | 0.442 |
| 120 | 0.436 | 0.431 | 0.426 | 0.421 | 0.416 | 0.411 | 0.406 | 0.401 | 0.396 | 0.392 |
| 130 | 0.387 | 0.383 | 0.378 | 0.374 | 0.369 | 0.365 | 0.361 | 0.357 | 0.352 | 0.348 |
| 140 | 0.344 | 0.340 | 0.337 | 0.333 | 0.329 | 0.325 | 0.322 | 0.318 | 0.314 | 0.311 |
| 150 | 0.308 | 0.304 | 0.301 | 0.297 | 0.294 | 0.291 | 0.288 | 0.285 | 0.282 | 0.279 |
| 160 | 0.276 | 0.273 | 0.270 | 0.267 | 0.264 | 0.262 | 0.259 | 0.256 | 0.253 | 0.251 |
| 170 | 0.248 | 0.246 | 0.243 | 0.241 | 0.238 | 0.236 | 0.234 | 0.229 | 0.227 | 0.226 |
| 180 | 0.225 | 0.222 | 0.220 | 0.218 | 0.216 | 0.214 | 0.212 | 0.210 | 0.208 | 0.206 |
| 190 | 0.204 | 0.202 | 0.200 | 0.198 | 0.196 | 0.195 | 0.193 | 0.191 | 0.189 | 0.188 |
| 200 | 0.186 | 0.184 | 0.183 | 0.181 | 0.179 | 0.178 | 0.176 | 0.175 | 0.173 | 0.172 |
| 210 | 0.170 | 0.169 | 0.167 | 0.166 | 0.164 | 0.163 | 0.162 | 0.160 | 0.159 | 0.158 |
| 220 | 0.156 | 0.155 | 0.154 | 0.152 | 0.151 | 0.150 | 0.149 | 0.147 | 0.146 | 0.145 |
| 230 | 0.144 | 0.143 | 0.142 | 0.141 | 0.139 | 0.138 | 0.137 | 0.136 | 0.135 | 0.134 |
| 240 | 0.133 | 0.132 | 0.131 | 0.130 | 0.129 | 0.128 | 0.127 | 0.126 | 0.125 | 0.124 |
| 250 | 0.123 | — | — | — | — | — | — | — | — | — |

Note: The values are calculated according to the formula in Appendix D. 5.

## D.3 Global stability factors of c-section of the axially loaded compression member

Stability factors $\varphi$ of c-section of the axially loaded compression member     Table D.3

| $\lambda/\varepsilon_k$ | 0 | 1 | 2 | 3 | 4 | 5 | 6 | 7 | 8 | 9 |
|---|---|---|---|---|---|---|---|---|---|---|
| 0   | 1.000 | 1.000 | 1.000 | 0.999 | 0.999 | 0.998 | 0.997 | 0.996 | 0.995 | 0.993 |
| 10  | 0.992 | 0.990 | 0.988 | 0.986 | 0.983 | 0.981 | 0.978 | 0.976 | 0.973 | 0.970 |
| 20  | 0.966 | 0.959 | 0.953 | 0.947 | 0.940 | 0.934 | 0.928 | 0.921 | 0.915 | 0.909 |
| 30  | 0.902 | 0.896 | 0.890 | 0.883 | 0.877 | 0.871 | 0.865 | 0.858 | 0.852 | 0.845 |
| 40  | 0.839 | 0.833 | 0.826 | 0.820 | 0.813 | 0.807 | 0.800 | 0.794 | 0.787 | 0.781 |
| 50  | 0.774 | 0.768 | 0.761 | 0.755 | 0.748 | 0.742 | 0.735 | 0.728 | 0.722 | 0.715 |
| 60  | 0.709 | 0.702 | 0.695 | 0.689 | 0.682 | 0.675 | 0.669 | 0.662 | 0.658 | 0.649 |
| 70  | 0.642 | 0.636 | 0.629 | 0.623 | 0.616 | 0.610 | 0.603 | 0.597 | 0.591 | 0.584 |
| 80  | 0.578 | 0.572 | 0.565 | 0.559 | 0.553 | 0.547 | 0.541 | 0.535 | 0.529 | 0.523 |
| 90  | 0.517 | 0.511 | 0.505 | 0.499 | 0.494 | 0.488 | 0.483 | 0.477 | 0.471 | 0.467 |
| 100 | 0.462 | 0.458 | 0.453 | 0.449 | 0.445 | 0.440 | 0.436 | 0.432 | 0.427 | 0.423 |
| 110 | 0.419 | 0.415 | 0.411 | 0.407 | 0.402 | 0.398 | 0.394 | 0.390 | 0.386 | 0.383 |
| 120 | 0.379 | 0.375 | 0.371 | 0.367 | 0.363 | 0.360 | 0.356 | 0.352 | 0.349 | 0.345 |
| 130 | 0.342 | 0.338 | 0.335 | 0.332 | 0.328 | 0.325 | 0.322 | 0.318 | 0.315 | 0.312 |
| 140 | 0.309 | 0.306 | 0.303 | 0.300 | 0.297 | 0.294 | 0.291 | 0.288 | 0.285 | 0.282 |
| 150 | 0.279 | 0.277 | 0.274 | 0.271 | 0.269 | 0.266 | 0.263 | 0.261 | 0.258 | 0.256 |
| 160 | 0.253 | 0.251 | 0.248 | 0.246 | 0.244 | 0.241 | 0.239 | 0.237 | 0.235 | 0.232 |
| 170 | 0.230 | 0.228 | 0.226 | 0.224 | 0.222 | 0.220 | 0.218 | 0.216 | 0.214 | 0.212 |
| 180 | 0.210 | 0.208 | 0.206 | 0.204 | 0.203 | 0.201 | 0.199 | 0.197 | 0.195 | 0.194 |
| 190 | 0.192 | 0.190 | 0.189 | 0.187 | 0.185 | 0.184 | 0.182 | 0.181 | 0.179 | 0.178 |
| 200 | 0.176 | 0.175 | 0.173 | 0.172 | 0.170 | 0.169 | 0.167 | 0.166 | 0.165 | 0.163 |
| 210 | 0.162 | 0.161 | 0.159 | 0.158 | 0.157 | 0.155 | 0.154 | 0.153 | 0.152 | 0.151 |
| 220 | 0.149 | 0.148 | 0.147 | 0.146 | 0.145 | 0.144 | 0.142 | 0.141 | 0.140 | 0.139 |
| 230 | 0.138 | 0.137 | 0.136 | 0.135 | 0.134 | 0.133 | 0.132 | 0.131 | 0.130 | 0.129 |
| 240 | 0.128 | 0.127 | 0.126 | 0.125 | 0.124 | 0.123 | 0.123 | 0.122 | 0.121 | 0.120 |
| 250 | 0.119 | — | — | — | — | — | — | — | — | — |

Note: The values are calculated according to the formula in Appendix D.5.

# APPENDIX D  GLOBAL STABILITY FACTORS OF AXIALLY LOADED COMPRESSION MEMBERS

## D.4 Global stability factors of d-section of the axially loaded compression member

Stability factors $\varphi$ of d-section of the axially loaded compression member  Table D.4

| $\lambda/\varepsilon_k$ | 0 | 1 | 2 | 3 | 4 | 5 | 6 | 7 | 8 | 9 |
|---|---|---|---|---|---|---|---|---|---|---|
| 0 | 1.000 | 1.000 | 0.999 | 0.999 | 0.998 | 0.996 | 0.994 | 0.992 | 0.990 | 0.987 |
| 10 | 0.984 | 0.981 | 0.978 | 0.974 | 0.969 | 0.965 | 0.960 | 0.955 | 0.949 | 0.944 |
| 20 | 0.937 | 0.927 | 0.918 | 0.909 | 0.900 | 0.891 | 0.883 | 0.874 | 0.865 | 9.857 |
| 30 | 0.848 | 0.840 | 0.831 | 0.823 | 0.815 | 0.807 | 0.798 | 0.790 | 0.782 | 0.774 |
| 40 | 0.766 | 0.758 | 0.751 | 0.743 | 0.735 | 0.727 | 0.720 | 0.712 | 0.705 | 0.697 |
| 50 | 0.690 | 0.682 | 0.675 | 0.668 | 0.660 | 0.653 | 0.646 | 0.639 | 0.632 | 0.625 |
| 60 | 0.618 | 0.611 | 0.605 | 0.598 | 0.591 | 0.585 | 0.578 | 0.571 | 0.565 | 0.559 |
| 70 | 0.552 | 0.546 | 0.540 | 0.534 | 0.528 | 0.521 | 0.516 | 0.510 | 0.504 | 0.498 |
| 80 | 0.492 | 0.487 | 0.481 | 0.476 | 0.470 | 0.465 | 0.469 | 0.454 | 0.449 | 0.444 |
| 90 | 0.439 | 0.434 | 0.429 | 0.424 | 0.419 | 0.414 | 0.409 | 0.405 | 0.401 | 0.397 |
| 100 | 0.393 | 0.390 | 0.386 | 0.383 | 0.380 | 0.376 | 0.373 | 0.369 | 0.366 | 0.363 |
| 110 | 0.359 | 0.356 | 0.353 | 0.350 | 0.346 | 0.343 | 0.340 | 0.337 | 0.334 | 0.331 |
| 120 | 0.328 | 0.325 | 0.322 | 0.319 | 0.316 | 0.313 | 0.310 | 0.307 | 0.304 | 0.301 |
| 130 | 0.298 | 0.296 | 0.293 | 0.290 | 0.288 | 0.285 | 0.282 | 0.280 | 0.277 | 0.275 |
| 140 | 0.272 | 0.270 | 0.267 | 0.265 | 0.262 | 0.260 | 0.257 | 0.255 | 0.253 | 0.250 |
| 150 | 0.248 | 0.246 | 0.244 | 0.242 | 0.239 | 0.237 | 0.235 | 0.233 | 0.231 | 0.229 |
| 160 | 0.227 | 0.225 | 0.223 | 0.221 | 0.219 | 0.217 | 0.215 | 0.213 | 0.211 | 0.210 |
| 170 | 0.208 | 0.206 | 0.204 | 0.202 | 0.201 | 0.199 | 0.197 | 0.196 | 0.194 | 0.192 |
| 180 | 0.191 | 0.189 | 0.187 | 0.186 | 0.184 | 0.183 | 0.181 | 0.180 | 0.178 | 0.177 |
| 190 | 0.175 | 0.174 | 0.173 | 0.171 | 0.170 | 0.168 | 0.167 | 0.166 | 0.164 | 0.163 |
| 200 | 0.162 | — | — | — | — | — | — | — | — | — |

Note: The values are calculated according to the formula in Appendix D.5.

## D.5 The calculation formula of the global stability factor of the axially loaded compression member

When the $\lambda/\varepsilon_k$ of member exceeds the scope of Table D.1~Table D.4, the stability factor of the axially loaded compression member should be calculated according to the following formula.

When $\lambda_n = \dfrac{\lambda}{\pi}\sqrt{f_y/E} \leqslant 0.215$,

$$\varphi = 1 - \alpha_1 \lambda_n^2 \tag{D-1}$$

# APPENDIX D  GLOBAL STABILITY FACTORS OF AXIALLY LOADED COMPRESSION MEMBERS

When $\lambda_n = \frac{\lambda}{\pi}\sqrt{f_y/E} > 0.215$,

$$\varphi = \frac{1}{2\lambda_n^2}\left[(\alpha_2 + \alpha_3\lambda_n + \lambda_n^2) - \sqrt{(\alpha_2 + \alpha_3\lambda_n + \lambda_n^2)^2 - 4\lambda_n^2}\right] \tag{D-2}$$

where $\alpha_1$, $\alpha_2$, and $\alpha_3$ are the coefficient, classified according to the section classification and adapted according to Table D.5.

Coefficient $\alpha_1$, $\alpha_2$, and $\alpha_3$ according to GB 50017—2017  Table D.5

| Section classification | | $\alpha_1$ | $\alpha_2$ | $\alpha_3$ |
|---|---|---|---|---|
| a | | 0.41 | 0.986 | 0.152 |
| b | | 0.65 | 0.965 | 0.300 |
| c | $\lambda_n \leqslant 1.05$ | 0.73 | 0.906 | 0.595 |
| c | $\lambda_n > 1.05$ | 0.73 | 1.216 | 0.302 |
| d | $\lambda_n \leqslant 1.05$ | 1.35 | 0.868 | 0.915 |
| d | $\lambda_n > 1.05$ | 1.35 | 1.375 | 0.432 |

# APPENDIX E  EFFECTIVE LENGTH FACTOR OF COLUMNS

## E.1 Effective length factor of frame column with sideway uninhibited

The effective length factor $\mu$ of frame columns with sideway uninhibited shall be taken according to Table E.1 and shall also meet the following requirements:

① When the beam is hinged to the column, take the beam's rotation stiffness as 0.

② For low-rise frame columns, when the column is hinged to the foundation, $K_2=0$ should be adopted. When the column is fixed to the foundation, $K_2=10$, and the flat plate support can be $K_2=0.1$.

**Effective length factor $\mu$ of frame column with sideway uninhibited**  Table E.1

| $K_2$ \ $K_1$ | 0 | 0.05 | 0.1 | 0.2 | 0.3 | 0.4 | 0.5 | 1 | 2 | 3 | 4 | 5 | ≥10 |
|---|---|---|---|---|---|---|---|---|---|---|---|---|---|
| 0 | ∞ | 6.02 | 4.45 | 3.42 | 3.01 | 2.78 | 2.64 | 2.33 | 2.17 | 2.11 | 2.08 | 2.07 | 2.03 |
| 0.05 | 6.02 | 4.16 | 3.47 | 2.86 | 2.58 | 2.42 | 2.31 | 2.07 | 1.94 | 1.90 | 1.87 | 1.86 | 1.83 |
| 0.1 | 4.46 | 3.47 | 3.01 | 2.56 | 2.33 | 2.20 | 2.11 | 1.90 | 1.79 | 1.75 | 1.73 | 1.72 | 1.70 |
| 0.2 | 3.42 | 2.86 | 2.36 | 2.23 | 2.05 | 1.94 | 1.87 | 1.70 | 1.60 | 1.57 | 1.55 | 1.54 | 1.52 |
| 0.3 | 3.01 | 2.58 | 2.33 | 2.05 | 1.90 | 1.80 | 1.74 | 1.58 | 1.49 | 1.46 | 1.45 | 1.44 | 1.42 |
| 0.4 | 2.78 | 2.42 | 2.20 | 1.94 | 1.80 | 1.71 | 1.65 | 1.50 | 1.42 | 1.39 | 1.37 | 1.37 | 1.35 |
| 0.5 | 2.64 | 2.31 | 2.11 | 1.87 | 1.74 | 1.65 | 1.59 | 1.45 | 1.37 | 1.34 | 1.32 | 1.32 | 1.30 |
| 1 | 2.33 | 2.07 | 1.90 | 1.70 | 1.58 | 1.50 | 1.45 | 1.32 | 1.24 | 1.21 | 1.20 | 1.19 | 1.17 |
| 2 | 2.17 | 1.94 | 1.79 | 1.60 | 1.49 | 1.42 | 1.37 | 1.24 | 1.16 | 1.14 | 1.12 | 1.12 | 1.10 |
| 3 | 2.11 | 1.90 | 1.75 | 1.57 | 1.46 | 1.39 | 1.34 | 1.21 | 1.14 | 1.11 | 1.10 | 1.09 | 1.07 |
| 4 | 2.08 | 1.87 | 1.73 | 1.55 | 1.45 | 1.37 | 1.32 | 1.20 | 1.12 | 1.10 | 1.08 | 1.08 | 1.06 |
| 5 | 2.07 | 1.86 | 1.72 | 1.54 | 1.44 | 1.37 | 1.32 | 1.19 | 1.12 | 1.09 | 1.08 | 1.07 | 1.05 |
| ≥10 | 2.03 | 1.83 | 1.70 | 1.52 | 1.42 | 1.35 | 1.30 | 1.17 | 1.10 | 1.07 | 1.06 | 1.05 | 1.03 |

③ When the axially compressive force $N_b$ of the beam rigidly connected to the column is large, the rotation stiffness reduction factor $\alpha_N$ of the beam should be calculated according to the following formula:

When the beam's far end is rigidly connected to the column:
$$\alpha_N = 1 - N_b/4N_{Eb} \tag{E-1}$$

When the beam's far end is hinged to the column:
$$\alpha_N = 1 - N_b/N_{Eb} \tag{E-2}$$

Whenthe beam's far end is fixed:

$$\alpha_N = 1 - N_b/2N_{Eb} \tag{E-3}$$
$$N_{Eb} = \pi^2 EI_b/l^2 \tag{E-4}$$

where $I_b$ = the moment of inertia of a beam (mm$^4$);

  $l$ = the length of a beam (mm).

## E.2 Effective length factor of frame column with sideway inhibited

The effective length factor $\mu$ of the frame column with sideway inhibited shall be taken according to Table E.2 and shall meet the following requirements:

① When the beam is hinged to the column, take the beam rotation stiffness as 0.

② For low-rise frame columns, when the column is hinged to the foundation, $K_2=0$ should be adopted. When the column is rigidly connected to the foundation, $K_2=10$, the flat plate support can be $K_2=0.1$.

Effective length factor $\mu$ of frame column with sideway inhibited     Table E.2

| $K_2$ \ $K_1$ | 0 | 0.05 | 0.1 | 0.2 | 0.3 | 0.4 | 0.5 | 1 | 2 | 3 | 4 | 5 | ≥10 |
|---|---|---|---|---|---|---|---|---|---|---|---|---|---|
| 0 | 1.000 | 0.990 | 0.981 | 0.964 | 0.949 | 0.935 | 0.922 | 0.875 | 0.820 | 0.791 | 0.773 | 0.760 | 0.732 |
| 0.05 | 0.990 | 0.981 | 0.971 | 0.955 | 0.940 | 0.926 | 0.914 | 0.867 | 0.814 | 0.784 | 0.766 | 0.754 | 0.726 |
| 0.1 | 0.981 | 0.971 | 0.962 | 0.946 | 0.931 | 0.918 | 0.906 | 0.860 | 0.807 | 0.778 | 0.760 | 0.748 | 0.721 |
| 0.2 | 0.964 | 0.955 | 0.946 | 0.930 | 0.916 | 0.903 | 0.891 | 0.846 | 0.795 | 0.767 | 0.749 | 0.737 | 0.711 |
| 0.3 | 0.949 | 0.940 | 0.931 | 0.916 | 0.902 | 0.889 | 0.878 | 0.834 | 0.784 | 0.756 | 0.739 | 0.728 | 0.701 |
| 0.4 | 0.935 | 0.926 | 0.918 | 0.903 | 0.889 | 0.877 | 0.866 | 0.823 | 0.774 | 0.747 | 0.730 | 0.719 | 0.693 |
| 0.5 | 0.922 | 0.914 | 0.906 | 0.891 | 0.878 | 0.866 | 0.855 | 0.813 | 0.765 | 0.738 | 0.721 | 0.710 | 0.685 |
| 1 | 0.875 | 0.867 | 0.860 | 0.846 | 0.834 | 0.823 | 0.813 | 0.774 | 0.729 | 0.704 | 0.688 | 0.677 | 0.654 |
| 2 | 0.820 | 0.814 | 0.807 | 0.795 | 0.784 | 0.774 | 0.765 | 0.729 | 0.686 | 0.663 | 0.646 | 0.638 | 0.615 |
| 3 | 0.791 | 0.784 | 0.778 | 0.767 | 0.756 | 0.747 | 0.738 | 0.704 | 0.663 | 0.640 | 0.625 | 0.616 | 0.593 |
| 4 | 0.773 | 0.766 | 0.760 | 0.749 | 0.739 | 0.730 | 0.721 | 0.688 | 0.648 | 0.625 | 0.611 | 0.601 | 0.580 |
| 5 | 0.760 | 0.754 | 0.748 | 0.737 | 0.728 | 0.719 | 0.710 | 0.677 | 0.638 | 0.616 | 0.601 | 0.592 | 0.570 |
| ≥10 | 0.732 | 0.726 | 0.721 | 0.711 | 0.701 | 0.693 | 0.685 | 0.654 | 0.615 | 0.593 | 0.580 | 0.570 | 0.549 |

③ When the axially compressive force $N_b$ of the beam rigidly connected to the column is large, the rotation stiffness reduction factor $\alpha_N$ of the beam should be calculated according to the following formula:

When the beam's far end is rigidly connected or hinged to the column:
$$\alpha_N = 1 - N_b/N_{Eb} \tag{E-5}$$

When the beam's far end is fixed:
$$\alpha_N = 1 - N_b/2N_{Eb} \tag{E-6}$$
$$N_{Eb} = \pi^2 EI_b/l^2 \tag{E-7}$$

where   $I_b$ = the Moment of inertia of a beam (mm$^4$);

  $l$ = the length of a beam (mm).

## E.3 Effective length factor $\mu_2$ of the lower section of a single-step column with the free upper end

The effective length factor $\mu_2$ of the lower section of a single-step column with a free

# APPENDIX E  EFFECTIVE LENGTH FACTOR OF COLUMNS

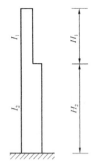

Figure E. 1  A single-step column

upper end (Figure E. 1) should be calculated as shown in Table E. 3.

$$K_1 = \frac{I_1}{I_2} \cdot \frac{H_2}{H_1} \quad \text{(E-8)}$$

$$\eta_1 = \frac{H_1}{H_2}\sqrt{\frac{N_1}{N_2} \cdot \frac{I_2}{I_1}} \quad \text{(E-9)}$$

where  $N_1$ = axial force of upper section column;
$N_2$ = axial force of the lower section column.

**Effective length factor of the lower section of a single-step column with a free upper end**  Table E. 3

| $K_1$ / $\eta_1$ | 0.06 | 0.08 | 0.10 | 0.12 | 0.14 | 0.16 | 0.18 | 0.20 | 0.22 | 0.24 | 0.26 | 0.28 | 0.3 | 0.4 | 0.5 | 0.6 | 0.7 | 0.8 |
|---|---|---|---|---|---|---|---|---|---|---|---|---|---|---|---|---|---|---|
| 0.2 | 2.00 | 2.01 | 2.01 | 2.01 | 2.01 | 2.01 | 2.01 | 2.02 | 2.02 | 2.02 | 2.02 | 2.02 | 2.02 | 2.03 | 2.04 | 2.05 | 2.06 | 2.07 |
| 0.3 | 2.01 | 2.02 | 2.02 | 2.02 | 2.03 | 2.03 | 2.03 | 2.04 | 2.04 | 2.05 | 2.05 | 2.05 | 2.06 | 2.08 | 2.10 | 2.12 | 2.13 | 2.15 |
| 0.4 | 2.02 | 2.03 | 2.04 | 2.04 | 2.05 | 2.06 | 2.07 | 2.07 | 2.08 | 2.09 | 2.09 | 2.10 | 2.11 | 2.14 | 2.18 | 2.21 | 2.25 | 2.28 |
| 0.5 | 2.04 | 2.05 | 2.06 | 2.07 | 2.09 | 2.10 | 2.11 | 2.12 | 2.13 | 2.15 | 2.16 | 2.17 | 2.18 | 2.24 | 2.29 | 2.35 | 2.40 | 2.45 |
| 0.6 | 2.06 | 2.08 | 2.10 | 2.12 | 2.14 | 2.16 | 2.18 | 2.19 | 2.21 | 2.23 | 2.25 | 2.26 | 2.28 | 2.36 | 2.44 | 2.52 | 2.59 | 2.66 |
| 0.7 | 2.10 | 2.13 | 2.16 | 2.18 | 2.21 | 2.24 | 2.26 | 2.29 | 2.31 | 2.34 | 2.36 | 2.38 | 2.41 | 2.52 | 2.62 | 2.72 | 2.81 | 2.90 |
| 0.8 | 2.15 | 2.20 | 2.24 | 2.27 | 2.31 | 2.34 | 2.38 | 2.41 | 2.44 | 2.47 | 2.50 | 2.53 | 2.56 | 2.70 | 2.82 | 2.94 | 3.06 | 3.16 |
| 0.9 | 2.24 | 2.29 | 2.35 | 2.39 | 2.44 | 2.48 | 2.52 | 2.56 | 2.60 | 2.63 | 2.67 | 2.71 | 2.74 | 2.90 | 3.05 | 3.19 | 3.32 | 3.44 |
| 1.0 | 2.36 | 2.43 | 2.48 | 2.54 | 2.59 | 2.64 | 2.69 | 2.73 | 2.77 | 2.82 | 2.86 | 2.90 | 2.94 | 3.12 | 3.29 | 3.45 | 3.59 | 3.74 |
| 1.2 | 2.69 | 2.76 | 2.83 | 2.89 | 2.95 | 3.01 | 3.07 | 3.12 | 3.17 | 3.22 | 3.27 | 3.32 | 3.37 | 3.59 | 3.80 | 3.99 | 4.17 | 4.34 |
| 1.4 | 3.07 | 3.14 | 3.22 | 3.29 | 3.36 | 3.42 | 3.48 | 3.55 | 3.61 | 3.66 | 3.72 | 3.78 | 3.83 | 4.09 | 4.33 | 4.56 | 4.77 | 4.97 |
| 1.6 | 3.47 | 3.55 | 3.63 | 3.71 | 3.78 | 3.85 | 3.92 | 3.99 | 4.07 | 4.12 | 4.18 | 4.25 | 4.31 | 4.61 | 4.88 | 5.14 | 5.38 | 5.62 |
| 1.8 | 3.88 | 3.97 | 4.05 | 4.13 | 4.21 | 4.29 | 4.37 | 4.44 | 4.52 | 4.59 | 4.66 | 4.73 | 4.80 | 5.13 | 5.44 | 5.73 | 6.00 | 6.26 |
| 2.0 | 4.29 | 4.39 | 4.48 | 4.57 | 4.65 | 4.74 | 4.82 | 4.90 | 4.99 | 5.07 | 5.14 | 5.22 | 5.30 | 5.66 | 6.00 | 6.32 | 6.63 | 6.92 |
| 2.2 | 4.71 | 4.81 | 4.91 | 5.00 | 5.10 | 5.19 | 5.28 | 5.37 | 5.46 | 5.54 | 5.63 | 5.71 | 5.80 | 6.19 | 6.57 | 6.92 | 7.26 | 7.58 |
| 2.4 | 5.13 | 5.24 | 5.34 | 5.44 | 5.54 | 5.64 | 5.74 | 5.84 | 5.93 | 6.03 | 6.12 | 6.21 | 6.30 | 6.73 | 7.14 | 7.52 | 7.89 | 8.24 |
| 2.6 | 5.55 | 5.66 | 5.77 | 5.88 | 5.99 | 6.10 | 6.20 | 6.31 | 6.41 | 6.51 | 6.61 | 6.71 | 6.80 | 7.27 | 7.71 | 8.13 | 8.52 | 8.90 |
| 2.8 | 5.97 | 6.09 | 6.21 | 6.33 | 6.44 | 6.55 | 6.67 | 6.78 | 6.89 | 6.99 | 7.10 | 7.21 | 7.31 | 7.81 | 8.28 | 8.73 | 9.16 | 9.57 |
| 3.0 | 6.39 | 6.52 | 6.64 | 6.77 | 6.89 | 7.01 | 7.13 | 7.25 | 7.37 | 7.48 | 7.59 | 7.71 | 7.82 | 8.35 | 8.86 | 9.34 | 9.80 | 10.24 |

## E.4 The effective length factor $\mu_2$ of the lower section of a single-step column with a movable upper end without rotation

The effective length factor $\mu_2$ of the lower section of a single-step column with a mov-able upper end without rotation (Figure E. 2) should be calculated as shown in Table E. 4.

$$K_1 = \frac{I_1}{I_2} \cdot \frac{H_2}{H_1} \quad \text{(E-10)}$$

$$\eta_1 = \frac{H_1}{H_2}\sqrt{\frac{N_1}{N_2} \cdot \frac{I_2}{I_1}} \quad \text{(E-11)}$$

where  $N_1$ = axial force of upper section column;
$N_2$ = axial force of the lower section column.

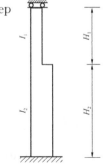

Figure E. 2  A single-step column

# APPENDIX E  EFFECTIVE LENGTH FACTOR OF COLUMNS

**Effective length factor of the lower section of a single-step column with a movable upper end without rotating**   Table E. 4

| $\eta_1$ \ $K_1$ | 0.06 | 0.08 | 0.10 | 0.12 | 0.14 | 0.16 | 0.18 | 0.20 | 0.22 | 0.24 | 0.26 | 0.28 | 0.3 | 0.4 | 0.5 | 0.6 | 0.7 | 0.8 |
|---|---|---|---|---|---|---|---|---|---|---|---|---|---|---|---|---|---|---|
| 0.2 | 1.96 | 1.94 | 1.93 | 1.91 | 1.90 | 1.89 | 1.88 | 1.86 | 1.85 | 1.84 | 1.83 | 1.82 | 1.81 | 1.76 | 1.72 | 1.68 | 1.65 | 1.62 |
| 0.3 | 1.96 | 1.94 | 1.93 | 1.92 | 1.91 | 1.89 | 1.88 | 1.87 | 1.86 | 1.85 | 1.84 | 1.83 | 1.82 | 1.77 | 1.73 | 1.70 | 1.66 | 1.63 |
| 0.4 | 1.96 | 1.95 | 1.94 | 1.92 | 1.91 | 1.90 | 1.89 | 1.88 | 1.87 | 1.86 | 1.85 | 1.84 | 1.83 | 1.79 | 1.75 | 1.72 | 1.68 | 1.66 |
| 0.5 | 1.96 | 1.95 | 1.94 | 1.93 | 1.92 | 1.91 | 1.90 | 1.89 | 1.88 | 1.87 | 1.86 | 1.85 | 1.85 | 1.81 | 1.77 | 1.74 | 1.71 | 1.69 |
| 0.6 | 1.97 | 1.96 | 1.95 | 1.94 | 1.93 | 1.92 | 1.91 | 1.90 | 1.90 | 1.89 | 1.88 | 1.87 | 1.87 | 1.83 | 1.80 | 1.78 | 1.75 | 1.73 |
| 0.7 | 1.97 | 1.97 | 1.96 | 1.95 | 1.94 | 1.94 | 1.93 | 1.92 | 1.92 | 1.91 | 1.90 | 1.90 | 1.89 | 1.86 | 1.84 | 1.82 | 1.80 | 1.78 |
| 0.8 | 1.98 | 1.98 | 1.97 | 1.96 | 1.96 | 1.95 | 1.95 | 1.94 | 1.94 | 1.93 | 1.93 | 1.93 | 1.92 | 1.90 | 1.88 | 1.87 | 1.86 | 1.84 |
| 0.9 | 1.99 | 1.99 | 1.98 | 1.98 | 1.98 | 1.97 | 1.97 | 1.97 | 1.97 | 1.96 | 1.96 | 1.96 | 1.96 | 1.95 | 1.94 | 1.93 | 1.92 | 1.92 |
| 1.0 | 2.00 | 2.00 | 2.00 | 2.00 | 2.00 | 2.00 | 2.00 | 2.00 | 2.00 | 2.00 | 2.00 | 2.00 | 2.00 | 2.00 | 2.00 | 2.00 | 2.00 | 2.00 |
| 1.2 | 2.03 | 2.04 | 2.04 | 2.05 | 2.06 | 2.07 | 2.07 | 2.08 | 2.08 | 2.09 | 2.10 | 2.10 | 2.11 | 2.13 | 2.15 | 2.17 | 2.18 | 2.20 |
| 1.4 | 2.07 | 2.09 | 2.11 | 2.12 | 2.14 | 2.16 | 2.17 | 2.18 | 2.20 | 2.21 | 2.22 | 2.23 | 2.24 | 2.29 | 2.33 | 2.37 | 2.40 | 2.42 |
| 1.6 | 2.13 | 2.16 | 2.19 | 2.22 | 2.25 | 2.27 | 2.30 | 2.32 | 2.34 | 2.36 | 2.37 | 2.39 | 2.41 | 2.48 | 2.54 | 2.59 | 2.63 | 2.67 |
| 1.8 | 2.22 | 2.27 | 2.31 | 2.35 | 2.39 | 2.42 | 2.45 | 2.48 | 2.50 | 2.53 | 2.55 | 2.57 | 2.59 | 2.69 | 2.76 | 2.83 | 2.88 | 2.93 |
| 2.0 | 2.35 | 2.41 | 2.46 | 2.50 | 2.55 | 2.59 | 2.62 | 2.66 | 2.69 | 2.72 | 2.75 | 2.77 | 2.80 | 2.91 | 3.00 | 3.08 | 3.14 | 3.20 |
| 2.2 | 2.51 | 2.57 | 2.63 | 2.68 | 2.73 | 2.77 | 2.81 | 2.85 | 2.89 | 2.92 | 2.95 | 2.98 | 3.01 | 3.14 | 3.25 | 3.33 | 3.41 | 3.47 |
| 2.4 | 2.68 | 2.75 | 2.81 | 2.87 | 2.92 | 2.97 | 3.01 | 3.05 | 3.09 | 3.13 | 3.17 | 3.20 | 3.24 | 3.38 | 3.50 | 3.59 | 3.68 | 3.75 |
| 2.6 | 2.87 | 2.94 | 3.00 | 3.06 | 3.12 | 3.17 | 3.22 | 3.27 | 3.31 | 3.35 | 3.39 | 3.43 | 3.46 | 3.62 | 3.75 | 3.86 | 3.95 | 4.03 |
| 2.8 | 3.06 | 3.14 | 3.20 | 3.27 | 3.33 | 3.38 | 3.43 | 3.48 | 3.53 | 3.58 | 3.62 | 3.66 | 3.70 | 3.87 | 4.01 | 4.13 | 4.23 | 4.32 |
| 3.0 | 3.26 | 3.34 | 3.41 | 3.47 | 3.54 | 3.60 | 3.65 | 3.70 | 3.75 | 3.80 | 3.85 | 3.89 | 3.93 | 4.12 | 4.27 | 4.40 | 4.51 | 4.61 |

## E.5  The effective length factor $\mu_3$ of the lower section of a double-step column with a free upper end

The effective length factor $\mu_3$ of the lower section of the double-step column with the free upper end (Figure E. 3) should be calculated according to the following formula or taken according to Table E. 5.

$$K_1 = \frac{I_1}{I_3} \cdot \frac{H_3}{H_1} \tag{E-12}$$

$$K_2 = \frac{I_2}{I_3} \cdot \frac{H_3}{H_2} \tag{E-13}$$

$$\eta_1 = \frac{H_1}{H_2}\sqrt{\frac{N_1}{N_3} \cdot \frac{I_3}{I_1}} \tag{E-14}$$

$$\eta_2 = \frac{H_2}{H_3}\sqrt{\frac{N_2}{N_3} \cdot \frac{I_3}{I_2}} \tag{E-15}$$

where  $N_1$ = axial force of the upper column;
$N_2$ = axial force of the lower middle column.
$N_3$ = axial force of the lower column.

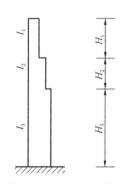

Figure E. 3  A double-step column

## APPENDIX E  EFFECTIVE LENGTH FACTOR OF COLUMNS

Table E.5  Effective length factor $\mu_3$ of the lower section of a double-step column with a free upper end

| $\eta_1$ | $\eta_2$ | $K_2$\$K_1$ | 0.2 | 0.3 | 0.4 | 0.5 | 0.6 | 0.7 | 0.8 | 0.9 | 1.0 | 1.1 | 1.2 | 0.2 | 0.3 | 0.4 | 0.5 | 0.6 | 0.7 | 0.8 | 0.9 | 1.0 | 1.1 | 1.2 |
|---|---|---|---|---|---|---|---|---|---|---|---|---|---|---|---|---|---|---|---|---|---|---|---|---|
| | | | | | | | 0.05 | | | | | | | | | | | 0.10 | | | | | | |
| 0.2 | 0.2 | | 2.02 | 2.03 | 2.04 | 2.05 | 2.05 | 2.06 | 2.07 | 2.08 | 2.09 | 2.10 | 2.10 | 2.03 | 2.03 | 2.04 | 2.05 | 2.06 | 2.07 | 2.08 | 2.08 | 2.09 | 2.10 | 2.11 |
| | 0.4 | | 2.08 | 2.11 | 2.15 | 2.19 | 2.22 | 2.25 | 2.29 | 2.32 | 2.35 | 2.39 | 2.42 | 2.09 | 2.12 | 2.16 | 2.19 | 2.23 | 2.26 | 2.29 | 2.33 | 2.36 | 2.39 | 2.42 |
| | 0.6 | | 2.20 | 2.29 | 2.37 | 2.45 | 2.52 | 2.60 | 2.67 | 2.73 | 2.80 | 2.87 | 2.93 | 2.21 | 2.30 | 2.38 | 2.46 | 2.53 | 2.60 | 2.67 | 2.74 | 2.81 | 2.87 | 2.93 |
| | 0.8 | | 2.42 | 2.57 | 2.71 | 2.83 | 2.95 | 3.06 | 3.17 | 3.27 | 3.37 | 3.47 | 3.56 | 2.44 | 2.58 | 2.71 | 2.84 | 2.96 | 3.07 | 3.17 | 3.28 | 3.37 | 3.47 | 3.56 |
| | 1.0 | | 2.75 | 2.95 | 3.13 | 3.30 | 3.45 | 3.60 | 3.74 | 3.87 | 4.00 | 4.13 | 4.25 | 2.76 | 2.96 | 3.14 | 3.30 | 3.46 | 3.60 | 3.74 | 3.88 | 4.01 | 4.13 | 4.25 |
| | 1.2 | | 3.13 | 3.38 | 3.60 | 3.80 | 4.00 | 4.18 | 4.35 | 4.51 | 4.67 | 4.82 | 4.97 | 3.15 | 3.39 | 3.61 | 3.81 | 4.00 | 4.18 | 4.35 | 4.52 | 4.68 | 4.83 | 4.98 |
| 0.4 | 0.2 | | 2.04 | 2.05 | 2.05 | 2.06 | 2.07 | 2.08 | 2.09 | 2.09 | 2.10 | 2.11 | 2.12 | 2.07 | 2.07 | 2.08 | 2.08 | 2.09 | 2.10 | 2.11 | 2.12 | 2.12 | 2.13 | 2.14 |
| | 0.4 | | 2.10 | 2.14 | 2.17 | 2.20 | 2.24 | 2.27 | 2.31 | 2.34 | 2.37 | 2.40 | 2.43 | 2.14 | 2.17 | 2.20 | 2.23 | 2.26 | 2.30 | 2.33 | 2.36 | 2.39 | 2.42 | 2.46 |
| | 0.6 | | 2.24 | 2.32 | 2.40 | 2.47 | 2.54 | 2.62 | 2.68 | 2.75 | 2.82 | 2.88 | 2.94 | 2.28 | 2.36 | 2.43 | 2.50 | 2.57 | 2.64 | 2.71 | 2.77 | 2.84 | 2.90 | 2.96 |
| | 0.8 | | 2.47 | 2.60 | 2.73 | 2.85 | 2.97 | 3.08 | 3.19 | 3.29 | 3.38 | 3.48 | 3.57 | 2.53 | 2.65 | 2.77 | 2.88 | 3.00 | 3.10 | 3.21 | 3.31 | 3.40 | 3.50 | 3.59 |
| | 1.0 | | 2.79 | 2.98 | 3.15 | 3.32 | 3.47 | 3.62 | 3.75 | 3.89 | 4.02 | 4.14 | 4.26 | 2.85 | 3.02 | 3.19 | 3.34 | 3.49 | 3.64 | 3.77 | 3.91 | 4.03 | 4.16 | 4.28 |
| | 1.2 | | 3.18 | 3.41 | 3.62 | 3.82 | 4.01 | 4.19 | 4.36 | 4.52 | 4.68 | 4.83 | 4.98 | 3.24 | 3.45 | 3.65 | 3.85 | 4.03 | 4.21 | 4.38 | 4.54 | 4.70 | 4.85 | 4.99 |
| 0.6 | 0.2 | | 2.09 | 2.09 | 2.10 | 2.10 | 2.11 | 2.12 | 2.12 | 2.13 | 2.14 | 2.15 | 2.15 | 2.22 | 2.19 | 2.18 | 2.17 | 2.18 | 2.18 | 2.19 | 2.19 | 2.20 | 2.20 | 2.21 |
| | 0.4 | | 2.17 | 2.19 | 2.22 | 2.25 | 2.28 | 2.31 | 2.34 | 2.38 | 2.41 | 2.44 | 2.47 | 2.31 | 2.30 | 2.31 | 2.33 | 2.35 | 2.38 | 2.41 | 2.44 | 2.47 | 2.49 | 2.52 |
| | 0.6 | | 2.32 | 2.38 | 2.45 | 2.52 | 2.59 | 2.66 | 2.72 | 2.79 | 2.85 | 2.91 | 2.97 | 2.48 | 2.49 | 2.54 | 2.60 | 2.66 | 2.72 | 2.78 | 2.84 | 2.90 | 2.96 | 3.02 |
| | 0.8 | | 2.56 | 2.67 | 2.79 | 2.90 | 3.01 | 3.11 | 3.22 | 3.32 | 3.41 | 3.50 | 3.60 | 2.72 | 2.78 | 2.87 | 2.97 | 3.07 | 3.17 | 3.27 | 3.36 | 3.46 | 3.55 | 3.64 |
| | 1.0 | | 2.88 | 3.04 | 3.20 | 3.36 | 3.50 | 3.65 | 3.78 | 3.91 | 4.04 | 4.16 | 4.26 | 3.04 | 3.15 | 3.28 | 3.42 | 3.56 | 3.70 | 3.83 | 3.95 | 4.08 | 4.20 | 4.31 |
| | 1.2 | | 3.26 | 3.46 | 3.66 | 3.86 | 4.04 | 4.22 | 4.38 | 4.55 | 4.70 | 4.85 | 5.00 | 3.40 | 3.56 | 3.74 | 3.91 | 4.09 | 4.26 | 4.42 | 4.58 | 4.73 | 4.88 | 5.03 |

APPENDIX E  EFFECTIVE LENGTH FACTOR OF COLUMNS

continued

| $K_1$ | | | | | 0.05 | | | | | | | | | | | | 0.10 | | | | | | |
|---|---|---|---|---|---|---|---|---|---|---|---|---|---|---|---|---|---|---|---|---|---|---|---|
| $\eta_1$ | $\eta_2$ / $K_2$ | 0.2 | 0.3 | 0.4 | 0.5 | 0.6 | 0.7 | 0.8 | 0.9 | 1.0 | 1.1 | 1.2 | 0.2 | 0.3 | 0.4 | 0.5 | 0.6 | 0.7 | 0.8 | 0.9 | 1.0 | 1.1 | 1.2 |
| 0.8 | 0.2 | 2.29 | 2.24 | 2.22 | 2.21 | 2.21 | 2.22 | 2.22 | 2.22 | 2.23 | 2.23 | 2.24 | 2.63 | 2.49 | 2.43 | 2.40 | 2.38 | 2.37 | 2.37 | 2.36 | 2.36 | 2.37 | 2.37 |
|  | 0.4 | 2.37 | 2.34 | 2.34 | 2.36 | 2.38 | 2.40 | 2.43 | 2.45 | 2.48 | 2.51 | 2.54 | 2.71 | 2.59 | 2.55 | 2.54 | 2.54 | 2.55 | 2.57 | 2.59 | 2.61 | 2.63 | 2.65 |
|  | 0.6 | 2.52 | 2.52 | 2.56 | 2.61 | 2.67 | 2.73 | 2.79 | 2.85 | 2.91 | 2.96 | 3.02 | 2.86 | 2.76 | 2.76 | 2.78 | 2.82 | 2.86 | 2.91 | 2.96 | 3.01 | 3.07 | 3.12 |
|  | 0.8 | 2.74 | 2.79 | 2.88 | 2.98 | 3.08 | 3.17 | 3.27 | 3.36 | 3.46 | 3.55 | 3.63 | 3.06 | 3.02 | 3.06 | 3.13 | 3.20 | 3.29 | 3.37 | 3.46 | 3.54 | 3.63 | 3.71 |
|  | 1.0 | 3.04 | 3.15 | 3.28 | 3.42 | 3.56 | 3.69 | 3.82 | 3.95 | 4.07 | 4.19 | 4.31 | 3.33 | 3.35 | 3.44 | 3.55 | 3.67 | 3.79 | 3.90 | 4.03 | 4.15 | 4.26 | 4.37 |
|  | 1.2 | 3.39 | 3.55 | 3.73 | 3.91 | 4.08 | 4.25 | 4.42 | 4.58 | 4.73 | 4.88 | 5.02 | 3.65 | 3.73 | 3.86 | 4.02 | 4.18 | 4.34 | 4.49 | 4.64 | 4.79 | 4.94 | 5.08 |
| 1.0 | 0.2 | 2.69 | 2.57 | 2.51 | 2.48 | 2.46 | 2.45 | 2.45 | 2.44 | 2.44 | 2.44 | 2.44 | 3.18 | 2.95 | 2.84 | 2.77 | 2.73 | 2.70 | 2.68 | 2.67 | 2.66 | 2.65 | 2.65 |
|  | 0.4 | 2.75 | 2.64 | 2.60 | 2.59 | 2.59 | 2.59 | 2.60 | 2.62 | 2.63 | 2.65 | 2.67 | 3.24 | 3.03 | 2.93 | 2.88 | 2.85 | 2.84 | 2.84 | 2.84 | 2.85 | 2.86 | 2.87 |
|  | 0.6 | 2.86 | 2.78 | 2.77 | 2.79 | 2.83 | 2.87 | 2.91 | 2.96 | 3.01 | 3.06 | 3.10 | 3.36 | 3.16 | 3.09 | 3.07 | 3.08 | 3.09 | 3.12 | 3.15 | 3.19 | 3.23 | 3.27 |
|  | 0.8 | 3.04 | 3.01 | 3.05 | 3.11 | 3.19 | 3.27 | 3.35 | 3.44 | 3.52 | 3.61 | 3.69 | 3.52 | 3.37 | 3.34 | 3.36 | 3.41 | 3.46 | 3.53 | 3.60 | 3.67 | 3.75 | 3.82 |
|  | 1.0 | 3.29 | 3.32 | 3.41 | 3.52 | 3.64 | 3.76 | 3.89 | 4.01 | 4.13 | 4.24 | 4.35 | 3.74 | 3.64 | 3.67 | 3.74 | 3.83 | 3.93 | 4.03 | 4.14 | 4.25 | 4.35 | 4.46 |
|  | 1.2 | 3.60 | 3.69 | 3.83 | 3.99 | 4.15 | 4.31 | 4.47 | 4.62 | 4.77 | 4.92 | 5.06 | 4.00 | 3.97 | 4.05 | 4.17 | 4.31 | 4.45 | 4.59 | 4.73 | 4.87 | 5.01 | 5.14 |
| 1.2 | 0.2 | 3.16 | 3.00 | 2.92 | 2.87 | 2.84 | 2.81 | 2.80 | 2.79 | 2.78 | 2.77 | 2.77 | 3.77 | 3.47 | 3.32 | 3.23 | 3.17 | 3.12 | 3.09 | 3.07 | 3.05 | 3.04 | 3.03 |
|  | 0.4 | 3.21 | 3.05 | 2.98 | 2.94 | 2.92 | 2.90 | 2.90 | 2.90 | 2.90 | 2.91 | 2.92 | 3.82 | 3.53 | 3.39 | 3.31 | 3.26 | 3.22 | 3.20 | 3.19 | 3.19 | 3.19 | 3.19 |
|  | 0.6 | 3.30 | 3.15 | 3.10 | 3.08 | 3.08 | 3.10 | 3.12 | 3.15 | 3.18 | 3.22 | 3.26 | 3.91 | 3.64 | 3.51 | 3.45 | 3.42 | 3.42 | 3.42 | 3.43 | 3.45 | 3.48 | 3.50 |
|  | 0.8 | 3.43 | 3.32 | 3.30 | 3.33 | 3.37 | 3.43 | 3.49 | 3.56 | 3.63 | 3.71 | 3.78 | 4.04 | 3.80 | 3.71 | 3.68 | 3.69 | 3.72 | 3.76 | 3.81 | 3.86 | 3.92 | 3.98 |
|  | 1.0 | 3.62 | 3.57 | 3.60 | 3.68 | 3.77 | 3.87 | 3.98 | 4.09 | 4.20 | 4.31 | 4.42 | 4.21 | 4.02 | 3.97 | 3.99 | 4.05 | 4.12 | 4.20 | 4.29 | 4.39 | 4.48 | 4.58 |
|  | 1.2 | 3.88 | 3.88 | 3.98 | 4.11 | 4.25 | 4.39 | 4.54 | 4.68 | 4.83 | 4.97 | 5.10 | 4.43 | 4.30 | 4.31 | 4.38 | 4.48 | 4.60 | 4.72 | 4.85 | 4.98 | 5.11 | 5.24 |
| 1.4 | 0.2 | 3.66 | 3.46 | 3.36 | 3.29 | 3.25 | 3.23 | 3.20 | 3.19 | 3.18 | 3.17 | 3.16 | 4.37 | 4.01 | 3.82 | 3.71 | 3.63 | 3.58 | 3.54 | 3.51 | 3.49 | 3.47 | 3.45 |
|  | 0.4 | 3.70 | 3.50 | 3.40 | 3.35 | 3.31 | 3.29 | 3.27 | 3.26 | 3.26 | 3.26 | 3.26 | 4.41 | 4.06 | 3.88 | 3.77 | 3.70 | 3.66 | 3.63 | 3.60 | 3.59 | 3.58 | 3.57 |
|  | 0.6 | 3.77 | 3.58 | 3.49 | 3.45 | 3.43 | 3.42 | 3.42 | 3.43 | 3.45 | 3.47 | 3.49 | 4.48 | 4.15 | 3.98 | 3.89 | 3.83 | 3.80 | 3.79 | 3.78 | 3.79 | 3.80 | 3.81 |
|  | 0.8 | 3.87 | 3.70 | 3.64 | 3.63 | 3.64 | 3.67 | 3.70 | 3.75 | 3.81 | 3.86 | 3.92 | 4.59 | 4.28 | 4.13 | 4.07 | 4.04 | 4.04 | 4.06 | 4.08 | 4.12 | 4.16 | 4.21 |
|  | 1.0 | 4.02 | 3.89 | 3.87 | 3.90 | 3.96 | 4.04 | 4.12 | 4.22 | 4.31 | 4.41 | 4.51 | 4.74 | 4.45 | 4.35 | 4.32 | 4.34 | 4.38 | 4.43 | 4.50 | 4.58 | 4.66 | 4.74 |
|  | 1.2 | 4.23 | 4.15 | 4.19 | 4.27 | 4.39 | 4.51 | 4.64 | 4.77 | 4.91 | 5.04 | 5.17 | 4.92 | 4.69 | 4.63 | 4.65 | 4.72 | 4.80 | 4.90 | 5.10 | 5.13 | 5.24 | 5.36 |

## APPENDIX E  EFFECTIVE LENGTH FACTOR OF COLUMNS

continued

| $\eta_1$ | $\eta_2$ | $K_1 \backslash K_2$ | 0.20 | | | | | | | | | | | 0.30 | | | | | | | | | | |
|---|---|---|---|---|---|---|---|---|---|---|---|---|---|---|---|---|---|---|---|---|---|---|---|---|
| | | | 0.2 | 0.3 | 0.4 | 0.5 | 0.6 | 0.7 | 0.8 | 0.9 | 1.0 | 1.1 | 1.2 | 0.2 | 0.3 | 0.4 | 0.5 | 0.6 | 0.7 | 0.8 | 0.9 | 1.0 | 1.1 | 1.2 |
| 0.2 | 0.2 | | 2.04 | 2.04 | 2.05 | 2.06 | 2.07 | 2.08 | 2.08 | 2.09 | 2.10 | 2.11 | 2.12 | 2.05 | 2.05 | 2.06 | 2.07 | 2.08 | 2.09 | 2.09 | 2.10 | 2.11 | 2.12 | 2.13 |
| | 0.4 | | 2.10 | 2.13 | 2.17 | 2.20 | 2.24 | 2.27 | 2.30 | 2.34 | 2.37 | 2.40 | 2.43 | 2.12 | 2.15 | 2.18 | 2.21 | 2.25 | 2.28 | 2.31 | 2.35 | 2.38 | 2.41 | 2.44 |
| | 0.6 | | 2.23 | 2.31 | 2.39 | 2.47 | 2.54 | 2.61 | 2.68 | 2.75 | 2.82 | 2.88 | 2.94 | 2.25 | 2.33 | 2.41 | 2.48 | 2.56 | 2.63 | 2.69 | 2.76 | 2.83 | 2.89 | 2.95 |
| | 0.8 | | 2.46 | 2.60 | 2.73 | 2.85 | 2.97 | 3.08 | 3.18 | 3.29 | 3.38 | 3.48 | 3.57 | 2.49 | 2.62 | 2.75 | 2.87 | 2.98 | 3.09 | 3.20 | 3.30 | 3.39 | 3.49 | 3.38 |
| | 1.0 | | 2.79 | 2.98 | 3.15 | 3.32 | 3.47 | 3.61 | 3.75 | 3.89 | 4.02 | 4.14 | 4.26 | 2.82 | 3.00 | 3.17 | 3.33 | 3.48 | 3.63 | 3.76 | 3.90 | 4.02 | 4.15 | 4.27 |
| | 1.2 | | 3.18 | 3.41 | 3.62 | 3.82 | 4.01 | 4.19 | 4.36 | 4.52 | 4.68 | 4.83 | 4.98 | 3.20 | 3.43 | 3.64 | 3.83 | 4.02 | 4.20 | 4.37 | 4.53 | 4.69 | 4.84 | 4.99 |
| 0.4 | 0.2 | | 2.15 | 2.13 | 2.13 | 2.14 | 2.14 | 2.15 | 2.15 | 2.16 | 2.17 | 2.17 | 2.18 | 2.26 | 2.21 | 2.20 | 2.19 | 2.19 | 2.20 | 2.20 | 2.21 | 2.21 | 2.22 | 2.23 |
| | 0.4 | | 2.24 | 2.24 | 2.26 | 2.29 | 2.32 | 2.35 | 2.38 | 2.41 | 2.44 | 2.47 | 2.50 | 2.36 | 2.33 | 2.33 | 2.35 | 2.38 | 2.40 | 2.43 | 2.46 | 2.49 | 2.51 | 2.54 |
| | 0.6 | | 2.40 | 2.44 | 2.50 | 2.56 | 2.63 | 2.69 | 2.76 | 2.82 | 2.88 | 2.94 | 3.00 | 2.54 | 2.54 | 2.58 | 2.63 | 2.69 | 2.75 | 2.81 | 2.87 | 2.93 | 2.99 | 3.04 |
| | 0.8 | | 2.66 | 2.74 | 2.84 | 2.95 | 3.05 | 3.15 | 3.25 | 3.35 | 3.44 | 3.53 | 3.62 | 2.79 | 2.83 | 2.91 | 3.01 | 3.10 | 3.20 | 3.30 | 3.39 | 3.48 | 3.57 | 3.46 |
| | 1.0 | | 2.98 | 3.12 | 3.25 | 3.40 | 3.54 | 3.68 | 3.81 | 3.94 | 4.07 | 4.19 | 4.30 | 3.11 | 3.20 | 3.32 | 3.46 | 3.59 | 3.72 | 3.85 | 3.98 | 4.10 | 4.22 | 4.33 |
| | 1.2 | | 3.35 | 3.53 | 3.71 | 3.90 | 4.08 | 4.25 | 4.41 | 4.57 | 4.73 | 4.87 | 5.02 | 3.47 | 3.60 | 3.77 | 3.95 | 4.12 | 4.28 | 4.45 | 4.60 | 4.75 | 4.90 | 5.04 |
| 0.6 | 0.2 | | 2.57 | 2.42 | 2.37 | 2.34 | 2.33 | 2.32 | 2.32 | 2.32 | 2.32 | 2.32 | 2.33 | 2.93 | 2.68 | 2.57 | 2.52 | 2.49 | 2.47 | 2.46 | 2.45 | 2.45 | 2.45 | 2.45 |
| | 0.4 | | 2.67 | 2.54 | 2.50 | 2.50 | 2.51 | 2.52 | 2.54 | 2.56 | 2.58 | 2.61 | 2.63 | 3.02 | 2.79 | 2.71 | 2.67 | 2.66 | 2.66 | 2.67 | 2.69 | 2.70 | 2.72 | 2.74 |
| | 0.6 | | 2.83 | 2.74 | 2.73 | 2.76 | 2.80 | 2.85 | 2.90 | 2.96 | 3.01 | 3.06 | 3.12 | 3.17 | 2.98 | 2.93 | 2.93 | 2.95 | 2.98 | 3.02 | 3.07 | 3.11 | 3.16 | 3.21 |
| | 0.8 | | 3.06 | 3.01 | 3.05 | 3.12 | 3.20 | 3.29 | 3.38 | 3.46 | 3.55 | 3.63 | 3.72 | 4.37 | 3.24 | 3.23 | 3.27 | 3.33 | 3.41 | 3.48 | 3.56 | 3.64 | 3.72 | 3.80 |
| | 1.0 | | 3.34 | 3.35 | 3.44 | 3.56 | 3.68 | 3.80 | 3.92 | 4.04 | 4.15 | 4.27 | 4.38 | 3.63 | 3.56 | 3.60 | 3.69 | 3.79 | 3.90 | 4.01 | 4.12 | 4.23 | 4.34 | 4.45 |
| | 1.2 | | 3.67 | 3.74 | 3.88 | 4.03 | 4.19 | 4.35 | 4.50 | 4.65 | 4.80 | 4.94 | 5.08 | 3.94 | 3.92 | 4.02 | 4.15 | 4.29 | 4.43 | 4.58 | 4.72 | 4.87 | 5.01 | 5.14 |

302

APPENDIX E  EFFECTIVE LENGTH FACTOR OF COLUMNS

continued

| $\eta_1$ | $K_2$ \ $K_1$ | \multicolumn{11}{c}{$\eta_2 = 0.20$} | \multicolumn{11}{c}{$\eta_2 = 0.30$} |
|---|---|---|---|---|---|---|---|---|---|---|---|---|---|---|---|---|---|---|---|---|---|---|---|
|  |  | 0.2 | 0.3 | 0.4 | 0.5 | 0.6 | 0.7 | 0.8 | 0.9 | 1.0 | 1.1 | 1.2 | 0.2 | 0.3 | 0.4 | 0.5 | 0.6 | 0.7 | 0.8 | 0.9 | 1.0 | 1.1 | 1.2 |
| 0.8 | 0.2 | 3.25 | 2.96 | 2.82 | 2.74 | 2.69 | 2.66 | 2.64 | 2.62 | 2.61 | 2.61 | 2.60 | 3.78 | 3.38 | 3.18 | 3.06 | 2.98 | 2.93 | 2.89 | 2.86 | 2.84 | 2.83 | 2.82 |
| 0.8 | 0.4 | 3.33 | 3.05 | 2.93 | 2.87 | 2.84 | 2.83 | 2.83 | 2.83 | 2.84 | 2.85 | 2.87 | 3.85 | 3.47 | 3.28 | 3.18 | 3.12 | 3.09 | 3.07 | 3.06 | 3.06 | 3.06 | 3.06 |
| 0.8 | 0.6 | 3.45 | 3.21 | 3.12 | 3.10 | 3.10 | 3.12 | 3.14 | 3.18 | 3.22 | 3.26 | 3.30 | 3.96 | 3.61 | 3.46 | 3.39 | 3.36 | 3.35 | 3.36 | 3.38 | 3.41 | 3.44 | 3.47 |
| 0.8 | 0.8 | 3.63 | 3.44 | 3.39 | 3.41 | 3.45 | 3.51 | 3.57 | 3.64 | 3.71 | 3.79 | 3.86 | 4.12 | 3.82 | 3.70 | 3.67 | 3.68 | 3.72 | 3.76 | 3.82 | 3.88 | 3.94 | 4.01 |
| 0.8 | 1.0 | 3.86 | 3.73 | 3.73 | 3.80 | 3.88 | 3.98 | 4.08 | 4.18 | 4.29 | 4.39 | 4.50 | 4.32 | 4.07 | 4.01 | 4.03 | 4.08 | 4.16 | 4.24 | 4.33 | 4.43 | 4.52 | 4.62 |
| 0.8 | 1.2 | 4.13 | 4.07 | 4.13 | 4.24 | 4.36 | 4.50 | 4.64 | 4.78 | 4.91 | 5.05 | 5.18 | 4.57 | 4.38 | 4.38 | 4.44 | 4.54 | 4.66 | 4.78 | 4.90 | 5.03 | 5.16 | 5.29 |
| 1.0 | 0.2 | 4.00 | 3.60 | 3.39 | 3.26 | 3.18 | 3.13 | 3.08 | 3.05 | 3.03 | 3.01 | 3.00 | 4.68 | 4.15 | 3.86 | 3.69 | 3.57 | 3.49 | 3.43 | 3.38 | 3.35 | 3.32 | 3.30 |
| 1.0 | 0.4 | 4.06 | 3.67 | 3.48 | 3.37 | 3.30 | 3.26 | 3.23 | 3.21 | 3.21 | 3.20 | 3.20 | 4.73 | 4.21 | 3.94 | 3.78 | 3.68 | 3.61 | 3.57 | 3.54 | 3.51 | 3.50 | 3.49 |
| 1.0 | 0.6 | 4.15 | 3.79 | 3.63 | 3.54 | 3.50 | 3.48 | 3.49 | 3.50 | 3.51 | 3.54 | 3.57 | 4.82 | 4.33 | 4.08 | 3.95 | 3.87 | 3.83 | 3.80 | 3.80 | 3.80 | 3.81 | 3.83 |
| 1.0 | 0.8 | 4.29 | 3.97 | 3.84 | 3.80 | 3.79 | 3.81 | 3.85 | 3.90 | 3.95 | 4.01 | 4.07 | 4.94 | 4.49 | 4.28 | 4.18 | 4.14 | 4.13 | 4.14 | 4.17 | 4.20 | 4.25 | 4.29 |
| 1.0 | 1.0 | 4.48 | 4.21 | 4.13 | 4.13 | 4.17 | 4.23 | 4.31 | 4.39 | 4.48 | 4.57 | 4.66 | 5.10 | 4.70 | 4.53 | 4.48 | 4.48 | 4.51 | 4.56 | 4.62 | 4.70 | 4.77 | 4.85 |
| 1.0 | 1.2 | 4.70 | 4.49 | 4.47 | 4.52 | 4.60 | 4.71 | 4.82 | 4.94 | 5.07 | 5.19 | 5.31 | 5.30 | 4.95 | 4.84 | 4.83 | 4.88 | 4.96 | 5.05 | 5.15 | 5.26 | 5.37 | 5.48 |
| 1.2 | 0.2 | 4.76 | 4.26 | 4.00 | 3.83 | 3.72 | 3.65 | 3.59 | 3.54 | 3.51 | 3.48 | 3.46 | 5.58 | 4.93 | 4.57 | 4.35 | 4.20 | 4.10 | 4.01 | 3.95 | 3.90 | 3.86 | 3.83 |
| 1.2 | 0.4 | 4.81 | 4.32 | 4.07 | 3.91 | 3.82 | 3.75 | 3.70 | 3.67 | 3.65 | 3.63 | 3.62 | 5.62 | 4.98 | 4.64 | 4.43 | 4.29 | 4.19 | 4.12 | 4.07 | 4.03 | 4.01 | 3.98 |
| 1.2 | 0.6 | 4.89 | 4.43 | 4.19 | 4.05 | 3.98 | 3.93 | 3.91 | 3.89 | 3.89 | 3.90 | 3.91 | 5.70 | 5.08 | 4.75 | 4.56 | 4.44 | 4.37 | 4.32 | 4.29 | 4.27 | 4.26 | 4.26 |
| 1.2 | 0.8 | 5.00 | 4.57 | 4.36 | 4.26 | 4.21 | 4.20 | 4.21 | 4.23 | 4.26 | 4.30 | 4.34 | 5.80 | 5.21 | 4.91 | 4.75 | 4.66 | 4.61 | 4.59 | 4.59 | 4.60 | 4.62 | 4.65 |
| 1.2 | 1.0 | 5.15 | 4.76 | 4.59 | 4.53 | 4.53 | 4.55 | 4.60 | 4.66 | 4.73 | 4.80 | 4.88 | 5.93 | 5.38 | 5.12 | 5.00 | 4.95 | 4.94 | 4.95 | 4.99 | 5.03 | 5.09 | 5.15 |
| 1.2 | 1.2 | 5.34 | 5.00 | 4.88 | 4.87 | 4.91 | 4.98 | 5.07 | 5.17 | 5.27 | 5.38 | 5.49 | 6.10 | 5.59 | 5.38 | 5.31 | 5.30 | 5.33 | 5.39 | 5.46 | 5.54 | 5.63 | 5.73 |
| 1.4 | 0.2 | 5.53 | 4.94 | 4.62 | 4.42 | 4.29 | 4.19 | 4.12 | 4.06 | 4.02 | 3.98 | 3.95 | 6.49 | 5.72 | 5.30 | 5.03 | 4.85 | 4.72 | 4.62 | 4.54 | 4.48 | 4.43 | 4.38 |
| 1.4 | 0.4 | 5.57 | 4.99 | 4.68 | 4.49 | 4.36 | 4.27 | 4.21 | 4.16 | 4.13 | 4.10 | 4.08 | 6.53 | 5.77 | 5.35 | 5.10 | 4.93 | 4.80 | 4.71 | 4.64 | 4.59 | 4.55 | 4.51 |
| 1.4 | 0.6 | 5.64 | 5.07 | 4.78 | 4.60 | 4.49 | 4.42 | 4.38 | 4.35 | 4.33 | 4.32 | 4.32 | 6.59 | 5.85 | 5.45 | 5.21 | 5.05 | 4.95 | 4.87 | 4.82 | 4.78 | 4.76 | 4.74 |
| 1.4 | 0.8 | 5.74 | 5.19 | 4.92 | 4.77 | 4.69 | 4.64 | 4.62 | 4.62 | 4.63 | 4.65 | 4.67 | 6.68 | 5.96 | 5.59 | 5.37 | 5.24 | 5.15 | 5.10 | 5.08 | 5.06 | 5.06 | 5.07 |
| 1.4 | 1.0 | 5.86 | 5.35 | 5.12 | 5.00 | 4.95 | 4.94 | 4.96 | 4.99 | 5.03 | 5.09 | 5.15 | 6.79 | 6.10 | 5.76 | 5.58 | 5.48 | 5.43 | 5.41 | 5.41 | 5.44 | 5.47 | 5.51 |
| 1.4 | 1.2 | 6.02 | 5.55 | 5.36 | 5.29 | 5.28 | 5.31 | 5.37 | 5.44 | 5.52 | 5.61 | 5.71 | 6.93 | 6.28 | 5.98 | 5.84 | 5.78 | 5.76 | 5.79 | 5.83 | 5.89 | 5.95 | 6.03 |

# APPENDIX E  EFFECTIVE LENGTH FACTOR OF COLUMNS

## E.6 The effective length factor $\mu_3$ of the lower section of a double-step column with a moveable upper end without rotation

The effective length factor $\mu_3$ of the lower section of the double-step column with the moveable upper end without rotation (Figure E.4) should be calculated according to the following formula or taken according to Table E.6.

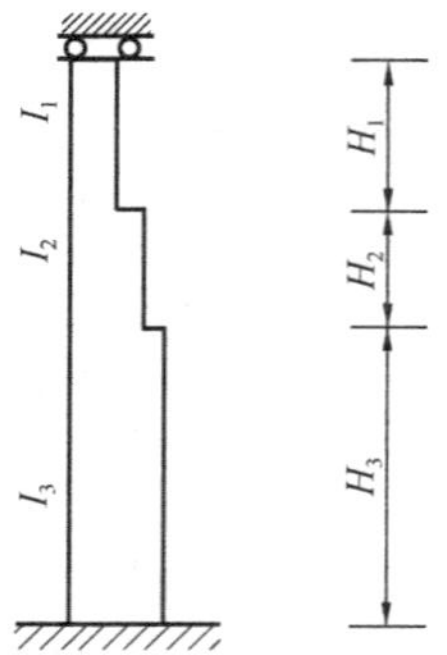

$$K_1 = \frac{I_1}{I_3} \cdot \frac{H_3}{H_1} \quad \text{(E-16)}$$

$$K_2 = \frac{I_2}{I_3} \cdot \frac{H_3}{H_2} \quad \text{(E-17)}$$

$$\eta_1 = \frac{H_1}{H_3}\sqrt{\frac{N_1}{N_3} \cdot \frac{I_3}{I_1}} \quad \text{(E-18)}$$

$$\eta_2 = \frac{H_2}{H_3}\sqrt{\frac{N_2}{N_3} \cdot \frac{I_3}{I_2}} \quad \text{(E-19)}$$

Figure E.4  A double-step column

**Effective length factor $\mu_3$ of the lower section of a double-step column with a movable upper end without rotating**   Table E.6

| $\eta_1$ | $\eta_2$ | $K_1$ \ $K_2$ | 0.2 | 0.3 | 0.4 | 0.5 | 0.6 | 0.7 | 0.8 | 0.9 | 1.0 | 1.1 | 1.2 | 0.2 | 0.3 | 0.4 | 0.5 | 0.6 | 0.7 | 0.8 | 0.9 | 1.0 | 1.1 | 1.2 |
|---|---|---|---|---|---|---|---|---|---|---|---|---|---|---|---|---|---|---|---|---|---|---|---|---|
| | | | 0.05 | | | | | | | | | | | 0.10 | | | | | | | | | | |
| 0.2 | | 0.2 | 1.99 | 1.99 | 2.00 | 2.00 | 2.01 | 2.02 | 2.02 | 2.03 | 2.04 | 2.05 | 2.06 | 1.96 | 1.96 | 1.97 | 1.97 | 1.98 | 1.98 | 1.99 | 2.00 | 2.00 | 2.01 | 2.02 |
| | | 0.4 | 2.03 | 2.06 | 2.09 | 2.12 | 2.16 | 2.19 | 2.22 | 2.25 | 2.29 | 2.32 | 2.35 | 2.00 | 2.02 | 2.05 | 2.08 | 2.11 | 2.14 | 2.17 | 2.20 | 2.23 | 2.26 | 2.29 |
| | | 0.6 | 2.12 | 2.20 | 2.28 | 2.36 | 2.43 | 2.50 | 2.57 | 2.64 | 2.71 | 2.77 | 2.83 | 2.07 | 2.14 | 2.22 | 2.29 | 2.36 | 2.43 | 2.50 | 2.56 | 2.63 | 2.69 | 2.75 |
| | | 0.8 | 2.28 | 2.43 | 2.57 | 2.70 | 2.82 | 2.94 | 3.04 | 3.15 | 3.25 | 3.34 | 3.43 | 2.20 | 2.35 | 2.48 | 2.61 | 2.73 | 2.84 | 2.94 | 3.05 | 3.14 | 3.24 | 3.33 |
| | | 1.0 | 2.53 | 2.76 | 2.96 | 3.13 | 3.29 | 3.44 | 3.59 | 3.72 | 3.85 | 3.98 | 4.10 | 2.41 | 2.64 | 2.83 | 3.01 | 3.17 | 3.32 | 3.46 | 3.59 | 3.72 | 3.85 | 3.97 |
| | | 1.2 | 2.86 | 3.15 | 3.39 | 3.61 | 3.80 | 3.99 | 4.16 | 4.33 | 4.49 | 4.64 | 4.79 | 2.70 | 2.99 | 3.23 | 3.45 | 3.65 | 3.84 | 4.01 | 4.18 | 4.34 | 4.49 | 4.64 |
| 0.4 | | 0.2 | 1.99 | 1.99 | 2.00 | 2.01 | 2.01 | 2.02 | 2.03 | 2.04 | 2.04 | 2.05 | 2.06 | 1.96 | 1.97 | 1.97 | 1.98 | 1.98 | 1.99 | 2.00 | 2.00 | 2.01 | 2.02 | 2.03 |
| | | 0.4 | 2.03 | 2.06 | 2.09 | 2.13 | 2.16 | 2.19 | 2.23 | 2.26 | 2.29 | 2.32 | 2.35 | 2.00 | 2.03 | 2.06 | 2.09 | 2.12 | 2.15 | 2.18 | 2.21 | 2.24 | 2.27 | 2.30 |
| | | 0.6 | 2.12 | 2.20 | 2.28 | 2.36 | 2.44 | 2.51 | 2.58 | 2.64 | 2.71 | 2.77 | 2.84 | 2.08 | 2.15 | 2.23 | 2.30 | 2.37 | 2.44 | 2.51 | 2.57 | 2.64 | 2.70 | 2.76 |
| | | 0.8 | 2.29 | 2.44 | 2.58 | 2.71 | 2.83 | 2.94 | 3.05 | 3.15 | 3.25 | 3.35 | 3.44 | 2.21 | 2.36 | 2.49 | 2.62 | 2.73 | 2.85 | 2.95 | 3.05 | 3.15 | 3.24 | 3.34 |
| | | 1.0 | 2.54 | 2.77 | 2.96 | 3.14 | 3.30 | 3.45 | 3.59 | 3.73 | 3.85 | 3.98 | 4.10 | 2.43 | 2.65 | 2.84 | 3.02 | 3.18 | 3.33 | 3.47 | 3.60 | 3.73 | 3.85 | 3.97 |
| | | 1.2 | 2.87 | 3.15 | 3.40 | 3.61 | 3.81 | 3.99 | 4.17 | 4.33 | 4.49 | 4.65 | 4.79 | 2.71 | 3.00 | 3.24 | 3.46 | 3.66 | 3.85 | 4.02 | 4.19 | 4.34 | 4.49 | 4.64 |
| 0.6 | | 0.2 | 1.99 | 1.98 | 2.00 | 2.01 | 2.02 | 2.03 | 2.04 | 2.04 | 2.05 | 2.06 | 2.07 | 1.97 | 1.98 | 1.98 | 1.99 | 2.00 | 2.00 | 2.01 | 2.02 | 2.02 | 2.03 | 2.04 |
| | | 0.4 | 2.04 | 2.07 | 2.10 | 2.14 | 2.17 | 2.20 | 2.23 | 2.27 | 2.30 | 2.33 | 2.36 | 2.01 | 2.04 | 2.07 | 2.10 | 2.13 | 2.16 | 2.19 | 2.22 | 2.26 | 2.29 | 2.32 |
| | | 0.6 | 2.13 | 2.21 | 2.29 | 2.37 | 2.45 | 2.52 | 2.59 | 2.65 | 2.72 | 2.78 | 2.84 | 2.09 | 2.17 | 2.24 | 2.32 | 2.39 | 2.46 | 2.52 | 2.59 | 2.65 | 2.71 | 2.77 |
| | | 0.8 | 2.30 | 2.45 | 2.59 | 2.72 | 2.84 | 2.95 | 3.06 | 3.16 | 3.26 | 3.35 | 3.44 | 2.23 | 2.38 | 2.51 | 2.64 | 2.75 | 2.86 | 2.97 | 3.07 | 3.16 | 3.26 | 3.35 |
| | | 1.0 | 2.56 | 2.78 | 2.97 | 3.15 | 3.31 | 3.46 | 3.60 | 3.73 | 3.86 | 3.99 | 4.11 | 2.45 | 2.68 | 2.86 | 3.03 | 3.19 | 3.34 | 3.48 | 3.61 | 3.74 | 3.86 | 3.98 |
| | | 1.2 | 2.89 | 3.17 | 3.41 | 3.62 | 3.82 | 4.00 | 4.17 | 4.34 | 4.50 | 4.65 | 4.80 | 2.74 | 3.02 | 3.26 | 3.48 | 3.67 | 3.86 | 4.03 | 4.20 | 4.35 | 4.50 | 4.65 |

## APPENDIX E  EFFECTIVE LENGTH FACTOR OF COLUMNS

continued

| $\eta_1$ | $\eta_2$ | $K_2$ | $K_1$=0.2 | 0.3 | 0.4 | 0.5 | 0.6 | 0.7 | 0.8 | 0.9 | 1.0 | 1.1 | 1.2 | $K_1$=0.2 | 0.3 | 0.4 | 0.5 | 0.6 | 0.7 | 0.8 | 0.9 | 1.0 | 1.1 | 1.2 |
|---|---|---|---|---|---|---|---|---|---|---|---|---|---|---|---|---|---|---|---|---|---|---|---|---|
|  |  |  | 0.05 |  |  |  |  |  |  |  |  |  |  | 0.10 |  |  |  |  |  |  |  |  |  |  |
| 0.8 |  | 0.2 | 2.00 | 2.01 | 2.02 | 2.02 | 2.03 | 2.04 | 2.05 | 2.05 | 2.06 | 2.07 | 2.08 | 1.99 | 1.99 | 2.00 | 2.01 | 2.01 | 2.02 | 2.03 | 2.04 | 2.04 | 2.05 | 2.06 |
|  |  | 0.4 | 2.05 | 2.08 | 2.12 | 2.15 | 2.18 | 2.21 | 2.25 | 2.28 | 2.31 | 2.34 | 2.37 | 2.03 | 2.06 | 2.09 | 2.12 | 2.15 | 2.19 | 2.22 | 2.25 | 2.28 | 2.31 | 2.34 |
|  |  | 0.6 | 2.15 | 2.23 | 2.31 | 2.39 | 2.46 | 2.53 | 2.60 | 2.67 | 2.73 | 2.79 | 2.85 | 2.12 | 2.19 | 2.27 | 2.34 | 2.41 | 2.48 | 2.55 | 2.61 | 2.67 | 2.73 | 2.79 |
|  |  | 0.8 | 2.32 | 2.47 | 2.61 | 2.73 | 2.85 | 2.96 | 3.07 | 3.17 | 3.27 | 3.36 | 3.45 | 2.27 | 2.41 | 2.54 | 2.66 | 2.78 | 2.89 | 2.99 | 3.09 | 3.18 | 3.28 | 3.37 |
|  |  | 1.0 | 2.59 | 2.80 | 2.99 | 3.16 | 3.32 | 3.47 | 3.61 | 3.74 | 3.87 | 3.99 | 4.11 | 2.49 | 2.70 | 2.89 | 3.06 | 3.21 | 3.36 | 3.50 | 3.63 | 3.76 | 3.88 | 4.00 |
|  |  | 1.2 | 2.92 | 3.19 | 3.42 | 3.63 | 3.83 | 4.01 | 4.18 | 4.35 | 4.51 | 4.66 | 4.81 | 2.78 | 3.05 | 3.29 | 3.50 | 3.69 | 3.88 | 4.05 | 4.21 | 4.37 | 4.52 | 4.66 |
| 1.0 |  | 0.2 | 2.02 | 2.02 | 2.03 | 2.04 | 2.05 | 2.05 | 2.06 | 2.07 | 2.08 | 2.09 | 2.09 | 2.01 | 2.02 | 2.03 | 2.04 | 2.04 | 2.05 | 2.06 | 2.07 | 2.08 | 2.08 | 2.09 |
|  |  | 0.4 | 2.07 | 2.10 | 2.14 | 2.17 | 2.20 | 2.23 | 2.26 | 2.30 | 2.33 | 2.36 | 2.39 | 2.06 | 2.10 | 2.13 | 2.16 | 2.19 | 2.22 | 2.25 | 2.28 | 2.31 | 2.34 | 2.37 |
|  |  | 0.6 | 2.17 | 2.26 | 2.33 | 2.41 | 2.48 | 2.55 | 2.62 | 2.68 | 2.75 | 2.81 | 2.87 | 2.16 | 2.24 | 2.31 | 2.38 | 2.45 | 2.51 | 2.58 | 2.64 | 2.70 | 2.76 | 2.82 |
|  |  | 0.8 | 2.36 | 2.50 | 2.63 | 2.76 | 2.87 | 2.98 | 3.08 | 3.19 | 3.28 | 3.38 | 3.47 | 2.32 | 2.46 | 2.58 | 2.70 | 2.81 | 2.92 | 3.02 | 3.12 | 3.21 | 3.30 | 3.39 |
|  |  | 1.0 | 2.62 | 2.83 | 3.01 | 3.18 | 3.34 | 3.48 | 3.62 | 3.75 | 3.88 | 4.01 | 4.12 | 2.55 | 2.75 | 2.93 | 3.09 | 3.25 | 3.39 | 3.53 | 3.66 | 3.78 | 3.90 | 4.02 |
|  |  | 1.2 | 2.95 | 3.21 | 3.44 | 3.65 | 3.82 | 4.02 | 4.20 | 4.36 | 4.52 | 4.67 | 4.81 | 2.84 | 3.10 | 3.32 | 3.53 | 3.72 | 3.90 | 4.07 | 4.23 | 4.39 | 4.54 | 4.68 |
| 1.2 |  | 0.2 | 2.04 | 2.05 | 2.06 | 2.06 | 2.07 | 2.08 | 2.09 | 2.09 | 2.10 | 2.11 | 2.12 | 2.07 | 2.08 | 2.08 | 2.09 | 2.09 | 2.10 | 2.11 | 2.11 | 2.12 | 2.13 | 2.13 |
|  |  | 0.4 | 2.10 | 2.13 | 2.17 | 2.20 | 2.23 | 2.26 | 2.29 | 2.32 | 2.35 | 2.38 | 2.41 | 2.13 | 2.16 | 2.18 | 2.21 | 2.24 | 2.27 | 2.30 | 2.33 | 2.35 | 2.38 | 2.41 |
|  |  | 0.6 | 2.22 | 2.29 | 2.37 | 2.44 | 2.51 | 2.58 | 2.64 | 2.71 | 2.77 | 2.83 | 2.89 | 2.24 | 2.30 | 2.37 | 2.43 | 2.50 | 2.56 | 2.63 | 2.68 | 2.74 | 2.80 | 2.86 |
|  |  | 0.8 | 2.41 | 2.54 | 2.67 | 2.78 | 2.90 | 3.00 | 3.11 | 3.20 | 3.30 | 3.39 | 3.48 | 2.41 | 2.53 | 2.64 | 2.75 | 2.86 | 2.96 | 3.06 | 3.15 | 3.24 | 3.33 | 3.42 |
|  |  | 1.0 | 2.68 | 2.87 | 3.04 | 3.21 | 3.36 | 3.50 | 3.64 | 3.77 | 3.90 | 4.02 | 4.14 | 2.64 | 2.82 | 2.98 | 3.14 | 3.29 | 3.43 | 3.56 | 3.69 | 3.81 | 3.93 | 4.04 |
|  |  | 1.2 | 3.00 | 3.25 | 3.47 | 3.67 | 3.86 | 4.04 | 4.21 | 4.37 | 4.53 | 4.68 | 4.83 | 2.92 | 3.16 | 3.37 | 3.57 | 3.76 | 3.93 | 4.10 | 4.26 | 4.41 | 4.56 | 4.70 |
| 1.4 |  | 0.2 | 2.10 | 2.10 | 2.10 | 2.11 | 2.11 | 2.12 | 2.13 | 2.13 | 2.14 | 2.15 | 2.15 | 2.20 | 2.18 | 2.17 | 2.17 | 2.17 | 2.18 | 2.18 | 2.19 | 2.19 | 2.20 | 2.20 |
|  |  | 0.4 | 2.17 | 2.19 | 2.21 | 2.24 | 2.27 | 2.30 | 2.33 | 2.36 | 2.39 | 2.41 | 2.44 | 2.26 | 2.26 | 2.27 | 2.29 | 2.32 | 2.34 | 2.37 | 2.39 | 2.42 | 2.44 | 2.47 |
|  |  | 0.6 | 2.29 | 2.35 | 2.41 | 2.48 | 2.55 | 2.61 | 2.67 | 2.74 | 2.80 | 2.86 | 2.91 | 2.37 | 2.41 | 2.46 | 2.51 | 2.57 | 2.63 | 2.68 | 2.74 | 2.80 | 2.85 | 2.91 |
|  |  | 0.8 | 2.48 | 2.60 | 2.71 | 2.82 | 2.93 | 3.03 | 3.13 | 3.23 | 3.32 | 3.41 | 3.50 | 2.53 | 2.62 | 2.72 | 2.82 | 2.92 | 3.01 | 3.11 | 3.20 | 3.29 | 3.37 | 3.46 |
|  |  | 1.0 | 2.74 | 2.92 | 3.08 | 3.24 | 3.39 | 3.53 | 3.66 | 3.79 | 3.92 | 4.04 | 4.15 | 2.75 | 2.90 | 3.05 | 3.20 | 3.34 | 3.47 | 3.60 | 3.72 | 3.84 | 3.96 | 4.07 |
|  |  | 1.2 | 3.06 | 3.29 | 3.50 | 3.70 | 3.89 | 4.06 | 4.23 | 4.39 | 4.55 | 4.70 | 4.84 | 3.02 | 3.23 | 3.43 | 3.62 | 3.80 | 3.97 | 4.13 | 4.29 | 4.44 | 4.59 | 4.73 |

# APPENDIX E  EFFECTIVE LENGTH FACTOR OF COLUMNS

*continued*

| $\eta_1$ | $\eta_2$ (K$_1$) | K$_2$ | 0.20 | | | | | | | | | | | 0.30 | | | | | | | | | | |
|---|---|---|---|---|---|---|---|---|---|---|---|---|---|---|---|---|---|---|---|---|---|---|---|---|
| | | | 0.2 | 0.3 | 0.4 | 0.5 | 0.6 | 0.7 | 0.8 | 0.9 | 1.0 | 1.1 | 1.2 | 0.2 | 0.3 | 0.4 | 0.5 | 0.6 | 0.7 | 0.8 | 0.9 | 1.0 | 1.1 | 1.2 |
| 0.2 | 0.2 | | 1.94 | 1.93 | 1.93 | 1.93 | 1.93 | 1.93 | 1.94 | 1.94 | 1.95 | 1.95 | 1.96 | 1.92 | 1.91 | 1.90 | 1.89 | 1.89 | 1.89 | 1.90 | 1.90 | 1.90 | 1.90 | 1.91 |
| | 0.4 | | 1.96 | 1.98 | 1.99 | 2.02 | 2.04 | 2.07 | 2.09 | 2.12 | 2.15 | 2.17 | 2.20 | 1.95 | 1.95 | 1.96 | 1.97 | 1.99 | 2.01 | 2.04 | 2.06 | 2.08 | 2.11 | 2.13 |
| | 0.6 | | 2.02 | 2.07 | 2.13 | 2.19 | 2.26 | 2.32 | 2.38 | 2.44 | 2.50 | 2.56 | 2.62 | 1.99 | 2.03 | 2.08 | 2.13 | 2.18 | 2.24 | 2.29 | 2.35 | 2.41 | 2.46 | 2.52 |
| | 0.8 | | 2.12 | 2.23 | 2.35 | 2.47 | 2.58 | 2.68 | 2.78 | 2.88 | 2.98 | 3.07 | 3.15 | 2.07 | 2.16 | 2.27 | 2.37 | 2.47 | 2.57 | 2.66 | 2.75 | 2.84 | 2.93 | 3.01 |
| | 1.0 | | 2.28 | 2.47 | 2.65 | 2.82 | 2.97 | 3.12 | 3.26 | 3.39 | 3.51 | 3.63 | 3.75 | 2.20 | 2.37 | 2.53 | 2.69 | 2.83 | 2.97 | 3.10 | 3.23 | 3.35 | 3.46 | 3.57 |
| | 1.2 | | 2.50 | 2.77 | 3.01 | 3.22 | 3.42 | 3.60 | 3.77 | 3.93 | 4.09 | 4.23 | 4.38 | 2.39 | 2.63 | 2.85 | 3.05 | 3.24 | 3.42 | 3.58 | 3.74 | 3.89 | 4.03 | 4.17 |
| 0.4 | 0.2 | | 1.93 | 1.93 | 1.93 | 1.93 | 1.94 | 1.94 | 1.95 | 1.95 | 1.96 | 1.96 | 1.97 | 1.92 | 1.91 | 1.91 | 1.90 | 1.90 | 1.91 | 1.91 | 1.91 | 1.92 | 1.92 | 1.92 |
| | 0.4 | | 1.97 | 1.98 | 2.00 | 2.03 | 2.05 | 2.08 | 2.11 | 2.13 | 2.16 | 2.19 | 2.22 | 1.95 | 1.96 | 1.97 | 1.99 | 2.01 | 2.03 | 2.05 | 2.08 | 2.10 | 2.12 | 2.15 |
| | 0.6 | | 2.03 | 2.08 | 2.14 | 2.21 | 2.27 | 2.33 | 2.40 | 2.46 | 2.52 | 2.58 | 2.63 | 2.00 | 2.04 | 2.09 | 2.14 | 2.20 | 2.26 | 2.31 | 2.37 | 2.42 | 2.48 | 2.53 |
| | 0.8 | | 2.13 | 2.25 | 2.37 | 2.48 | 2.59 | 2.70 | 2.80 | 2.90 | 2.99 | 3.08 | 3.17 | 2.08 | 2.18 | 2.28 | 2.39 | 2.49 | 2.59 | 2.68 | 2.77 | 2.86 | 2.95 | 3.03 |
| | 1.0 | | 2.29 | 2.49 | 2.67 | 2.83 | 2.99 | 3.13 | 3.27 | 3.40 | 3.53 | 3.64 | 3.76 | 2.22 | 2.39 | 2.55 | 2.71 | 2.85 | 2.99 | 3.12 | 3.24 | 3.36 | 3.48 | 3.59 |
| | 1.2 | | 2.52 | 2.79 | 3.02 | 3.23 | 3.43 | 3.61 | 3.78 | 3.94 | 4.10 | 4.24 | 4.39 | 2.41 | 2.65 | 2.87 | 3.07 | 3.26 | 3.43 | 3.60 | 3.75 | 3.90 | 4.04 | 4.18 |
| 0.6 | 0.2 | | 1.95 | 1.95 | 1.95 | 1.95 | 1.96 | 1.96 | 1.97 | 1.97 | 1.98 | 1.98 | 1.99 | 1.93 | 1.93 | 1.92 | 1.92 | 1.93 | 1.93 | 1.93 | 1.94 | 1.94 | 1.95 | 1.95 |
| | 0.4 | | 1.98 | 2.00 | 2.02 | 2.05 | 2.08 | 2.10 | 2.13 | 2.16 | 2.19 | 2.21 | 2.24 | 1.96 | 1.97 | 1.99 | 2.01 | 2.03 | 2.06 | 2.08 | 2.11 | 2.13 | 2.16 | 2.18 |
| | 0.6 | | 2.04 | 2.10 | 2.17 | 2.23 | 2.30 | 2.36 | 2.42 | 2.48 | 2.54 | 2.60 | 2.66 | 2.02 | 2.06 | 2.12 | 2.17 | 2.23 | 2.29 | 2.35 | 2.40 | 2.46 | 2.51 | 2.57 |
| | 0.8 | | 2.15 | 2.27 | 2.39 | 2.51 | 2.62 | 2.72 | 2.82 | 2.92 | 3.01 | 3.10 | 3.19 | 2.11 | 2.21 | 2.32 | 2.42 | 2.52 | 2.62 | 2.71 | 2.80 | 2.89 | 2.98 | 3.06 |
| | 1.0 | | 2.32 | 2.52 | 2.70 | 2.86 | 3.01 | 3.16 | 3.29 | 3.42 | 3.55 | 3.66 | 3.78 | 2.25 | 2.42 | 2.59 | 2.74 | 2.88 | 3.02 | 3.15 | 3.27 | 3.39 | 3.50 | 3.61 |
| | 1.2 | | 2.55 | 2.82 | 3.05 | 3.26 | 3.45 | 3.63 | 3.80 | 3.96 | 4.11 | 4.26 | 4.40 | 2.44 | 2.69 | 2.91 | 3.11 | 3.29 | 3.46 | 3.62 | 3.78 | 3.93 | 4.07 | 4.20 |

APPENDIX E  EFFECTIVE LENGTH FACTOR OF COLUMNS

*continued*

| $\eta_1$ | $K_1$ $K_2$ $\eta_2$ | 0.20 | | | | | | | | | | | 0.30 | | | | | | | | | | | |
|---|---|---|---|---|---|---|---|---|---|---|---|---|---|---|---|---|---|---|---|---|---|---|---|---|
| | | 0.2 | 0.3 | 0.4 | 0.5 | 0.6 | 0.7 | 0.8 | 0.9 | 1.0 | 1.1 | 1.2 | 0.2 | 0.3 | 0.4 | 0.5 | 0.6 | 0.7 | 0.8 | 0.9 | 1.0 | 1.1 | 1.2 |
| 0.8 | 0.2 | 1.97 | 1.97 | 1.98 | 1.98 | 1.99 | 1.99 | 2.00 | 2.01 | 2.01 | 2.02 | 2.03 | 1.96 | 1.95 | 1.96 | 1.96 | 1.97 | 1.97 | 1.98 | 1.98 | 1.99 | 1.99 | 2.00 |
| | 0.4 | 2.00 | 2.03 | 2.06 | 2.08 | 2.11 | 2.14 | 2.17 | 2.20 | 2.22 | 2.25 | 2.28 | 1.99 | 2.01 | 2.03 | 2.05 | 2.08 | 2.10 | 2.13 | 2.15 | 2.18 | 2.21 | 2.23 |
| | 0.6 | 2.08 | 2.14 | 2.21 | 2.27 | 2.34 | 2.40 | 2.46 | 2.52 | 2.58 | 2.64 | 2.69 | 2.05 | 2.10 | 2.16 | 2.22 | 2.28 | 2.34 | 2.40 | 2.45 | 2.51 | 2.56 | 2.81 |
| | 0.8 | 2.19 | 2.32 | 2.44 | 2.55 | 2.66 | 2.76 | 2.86 | 2.96 | 3.05 | 3.13 | 3.22 | 2.15 | 2.26 | 2.37 | 2.47 | 2.57 | 2.67 | 2.76 | 2.85 | 2.94 | 3.02 | 3.10 |
| | 1.0 | 2.37 | 2.57 | 2.74 | 2.90 | 3.05 | 3.19 | 3.33 | 3.45 | 3.58 | 3.69 | 3.81 | 2.30 | 2.48 | 2.64 | 2.79 | 2.93 | 3.07 | 3.19 | 3.31 | 3.43 | 3.54 | 3.65 |
| | 1.2 | 2.61 | 2.87 | 3.09 | 3.30 | 3.49 | 3.66 | 3.83 | 3.99 | 4.14 | 4.29 | 4.42 | 2.50 | 2.74 | 2.96 | 3.15 | 3.33 | 3.50 | 3.66 | 3.81 | 3.96 | 4.10 | 4.23 |
| 1.0 | 0.2 | 2.01 | 2.02 | 2.03 | 2.03 | 2.04 | 2.05 | 2.05 | 2.06 | 2.07 | 2.07 | 2.08 | 2.01 | 2.02 | 2.02 | 2.03 | 2.04 | 2.04 | 2.05 | 2.06 | 2.06 | 2.07 | 2.07 |
| | 0.4 | 2.06 | 2.09 | 2.11 | 2.14 | 2.17 | 2.20 | 2.23 | 2.25 | 2.28 | 2.31 | 2.33 | 2.05 | 2.08 | 2.10 | 2.13 | 2.16 | 2.18 | 2.21 | 2.23 | 2.26 | 2.28 | 2.31 |
| | 0.6 | 2.14 | 2.21 | 2.27 | 2.34 | 2.40 | 2.46 | 2.52 | 2.58 | 2.63 | 2.69 | 2.74 | 2.13 | 2.19 | 2.25 | 2.30 | 2.36 | 2.42 | 2.47 | 2.53 | 2.58 | 2.63 | 2.68 |
| | 0.8 | 2.27 | 2.39 | 2.51 | 2.62 | 2.72 | 2.82 | 2.91 | 3.00 | 3.09 | 3.18 | 3.26 | 2.24 | 2.35 | 2.45 | 2.55 | 2.65 | 2.74 | 2.83 | 2.92 | 3.00 | 3.08 | 3.16 |
| | 1.0 | 2.46 | 2.64 | 2.81 | 2.96 | 3.10 | 3.24 | 3.37 | 3.50 | 3.61 | 3.73 | 3.84 | 2.40 | 2.57 | 2.72 | 2.86 | 3.00 | 3.13 | 3.25 | 3.37 | 3.48 | 3.59 | 3.70 |
| | 1.2 | 2.69 | 2.94 | 3.15 | 3.35 | 3.53 | 3.71 | 3.87 | 4.02 | 4.17 | 4.32 | 4.46 | 2.60 | 2.83 | 3.03 | 3.22 | 3.39 | 3.56 | 3.71 | 3.86 | 4.01 | 4.14 | 4.28 |
| 1.2 | 0.2 | 2.13 | 2.12 | 2.12 | 2.13 | 2.13 | 2.14 | 2.14 | 2.15 | 2.15 | 2.16 | 2.16 | 2.17 | 2.16 | 2.16 | 2.16 | 2.16 | 2.16 | 2.17 | 2.17 | 2.18 | 2.18 | 2.19 |
| | 0.4 | 2.18 | 2.19 | 2.21 | 2.24 | 2.26 | 2.29 | 2.31 | 2.34 | 2.36 | 2.38 | 2.41 | 2.22 | 2.22 | 2.24 | 2.26 | 2.28 | 2.30 | 2.32 | 2.34 | 2.36 | 2.39 | 2.41 |
| | 0.6 | 2.27 | 2.32 | 2.37 | 2.43 | 2.49 | 2.54 | 2.60 | 2.65 | 2.70 | 2.76 | 2.81 | 2.29 | 2.33 | 2.38 | 2.43 | 2.48 | 2.53 | 2.58 | 2.62 | 2.67 | 2.72 | 2.77 |
| | 0.8 | 2.41 | 2.50 | 2.60 | 2.70 | 2.80 | 2.89 | 2.98 | 3.07 | 3.15 | 3.23 | 3.32 | 2.41 | 2.49 | 2.58 | 2.67 | 2.75 | 2.84 | 2.92 | 3.00 | 3.08 | 3.16 | 3.23 |
| | 1.0 | 2.59 | 2.74 | 2.89 | 3.04 | 3.17 | 3.30 | 3.43 | 3.55 | 3.66 | 3.78 | 3.89 | 2.56 | 2.69 | 2.83 | 2.96 | 3.09 | 3.21 | 3.33 | 3.44 | 3.55 | 3.66 | 3.76 |
| | 1.2 | 2.81 | 3.03 | 3.23 | 3.42 | 3.59 | 3.76 | 3.92 | 4.07 | 4.22 | 4.36 | 4.49 | 2.74 | 2.94 | 3.13 | 3.30 | 3.47 | 3.63 | 3.78 | 3.92 | 4.06 | 4.20 | 4.33 |
| 1.4 | 0.2 | 2.35 | 2.31 | 2.29 | 2.28 | 2.27 | 2.27 | 2.27 | 2.27 | 2.27 | 2.28 | 2.28 | 2.45 | 2.40 | 2.37 | 2.35 | 2.35 | 2.34 | 2.34 | 2.34 | 2.34 | 2.34 | 2.34 |
| | 0.4 | 2.40 | 2.37 | 2.37 | 2.38 | 2.39 | 2.41 | 2.43 | 2.45 | 2.47 | 2.49 | 2.51 | 2.48 | 2.45 | 2.44 | 2.44 | 2.45 | 2.46 | 2.48 | 2.49 | 2.51 | 2.53 | 2.55 |
| | 0.6 | 2.48 | 2.49 | 2.52 | 2.56 | 2.61 | 2.65 | 2.70 | 2.75 | 2.80 | 2.85 | 2.89 | 2.55 | 2.54 | 2.56 | 2.60 | 2.63 | 2.67 | 2.71 | 2.75 | 2.80 | 2.84 | 2.88 |
| | 0.8 | 2.60 | 2.66 | 2.73 | 2.82 | 2.90 | 2.98 | 3.07 | 3.15 | 3.23 | 3.31 | 3.38 | 2.64 | 2.68 | 2.74 | 2.81 | 2.89 | 2.96 | 3.04 | 3.11 | 3.18 | 3.25 | 3.33 |
| | 1.0 | 2.77 | 2.88 | 3.01 | 3.14 | 3.26 | 3.38 | 3.50 | 3.62 | 3.73 | 3.84 | 3.94 | 2.77 | 2.87 | 2.98 | 3.09 | 3.20 | 3.32 | 3.43 | 3.53 | 3.64 | 3.74 | 3.84 |
| | 1.2 | 2.97 | 3.15 | 3.33 | 3.50 | 3.67 | 3.83 | 3.98 | 4.13 | 4.27 | 4.41 | 4.54 | 2.94 | 3.09 | 3.26 | 3.41 | 3.57 | 3.72 | 3.86 | 4.00 | 4.13 | 4.26 | 4.39 |

# APPENDIX F  TYPICAL STEEL SHAPES

## Universal I-shape                                                          Table F.1

Symbol: $h$—Depth
$b$—Flange width
$t_w$—Web thickness
$t$—Mean flange thickness

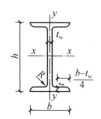

$I$—Moment of inertia
$W$—Section modulus
$i$—Radius of gyration
$S_x$—Area moment of half-section
$R$—Fillet radius

| Type | Section dimension (mm) | | | | | | Sectional area ($cm^2$) | Theoretical weight (kg/m) | Moment of inertia ($cm^4$) | | The radius of inertia (cm) | | Section modulus ($cm^3$) | |
|---|---|---|---|---|---|---|---|---|---|---|---|---|---|---|
| | $h$ | $b$ | $t_w$ | $t$ | $R$ | $r_1$ | | | $I_x$ | $I_y$ | $i_x$ | $i_y$ | $W_x$ | $W_y$ |
| 10 | 100 | 68 | 4.5 | 7.6 | 6.5 | 3.3 | 14.345 | 11.261 | 245 | 33.0 | 4.14 | 1.52 | 49.0 | 9.72 |
| 12 | 120 | 74 | 5.0 | 8.4 | 7.0 | 3.5 | 17.818 | 13.987 | 436 | 46.9 | 4.95 | 1.62 | 72.7 | 12.7 |
| 12.6 | 126 | 74 | 5.0 | 8.4 | 7.0 | 3.5 | 18.118 | 14.223 | 488 | 46.9 | 5.20 | 1.61 | 77.5 | 12.7 |
| 14 | 140 | 80 | 5.5 | 9.1 | 7.5 | 3.8 | 21.516 | 16.890 | 712 | 64.4 | 5.76 | 1.73 | 102 | 16.1 |
| 16 | 160 | 88 | 6.0 | 9.9 | 8.0 | 4.0 | 26.131 | 20.513 | 1130 | 93.1 | 6.58 | 1.89 | 141 | 21.2 |
| 18 | 180 | 94 | 6.5 | 10.7 | 8.5 | 4.3 | 30.756 | 24.143 | 1660 | 122 | 7.36 | 2.00 | 185 | 26.0 |
| 20a | 200 | 100 | 7.0 | 11.4 | 9.0 | 4.5 | 35.578 | 27.929 | 2370 | 158 | 8.15 | 2.12 | 237 | 31.5 |
| 20b | 200 | 102 | 9.0 | 11.4 | 9.0 | 4.5 | 39.578 | 31.069 | 2500 | 169 | 7.96 | 2.06 | 250 | 33.1 |
| 22a | 220 | 110 | 7.5 | 12.3 | 9.5 | 4.8 | 42.128 | 33.070 | 3400 | 225 | 8.99 | 2.31 | 309 | 40.9 |
| 22b | 220 | 112 | 9.5 | 12.3 | 9.5 | 4.8 | 46.528 | 36.524 | 3570 | 239 | 8.78 | 2.27 | 325 | 42.7 |
| 24a | 240 | 116 | 8.0 | 13.0 | 10.0 | 5.0 | 47.741 | 37.477 | 4570 | 280 | 9.77 | 2.42 | 381 | 48.4 |
| 24b | 240 | 118 | 10.0 | 13.0 | 10.0 | 5.0 | 52.541 | 41.245 | 4800 | 297 | 9.57 | 2.38 | 400 | 50.4 |
| 25a | 250 | 116 | 8.0 | 13.0 | 10.0 | 5.0 | 48.541 | 38.105 | 5020 | 280 | 10.2 | 2.40 | 402 | 48.3 |
| 25b | 250 | 118 | 10.0 | 13.0 | 10.0 | 5.0 | 53.541 | 42.030 | 5280 | 309 | 9.94 | 2.40 | 423 | 52.4 |
| 27a | 270 | 122 | 8.5 | 13.7 | 10.5 | 5.3 | 54.554 | 42.825 | 6550 | 345 | 10.9 | 2.51 | 485 | 56.6 |
| 27b | 270 | 124 | 10.5 | 13.7 | 10.5 | 5.3 | 59.954 | 47.064 | 6870 | 366 | 10.7 | 2.47 | 509 | 58.9 |
| 28a | 280 | 122 | 8.5 | 13.7 | 10.5 | 5.3 | 55.404 | 43.492 | 7110 | 345 | 11.3 | 2.50 | 508 | 56.6 |
| 28b | 280 | 124 | 10.5 | 13.7 | 10.5 | 5.3 | 61.004 | 47.888 | 7480 | 379 | 11.1 | 2.49 | 534 | 61.2 |

APPENDIX F  TYPICAL STEEL SHAPES

continued

| Type | Section dimension(mm) | | | | | | Sectional area (cm²) | Theor-etical weight (kg/m) | Moment of inertia(cm⁴) | | The radius of inertia(cm) | | Section modulus (cm³) | |
|---|---|---|---|---|---|---|---|---|---|---|---|---|---|---|
| | $h$ | $b$ | $t_w$ | $t$ | $R$ | $r_1$ | | | $I_x$ | $I_y$ | $i_x$ | $i_y$ | $W_x$ | $W_y$ |
| 30a | 300 | 126 | 9.0 | 14.4 | 11.0 | 5.5 | 61.254 | 48.084 | 8950 | 400 | 12.1 | 2.55 | 597 | 63.5 |
| 30b | 300 | 128 | 11.0 | 14.4 | 11.0 | 5.5 | 67.254 | 52.794 | 9400 | 422 | 11.8 | 2.50 | 627 | 65.9 |
| 30c | 300 | 130 | 13.0 | 14.4 | 11.0 | 5.5 | 73.254 | 57.504 | 9850 | 445 | 11.6 | 2.46 | 657 | 68.5 |
| 32a | 320 | 130 | 9.5 | 15.0 | 11.5 | 5.8 | 67.156 | 52.717 | 11100 | 460 | 12.8 | 2.62 | 692 | 70.8 |
| 32b | 320 | 132 | 11.5 | 15.0 | 11.5 | 5.8 | 73.556 | 57.741 | 11600 | 502 | 12.6 | 2.61 | 726 | 76.0 |
| 32c | 320 | 134 | 13.5 | 15.0 | 11.5 | 5.8 | 79.956 | 62.765 | 12200 | 544 | 12.3 | 2.61 | 760 | 81.2 |
| 36a | 360 | 136 | 10.0 | 15.8 | 12.0 | 6.0 | 76.480 | 60.037 | 15800 | 552 | 14.4 | 2.69 | 875 | 81.2 |
| 36b | 360 | 138 | 12.0 | 15.8 | 12.0 | 6.0 | 83.680 | 65.689 | 16500 | 582 | 14.1 | 2.64 | 919 | 84.3 |
| 36c | 360 | 140 | 14.0 | 15.8 | 12.0 | 6.0 | 90.880 | 71.341 | 17300 | 612 | 13.8 | 2.60 | 962 | 87.4 |
| 40a | 400 | 142 | 10.5 | 16.5 | 12.5 | 6.3 | 86.112 | 67.598 | 21700 | 660 | 15.9 | 2.77 | 1090 | 93.2 |
| 40b | 400 | 144 | 12.5 | 16.5 | 12.5 | 6.3 | 94.112 | 73.878 | 22800 | 692 | 15.6 | 2.71 | 1140 | 96.2 |
| 40c | 400 | 146 | 14.5 | 16.5 | 12.5 | 6.3 | 102.112 | 80.158 | 23900 | 727 | 15.2 | 2.65 | 1190 | 99.6 |
| 45a | 450 | 150 | 11.5 | 18.0 | 13.5 | 6.8 | 102.446 | 80.420 | 32200 | 855 | 17.7 | 2.89 | 1430 | 114 |
| 45b | 450 | 152 | 13.5 | 18.0 | 13.5 | 6.8 | 111.446 | 87.485 | 33800 | 894 | 17.4 | 2.84 | 1500 | 118 |
| 45c | 450 | 154 | 15.5 | 18.0 | 13.5 | 6.8 | 120.446 | 94.550 | 35300 | 938 | 17.1 | 2.79 | 1570 | 122 |
| 50a | 500 | 158 | 12.0 | 20.0 | 14.0 | 7.0 | 119.304 | 93.654 | 46500 | 1120 | 19.7 | 3.07 | 1860 | 142 |
| 50b | 500 | 160 | 14.0 | 20.0 | 14.0 | 7.0 | 129.304 | 101.504 | 48600 | 1170 | 19.4 | 3.01 | 1940 | 146 |
| 50c | 500 | 162 | 16.0 | 20.0 | 14.0 | 7.0 | 139.304 | 109.354 | 50600 | 1220 | 19.0 | 2.96 | 2080 | 151 |
| 55a | 550 | 166 | 12.5 | 21.0 | 14.5 | 7.3 | 134.185 | 105.335 | 62900 | 1370 | 21.6 | 3.19 | 2290 | 164 |
| 55b | 550 | 168 | 14.5 | 21.0 | 14.5 | 7.3 | 145.185 | 113.970 | 65600 | 1420 | 21.2 | 3.14 | 2390 | 170 |
| 55c | 550 | 170 | 16.5 | 21.0 | 14.5 | 7.3 | 156.185 | 122.605 | 68400 | 1480 | 20.9 | 3.08 | 2490 | 175 |
| 56a | 560 | 166 | 12.5 | 21.0 | 14.5 | 7.3 | 135.435 | 106.316 | 65600 | 1370 | 22.0 | 3.18 | 2340 | 165 |
| 56b | 560 | 168 | 14.5 | 21.0 | 14.5 | 7.3 | 146.635 | 115.108 | 68500 | 1490 | 21.6 | 3.16 | 2450 | 174 |
| 56c | 560 | 170 | 16.5 | 21.0 | 14.5 | 7.3 | 157.835 | 123.900 | 71400 | 1560 | 21.3 | 3.16 | 2550 | 183 |
| 63a | 630 | 176 | 13.0 | 22.0 | 15.0 | 7.5 | 154.658 | 121.407 | 93900 | 1700 | 24.5 | 3.31 | 2980 | 193 |
| 63b | 630 | 178 | 15.0 | 22.0 | 15.0 | 7.5 | 167.258 | 131.298 | 98100 | 1810 | 24.2 | 3.29 | 3160 | 204 |
| 63c | 630 | 180 | 17.0 | 22.0 | 15.0 | 7.5 | 179.858 | 141.189 | 102000 | 1920 | 23.8 | 3.27 | 3300 | 214 |

Note: The data of $R$ and $r_1$ are used for design, not as delivery conditions.

# APPENDIX F  TYPICAL STEEL SHAPES

**H-shape**  Table F. 2

Symbol: $h$ — Depth
$b$ — Flange width
$t_1$ — Web thickness
$t_2$ — Flange thickness
$I$ — Moment of inertia
$W$ — Section modulus
$i$ — Radius of gyration
$S_x$ — Area moment of half-section

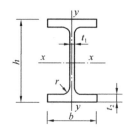

| Type | Depth×Width (mm×mm) | Section dimension (mm) | | | | | Sectional area (cm²) | Theoretical weight (kg/m) | Moment of inertia (cm⁴) | | Radius of gyration (cm) | | Section modulus (cm³) | |
|---|---|---|---|---|---|---|---|---|---|---|---|---|---|---|
| | | $h$ | $b$ | $t_1$ | $t_2$ | $r$ | | | $I_x$ | $I_y$ | $i_x$ | $i_y$ | $W_x$ | $W_y$ |
| HW | 100×100 | 100 | 100 | 6 | 8 | 8 | 21.59 | 16.9 | 386 | 134 | 4.23 | 2.49 | 77.1 | 26.7 |
| | 125×125 | 125 | 125 | 6.5 | 9 | 8 | 30.00 | 23.6 | 843 | 293 | 5.30 | 3.13 | 135 | 46.9 |
| | 150×150 | 150 | 150 | 7 | 10 | 8 | 39.65 | 31.1 | 1620 | 563 | 6.39 | 3.77 | 216 | 75.1 |
| | 175×175 | 175 | 175 | 7.5 | 11 | 13 | 51.43 | 40.4 | 2918 | 983 | 7.53 | 4.37 | 334 | 112 |
| | 200×200 | 200 | 200 | 8 | 12 | 13 | 63.53 | 49.9 | 4717 | 1601 | 8.62 | 5.02 | 472 | 160 |
| | | 200 | 204 | 12 | 12 | 13 | 71.53 | 56.2 | 4984 | 1701 | 8.35 | 4.88 | 498 | 167 |
| | 250×250 | 244 | 252 | 11 | 11 | 13 | 81.31 | 63.8 | 8573 | 2937 | 10.27 | 6.01 | 703 | 233 |
| | | 250 | 250 | 9 | 14 | 13 | 91.43 | 71.8 | 10689 | 3648 | 10.81 | 6.32 | 855 | 292 |
| | | 250 | 255 | 14 | 14 | 13 | 103.93 | 81.6 | 11340 | 3875 | 10.45 | 6.11 | 907 | 304 |
| | 300×300 | 294 | 302 | 12 | 12 | 13 | 106.33 | 83.5 | 16384 | 5513 | 12.41 | 7.20 | 1115 | 365 |
| | | 300 | 300 | 10 | 15 | 13 | 118.45 | 93.0 | 20010 | 6753 | 13.00 | 7.55 | 1334 | 450 |
| | | 300 | 305 | 15 | 15 | 13 | 133.45 | 104.8 | 21135 | 7102 | 12.58 | 7.29 | 1409 | 466 |
| | 350×350 | 338 | 351 | 13 | 13 | 13 | 133.27 | 104.6 | 27352 | 9376 | 14.33 | 8.39 | 1618 | 534 |
| | | 344 | 348 | 10 | 16 | 13 | 144.01 | 113.0 | 32545 | 11242 | 15.03 | 8.84 | 1892 | 646 |
| | | 344 | 354 | 16 | 16 | 13 | 164.65 | 129.3 | 34581 | 11841 | 14.49 | 8.48 | 2011 | 669 |
| | | 350 | 350 | 12 | 19 | 13 | 171.89 | 134.9 | 39637 | 13582 | 15.19 | 8.89 | 2265 | 776 |
| | | 350 | 357 | 19 | 19 | 13 | 196.39 | 154.2 | 42138 | 14427 | 14.65 | 8.57 | 2408 | 808 |
| | 400×400 | 388 | 402 | 15 | 15 | 22 | 178.45 | 140.1 | 48040 | 16255 | 16.41 | 9.54 | 2476 | 809 |
| | | 394 | 398 | 11 | 18 | 22 | 186.81 | 146.6 | 55597 | 18920 | 17.25 | 10.06 | 2822 | 951 |
| | | 394 | 405 | 18 | 18 | 22 | 214.39 | 168.3 | 59165 | 19951 | 16.61 | 9.65 | 3003 | 985 |
| | | 400 | 400 | 13 | 21 | 22 | 218.69 | 171.7 | 66455 | 22410 | 17.43 | 10.12 | 3323 | 1120 |
| | | 400 | 408 | 21 | 21 | 22 | 250.69 | 196.8 | 70722 | 23804 | 16.80 | 9.74 | 3536 | 1167 |
| | | 414 | 405 | 18 | 28 | 22 | 295.39 | 231.9 | 93518 | 31022 | 17.79 | 10.25 | 4518 | 1532 |
| | | 428 | 407 | 20 | 35 | 22 | 360.65 | 283.1 | 12089 | 39357 | 18.31 | 10.45 | 5649 | 1934 |
| | | 458 | 417 | 30 | 50 | 22 | 528.55 | 414.9 | 19093 | 60516 | 19.01 | 10.70 | 8338 | 2902 |
| | | *498 | 432 | 45 | 70 | 22 | 770.05 | 604.5 | 30473 | 94346 | 19.89 | 11.07 | 12238 | 4368 |
| | *500×500 | 492 | 465 | 15 | 20 | 22 | 257.95 | 202.5 | 115559 | 33531 | 21.17 | 11.40 | 4698 | 1442 |
| | | 502 | 465 | 15 | 25 | 22 | 304.45 | 239.0 | 145012 | 41910 | 21.82 | 11.73 | 5777 | 1803 |
| | | 502 | 470 | 20 | 25 | 22 | 329.55 | 258.7 | 150283 | 43295 | 21.35 | 11.46 | 5987 | 1842 |

APPENDIX F  TYPICAL STEEL SHAPES

continued

| Type | Depth×Width (mm×mm) | Section dimension(mm) | | | | | Sectional area ($cm^2$) | Theoretical weight (kg/m) | Moment of inertia($cm^4$) | | Radius of gyration(cm) | | Section modulus ($cm^3$) | |
|---|---|---|---|---|---|---|---|---|---|---|---|---|---|---|
| | | $h$ | $b$ | $t_1$ | $t_2$ | $r$ | | | $I_x$ | $I_y$ | $i_x$ | $i_y$ | $W_x$ | $W_y$ |
| HM | 150×100 | 148 | 100 | 6 | 9 | 8 | 26.35 | 20.7 | 995.3 | 150.3 | 6.15 | 2.39 | 134.5 | 30.1 |
| | 200×150 | 194 | 150 | 6 | 9 | 8 | 38.11 | 29.9 | 2586 | 506.6 | 8.24 | 3.65 | 266.6 | 67.6 |
| | 250×175 | 244 | 175 | 7 | 11 | 13 | 55.49 | 43.6 | 5908 | 983.5 | 10.32 | 4.21 | 484.3 | 112.4 |
| | 300×200 | 294 | 200 | 8 | 12 | 13 | 71.05 | 55.8 | 10858 | 1602 | 12.36 | 4.75 | 738.6 | 160.2 |
| | 350×250 | 340 | 250 | 9 | 14 | 13 | 99.53 | 78.1 | 20867 | 3648 | 14.48 | 6.05 | 1227 | 291.9 |
| | 400×300 | 390 | 300 | 10 | 16 | 13 | 133.25 | 104.6 | 37363 | 7203 | 16.75 | 7.35 | 1916 | 480.2 |
| | 450×300 | 440 | 300 | 11 | 18 | 13 | 153.89 | 120.8 | 54067 | 8105 | 18.74 | 7.26 | 2458 | 540.3 |
| | 500×300 | 482 | 300 | 11 | 15 | 13 | 141.17 | 110.8 | 57212 | 6756 | 20.13 | 6.92 | 2374 | 450.4 |
| | | 488 | 300 | 11 | 18 | 13 | 159.17 | 124.9 | 67916 | 8106 | 20.66 | 7.14 | 2783 | 540.4 |
| | 550×300 | 544 | 300 | 11 | 15 | 13 | 147.99 | 116.2 | 74874 | 6756 | 22.49 | 6.76 | 2753 | 450.4 |
| | | 550 | 300 | 11 | 18 | 13 | 165.99 | 130.3 | 88470 | 8106 | 23.09 | 6.99 | 3217 | 540.4 |
| | 600×300 | 582 | 300 | 12 | 17 | 13 | 169.21 | 132.8 | 97287 | 7659 | 23.98 | 6.73 | 3343 | 510.6 |
| | | 588 | 300 | 12 | 20 | 13 | 187.21 | 147.0 | 112827 | 9009 | 24.55 | 6.94 | 3838 | 600.6 |
| | | 594 | 302 | 14 | 23 | 13 | 217.09 | 170.4 | 132179 | 10572 | 24.68 | 6.98 | 4450 | 700.1 |
| HN | 100×50 | 100 | 50 | 5 | 7 | 8 | 11.85 | 9.3 | 191.0 | 14.7 | 4.02 | 1.11 | 38.2 | 5.9 |
| | 125×60 | 125 | 60 | 6 | 8 | 8 | 16.69 | 13.1 | 407.7 | 29.1 | 4.94 | 1.32 | 65.2 | 9.7 |
| | 150×75 | 150 | 75 | 5 | 7 | 8 | 17.85 | 14.0 | 645.7 | 49.4 | 6.01 | 1.66 | 86.1 | 13.2 |
| | 175×90 | 175 | 90 | 5 | 8 | 8 | 22.90 | 18.0 | 1174 | 97.4 | 7.16 | 2.06 | 134.2 | 21.6 |
| | 200×100 | 198 | 99 | 4.5 | 7 | 8 | 22.69 | 17.8 | 1484 | 113.4 | 8.09 | 2.24 | 149.9 | 22.9 |
| | | 200 | 100 | 5.5 | 8 | 8 | 26.67 | 20.9 | 1753 | 133.7 | 8.11 | 2.24 | 175.3 | 26.7 |
| | 250×125 | 248 | 124 | 5 | 8 | 8 | 31.99 | 25.1 | 3346 | 254.5 | 10.23 | 2.82 | 269.8 | 41.1 |
| | | 250 | 125 | 6 | 9 | 8 | 36.97 | 29.0 | 3868 | 293.5 | 10.23 | 2.82 | 309.4 | 47.0 |
| | 300×150 | 298 | 149 | 5.5 | 8 | 13 | 40.80 | 32.0 | 5911 | 441.7 | 12.04 | 3.29 | 396.7 | 59.3 |
| | | 300 | 150 | 6.5 | 9 | 13 | 46.78 | 36.7 | 6829 | 507.2 | 12.08 | 3.29 | 455.3 | 67.6 |
| | 350×175 | 346 | 174 | 6 | 9 | 13 | 52.45 | 41.2 | 10456 | 791.1 | 14.12 | 3.88 | 604.4 | 90.9 |
| | | 350 | 175 | 7 | 11 | 13 | 62.91 | 49.4 | 12980 | 983.8 | 14.36 | 3.95 | 741.7 | 112.4 |
| | 400×150 | 400 | 150 | 8 | 13 | 13 | 70.37 | 55.2 | 17906 | 733.2 | 15.95 | 3.23 | 895.3 | 97.8 |
| | 400×200 | 396 | 199 | 7 | 11 | 13 | 71.41 | 56.1 | 19023 | 1446 | 16.32 | 4.50 | 960.8 | 145.3 |
| | | 400 | 200 | 8 | 13 | 13 | 83.37 | 65.4 | 22775 | 1735 | 16.53 | 4.56 | 1139 | 173.5 |
| | 450×200 | 446 | 199 | 8 | 12 | 13 | 82.97 | 65.1 | 27146 | 1578 | 18.09 | 4.36 | 1217 | 158.6 |
| | | 450 | 200 | 9 | 14 | 13 | 95.43 | 74.9 | 31973 | 1870 | 18.30 | 4.43 | 1421 | 187.0 |
| | 500×200 | 496 | 199 | 9 | 14 | 13 | 99.29 | 77.9 | 39628 | 1842 | 19.98 | 4.31 | 1598 | 185.1 |
| | | 500 | 200 | 10 | 16 | 13 | 112.25 | 88.1 | 45685 | 2138 | 20.17 | 4.36 | 1827 | 213.8 |
| | | 506 | 201 | 11 | 19 | 13 | 129.31 | 101.5 | 54478 | 2577 | 20.53 | 4.46 | 2153 | 256.4 |
| | 550×200 | 546 | 199 | 9 | 14 | 13 | 103.79 | 81.5 | 49245 | 1842 | 21.78 | 4.21 | 1804 | 185.2 |
| | | 550 | 200 | 10 | 16 | 13 | 117.25 | 92.0 | 56695 | 2138 | 21.99 | 4.27 | 2062 | 213.8 |
| | 600×200 | 596 | 199 | 10 | 15 | 13 | 117.75 | 92.4 | 64739 | 1975 | 23.45 | 4.10 | 2172 | 198.5 |
| | | 600 | 200 | 11 | 17 | 13 | 131.71 | 103.4 | 73749 | 2273 | 23.66 | 4.15 | 2458 | 227.3 |
| | | 606 | 201 | 12 | 20 | 13 | 149.77 | 117.6 | 86656 | 2716 | 24.05 | 4.26 | 2860 | 270.2 |
| | 650×300 | 646 | 299 | 10 | 15 | 13 | 152.75 | 119.9 | 107794 | 6688 | 26.56 | 6.62 | 3337 | 447.4 |
| | | 650 | 300 | 11 | 17 | 13 | 171.21 | 134.4 | 122739 | 7657 | 26.77 | 6.69 | 3777 | 510.5 |
| | | 656 | 301 | 12 | 20 | 13 | 195.77 | 153.7 | 144433 | 9100 | 27.16 | 6.82 | 4403 | 604.6 |
| | 700×300 | 692 | 300 | 13 | 20 | 18 | 207.54 | 162.9 | 164101 | 9014 | 28.12 | 6.59 | 4743 | 600.9 |
| | | 700 | 300 | 13 | 24 | 18 | 231.54 | 181.8 | 193622 | 10814 | 28.92 | 6.83 | 5532 | 720.9 |
| | 750×300 | 734 | 299 | 12 | 16 | 18 | 182.70 | 143.4 | 155539 | 7140 | 29.18 | 6.25 | 4238 | 477.6 |
| | | 742 | 300 | 13 | 20 | 18 | 214.04 | 168.0 | 191989 | 9015 | 29.95 | 6.49 | 5175 | 601.0 |
| | | 750 | 300 | 13 | 24 | 18 | 238.04 | 186.9 | 225863 | 10815 | 30.80 | 6.74 | 6023 | 721.0 |
| | | 758 | 303 | 16 | 28 | 18 | 284.78 | 223.6 | 271350 | 13008 | 30.87 | 6.76 | 7160 | 858.6 |

# APPENDIX F  TYPICAL STEEL SHAPES

continued

| Type | Depth×Width (mm×mm) | Section dimension(mm) | | | | | Sectional area (cm²) | Theoretical weight (kg/m) | Moment of inertia(cm⁴) | | Radius of gyration(cm) | | Section modulus (cm³) | |
|---|---|---|---|---|---|---|---|---|---|---|---|---|---|---|
| | | $h$ | $b$ | $t_1$ | $t_2$ | $r$ | | | $I_x$ | $I_y$ | $i_x$ | $i_y$ | $W_x$ | $W_y$ |
| HN | 800×300 | 792 | 300 | 14 | 22 | 18 | 239.50 | 188.0 | 242399 | 9919 | 31.81 | 6.44 | 6121 | 661.3 |
| | | 800 | 300 | 14 | 26 | 18 | 263.50 | 206.8 | 280925 | 11719 | 32.65 | 6.67 | 7023 | 781.3 |
| | 850×300 | 834 | 298 | 14 | 19 | 18 | 227.46 | 178.6 | 243858 | 8400 | 32.74 | 6.08 | 5848 | 563.8 |
| | | 842 | 299 | 15 | 23 | 18 | 259.72 | 203.9 | 291216 | 10271 | 33.49 | 6.29 | 6917 | 687.0 |
| | | 850 | 300 | 16 | 27 | 18 | 292.14 | 229.3 | 339670 | 12179 | 34.10 | 6.46 | 7992 | 812.0 |
| | | 858 | 301 | 17 | 31 | 18 | 324.72 | 254.9 | 389234 | 14125 | 34.62 | 6.60 | 9073 | 938.5 |
| | 900×300 | 890 | 299 | 15 | 23 | 18 | 266.92 | 209.5 | 330588 | 10273 | 35.19 | 6.20 | 7429 | 687.1 |
| | | 900 | 300 | 16 | 28 | 18 | 305.82 | 240.1 | 397241 | 12631 | 36.04 | 6.43 | 8828 | 842.1 |
| | | 912 | 302 | 18 | 34 | 18 | 360.06 | 282.6 | 484615 | 15652 | 36.69 | 6.59 | 10628 | 1037 |
| | 1000×300 | 970 | 297 | 16 | 21 | 18 | 276.00 | 216.7 | 382977 | 9203 | 37.25 | 5.77 | 7896 | 619.7 |
| | | 980 | 298 | 17 | 26 | 18 | 315.50 | 247.7 | 462157 | 11508 | 38.27 | 6.04 | 9432 | 772.3 |
| | | 990 | 298 | 17 | 31 | 18 | 345.30 | 271.1 | 535201 | 13713 | 39.37 | 6.30 | 10812 | 920.3 |
| | | 1000 | 300 | 19 | 36 | 18 | 395.10 | 310.2 | 626396 | 16256 | 39.82 | 6.41 | 12528 | 1084 |
| | | 1008 | 302 | 21 | 40 | 18 | 439.26 | 344.8 | 704572 | 18437 | 40.05 | 6.48 | 13980 | 1221 |
| HT | 100×50 | 95 | 48 | 3.2 | 4.5 | 8 | 7.62 | 6.0 | 109.7 | 8.4 | 3.79 | 1.05 | 23.1 | 3.5 |
| | | 97 | 49 | 4 | 5.5 | 8 | 9.38 | 7.4 | 141.8 | 10.9 | 3.89 | 1.08 | 29.2 | 4.4 |
| | 100×100 | 96 | 99 | 4.5 | 6 | 8 | 16.21 | 12.7 | 272.7 | 97.1 | 4.10 | 2.45 | 56.8 | 19.6 |
| | 125×60 | 118 | 58 | 3.2 | 4.5 | 8 | 9.26 | 7.3 | 202.4 | 14.7 | 4.68 | 1.26 | 34.3 | 5.1 |
| | | 120 | 59 | 4 | 5.5 | 8 | 11.40 | 8.9 | 259.7 | 18.9 | 4.77 | 1.29 | 43.3 | 6.4 |
| | 125×125 | 119 | 123 | 4.5 | 6 | 8 | 20.12 | 15.8 | 523.6 | 186.2 | 5.10 | 3.04 | 88.0 | 30.3 |
| | 150×75 | 145 | 73 | 3.2 | 4.5 | 8 | 11.47 | 9.0 | 383.2 | 29.3 | 5.78 | 1.60 | 52.9 | 8.0 |
| | | 147 | 74 | 4 | 5.5 | 8 | 14.13 | 11.1 | 488.0 | 37.3 | 5.88 | 1.62 | 66.4 | 10.1 |
| | 150×100 | 139 | 97 | 3.2 | 4.5 | 8 | 13.44 | 10.5 | 447.3 | 68.5 | 5.77 | 2.26 | 64.4 | 14.1 |
| | | 142 | 99 | 4.5 | 6 | 8 | 18.28 | 14.3 | 632.7 | 97.2 | 5.88 | 2.31 | 89.1 | 19.6 |
| | 150×150 | 144 | 148 | 5 | 7 | 8 | 27.77 | 21.8 | 1070 | 378.4 | 6.21 | 3.69 | 148.6 | 51.1 |
| | | 147 | 149 | 6 | 8.5 | 8 | 33.68 | 26.4 | 1338 | 468.9 | 6.30 | 3.73 | 182.1 | 62.9 |
| | 175×90 | 168 | 88 | 3.2 | 4.5 | 8 | 13.56 | 10.6 | 619.6 | 51.2 | 6.76 | 1.94 | 73.8 | 11.6 |
| | | 171 | 89 | 4 | 6 | 8 | 17.59 | 13.8 | 852.1 | 70.6 | 6.96 | 2.00 | 99.7 | 15.9 |
| | 175×175 | 167 | 173 | 5 | 7 | 13 | 33.32 | 26.2 | 1731 | 604.5 | 7.21 | 4.26 | 207.2 | 69.9 |
| | | 172 | 175 | 6.5 | 9.5 | 13 | 44.65 | 35.0 | 2466 | 849.2 | 7.43 | 4.36 | 286.8 | 97.1 |
| | 200×100 | 193 | 98 | 3.2 | 4.5 | 8 | 15.26 | 12.0 | 921.0 | 70.7 | 7.77 | 2.15 | 95.4 | 14.4 |
| | | 196 | 99 | 4 | 6 | 8 | 19.79 | 15.5 | 1260 | 97.2 | 7.98 | 2.22 | 128.6 | 19.6 |
| | 200×150 | 188 | 149 | 4.5 | 6 | 8 | 26.35 | 20.7 | 1669 | 331.0 | 7.96 | 3.54 | 177.6 | 44.4 |
| | 200×200 | 192 | 198 | 6 | 8 | 13 | 43.69 | 34.3 | 2984 | 1036 | 8.26 | 4.87 | 310.8 | 104.6 |
| | 250×125 | 244 | 124 | 4.5 | 6 | 8 | 25.87 | 20.3 | 2529 | 190.9 | 9.89 | 2.72 | 207.3 | 30.8 |
| | 250×175 | 238 | 173 | 4.5 | 8 | 13 | 39.12 | 30.7 | 4045 | 690.8 | 10.17 | 4.20 | 339.9 | 79.9 |
| | 300×150 | 294 | 148 | 4.5 | 6 | 13 | 31.90 | 25.0 | 4342 | 324.6 | 11.67 | 3.19 | 295.4 | 43.9 |
| | 300×200 | 286 | 198 | 6 | 8 | 13 | 49.33 | 38.7 | 7000 | 1036 | 11.91 | 4.58 | 489.5 | 104.6 |
| | 350×175 | 340 | 173 | 4.5 | 6 | 13 | 36.97 | 29.0 | 6823 | 518.3 | 13.58 | 3.74 | 401.3 | 59.9 |
| | 400×150 | 390 | 148 | 6 | 8 | 13 | 47.57 | 37.3 | 10900 | 433.2 | 15.14 | 3.02 | 559.0 | 58.5 |
| | 400×200 | 390 | 198 | 6 | 8 | 13 | 55.57 | 43.6 | 13819 | 1036 | 15.77 | 4.32 | 708.7 | 104.6 |

Note: 1. Products of the same model have the same inner dimensions and height;
2. The specifications shown by * mean that production is temporarily unavailable.

# APPENDIX F  TYPICAL STEEL SHAPES

**Structural Tee**  Table F.3

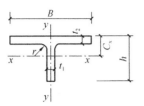

Symbol: $h$ — Depth
$B$ — Flange width
$t_1$ — Web thickness
$t_2$ — Flange thickness
$r$ — Fillet radius
$C_x$ — Gravity center

| Type | Depth×Width (mm×mm) | Section dimension (mm) | | | | | Sectional area (cm²) | Theoretical weight (kg/m) | Section characteristics | | | | | | Gravity center (cm) |
|---|---|---|---|---|---|---|---|---|---|---|---|---|---|---|---|
| | | | | | | | | | Moment of inertia (cm⁴) | | Radius of gyration (cm) | | Section modulus (cm³) | | |
| | | $h$ | $B$ | $t_1$ | $t_2$ | $r$ | | | $I_x$ | $I_y$ | $i_x$ | $i_y$ | $W_x$ | $W_y$ | $C_x$ |
| TW | 50×100 | 50 | 100 | 6 | 8 | 8 | 10.79 | 8.47 | 16.7 | 67.7 | 1.23 | 2.49 | 4.2 | 13.5 | 1.00 |
| | 62.5×125 | 62.5 | 125 | 6.5 | 9 | 8 | 15.00 | 11.8 | 35.2 | 147.1 | 1.53 | 3.13 | 6.9 | 23.5 | 1.19 |
| | 75×150 | 75 | 150 | 7 | 10 | 8 | 19.82 | 15.6 | 66.6 | 281.9 | 1.83 | 3.77 | 10.9 | 37.6 | 1.37 |
| | 87.5×175 | 87.5 | 175 | 7.5 | 11 | 13 | 25.71 | 20.2 | 115.8 | 494.4 | 2.12 | 4.38 | 16.1 | 56.5 | 1.55 |
| | 100×200 | 100 | 200 | 8 | 12 | 13 | 31.77 | 24.9 | 185.6 | 803.3 | 2.42 | 5.03 | 22.4 | 80.3 | 1.73 |
| | | 100 | 204 | 12 | 12 | 13 | 35.77 | 28.1 | 256.3 | 853.6 | 2.68 | 4.89 | 32.4 | 83.7 | 2.09 |
| | 125×250 | 125 | 250 | 9 | 14 | 13 | 45.72 | 35.9 | 413.0 | 1827 | 3.01 | 6.32 | 39.6 | 146.1 | 2.08 |
| | | 125 | 255 | 14 | 14 | 13 | 51.97 | 40.8 | 589.3 | 1941 | 3.37 | 6.11 | 59.4 | 152.2 | 2.58 |
| | 150×300 | 147 | 302 | 12 | 12 | 13 | 53.17 | 41.7 | 855.8 | 2760 | 4.01 | 7.20 | 72.2 | 182.8 | 2.85 |
| | | 150 | 300 | 10 | 15 | 13 | 59.23 | 46.5 | 798.7 | 3379 | 3.67 | 7.55 | 63.8 | 225.3 | 2.47 |
| | | 150 | 305 | 15 | 15 | 13 | 66.73 | 52.4 | 1107 | 3554 | 4.07 | 7.30 | 92.6 | 233.1 | 3.04 |
| | 175×350 | 172 | 348 | 10 | 16 | 13 | 72.01 | 56.5 | 1231 | 5624 | 4.13 | 8.84 | 84.7 | 323.2 | 2.67 |
| | | 175 | 350 | 12 | 19 | 13 | 85.95 | 67.5 | 1520 | 6794 | 4.21 | 8.89 | 103.9 | 388.2 | 2.87 |
| | 200×400 | 194 | 402 | 15 | 15 | 22 | 89.23 | 70.0 | 2479 | 8150 | 5.27 | 9.56 | 157.9 | 405.5 | 3.70 |
| | | 197 | 398 | 11 | 18 | 22 | 93.41 | 73.3 | 2052 | 9481 | 4.69 | 10.07 | 122.9 | 476.4 | 3.01 |
| | | 200 | 400 | 13 | 21 | 22 | 109.35 | 85.8 | 2483 | 11220 | 4.77 | 10.13 | 147.9 | 561.3 | 3.21 |
| | | 200 | 408 | 21 | 21 | 22 | 125.35 | 98.4 | 3654 | 11920 | 5.40 | 9.75 | 229.4 | 584.7 | 4.07 |
| | | 207 | 405 | 18 | 28 | 22 | 147.70 | 115.9 | 3634 | 15530 | 4.96 | 10.26 | 213.6 | 767.2 | 3.68 |
| | | 214 | 407 | 20 | 35 | 22 | 180.33 | 141.6 | 4393 | 19700 | 4.94 | 10.45 | 251.0 | 968.2 | 3.90 |
| TM | 75×100 | 74 | 100 | 6 | 9 | 8 | 13.17 | 10.3 | 51.7 | 75.6 | 1.98 | 2.39 | 8.9 | 15.1 | 1.56 |
| | 100×150 | 97 | 150 | 6 | 9 | 8 | 19.05 | 15.0 | 124.4 | 253.7 | 2.56 | 3.65 | 15.8 | 33.8 | 1.80 |
| | 125×175 | 122 | 175 | 7 | 11 | 13 | 27.75 | 21.8 | 288.3 | 494.4 | 3.22 | 4.22 | 29.1 | 56.5 | 2.28 |
| | 150×200 | 147 | 200 | 8 | 12 | 13 | 35.53 | 27.9 | 570.0 | 803.5 | 4.01 | 4.76 | 48.1 | 80.3 | 2.85 |
| | 175×250 | 170 | 250 | 9 | 14 | 13 | 49.77 | 39.1 | 1016 | 1827 | 4.52 | 6.06 | 73.1 | 146.1 | 3.11 |
| | 200×300 | 195 | 300 | 10 | 16 | 13 | 66.62 | 52.3 | 1730 | 3600 | 5.09 | 7.35 | 108 | 240 | 3.43 |
| | 225×300 | 220 | 300 | 11 | 18 | 13 | 76.94 | 60.4 | 2680 | 4050 | 5.89 | 7.25 | 150 | 270 | 4.09 |
| | 250×300 | 241 | 300 | 11 | 15 | 13 | 70.58 | 55.4 | 3400 | 3380 | 6.93 | 6.91 | 178 | 225 | 5.00 |
| | | 244 | 300 | 11 | 18 | 13 | 79.58 | 62.5 | 3610 | 4050 | 6.73 | 7.13 | 184 | 270 | 4.72 |
| | 275×300 | 272 | 300 | 11 | 15 | 13 | 73.99 | 58.1 | 4790 | 3380 | 8.04 | 6.75 | 225 | 225 | 5.96 |
| | | 275 | 300 | 11 | 18 | 13 | 82.99 | 65.2 | 5090 | 4050 | 7.82 | 6.98 | 232 | 270 | 5.59 |
| | 300×300 | 291 | 300 | 12 | 17 | 13 | 84.60 | 66.4 | 6320 | 3830 | 8.64 | 6.72 | 280 | 255 | 6.51 |
| | | 294 | 300 | 12 | 20 | 13 | 93.60 | 73.5 | 6680 | 4500 | 8.44 | 6.93 | 288 | 300 | 6.17 |
| | | 297 | 302 | 14 | 23 | 13 | 108.5 | 85.2 | 7890 | 5290 | 8.52 | 6.97 | 339 | 350 | 6.41 |

## APPENDIX F  TYPICAL STEEL SHAPES

continued

| Type | Depth×Width (mm×mm) | Section dimension (mm) | | | | | Sectional area ($cm^2$) | Theoretical weight (kg/m) | Moment of inertia ($cm^4$) | | Radius of gyration (cm) | | Section modulus ($cm^3$) | | Gravity center (cm) |
|---|---|---|---|---|---|---|---|---|---|---|---|---|---|---|---|
| | | $h$ | $B$ | $t_1$ | $t_2$ | $r$ | | | $I_x$ | $I_y$ | $i_x$ | $i_y$ | $W_x$ | $W_y$ | $C_x$ |
| TN | 50×50 | 50 | 50 | 5 | 7 | 8 | 5.92 | 4.7 | 11.9 | 7.8 | 1.42 | 1.14 | 3.2 | 3.1 | 1.28 |
| | 62.5×60 | 62.5 | 60 | 6 | 8 | 8 | 8.34 | 6.6 | 27.5 | 14.9 | 1.81 | 1.34 | 6.0 | 5.0 | 1.64 |
| | 75×75 | 75 | 75 | 5 | 7 | 8 | 8.92 | 7.0 | 42.4 | 25.1 | 2.18 | 1.68 | 7.4 | 6.7 | 1.79 |
| | 87.5×90 | 87.5 | 90 | 5 | 8 | 8 | 11.45 | 9.0 | 70.5 | 49.1 | 2.48 | 2.07 | 10.3 | 10.9 | 1.93 |
| | 100×100 | 99 | 99 | 4.5 | 7 | 8 | 11.34 | 8.9 | 93.1 | 57.1 | 2.87 | 2.24 | 12.0 | 11.5 | 2.17 |
| | | 100 | 100 | 5.5 | 8 | 8 | 13.33 | 10.5 | 113.9 | 67.2 | 2.92 | 2.25 | 14.8 | 13.4 | 2.31 |
| | 125×125 | 124 | 124 | 5 | 8 | 8 | 15.99 | 12.6 | 206.7 | 127.6 | 3.59 | 2.82 | 21.2 | 20.6 | 2.66 |
| | | 125 | 125 | 6 | 9 | 8 | 18.48 | 14.5 | 247.5 | 147.1 | 3.66 | 2.82 | 25.5 | 23.5 | 2.81 |
| | 150×150 | 149 | 149 | 5.5 | 8 | 13 | 20.40 | 16.0 | 390.4 | 223.3 | 4.37 | 3.31 | 33.5 | 30.0 | 3.26 |
| | | 150 | 150 | 6.5 | 9 | 13 | 23.39 | 18.4 | 460.4 | 256.1 | 4.44 | 3.31 | 39.7 | 34.2 | 3.41 |
| | 175×175 | 173 | 174 | 6 | 9 | 13 | 26.23 | 20.6 | 674.7 | 398.0 | 5.07 | 3.90 | 49.7 | 45.8 | 3.72 |
| | | 175 | 175 | 7 | 11 | 13 | 31.46 | 24.7 | 811.1 | 494.5 | 5.08 | 3.96 | 59.0 | 56.5 | 3.76 |
| | 200×200 | 198 | 199 | 7 | 11 | 13 | 35.71 | 28.0 | 1188 | 725.7 | 5.77 | 4.51 | 76.2 | 72.9 | 4.20 |
| | | 200 | 200 | 8 | 13 | 13 | 41.69 | 32.7 | 1392 | 870.3 | 5.78 | 4.57 | 88.4 | 87.0 | 4.26 |
| | 225×200 | 223 | 199 | 8 | 12 | 13 | 41.49 | 32.6 | 1863 | 791.8 | 6.70 | 4.37 | 108.7 | 79.6 | 5.15 |
| | | 225 | 200 | 9 | 14 | 13 | 47.72 | 37.5 | 2148 | 937.6 | 6.71 | 4.43 | 124.1 | 93.8 | 5.19 |
| | 250×200 | 248 | 199 | 9 | 14 | 13 | 49.65 | 39.0 | 2820 | 923.8 | 7.54 | 4.31 | 149.8 | 92.8 | 5.97 |
| | | 250 | 200 | 10 | 16 | 13 | 56.13 | 44.1 | 3201 | 1072 | 7.55 | 4.37 | 168.7 | 107.2 | 6.03 |
| | | 253 | 201 | 11 | 19 | 13 | 64.66 | 50.8 | 3666 | 1292 | 7.53 | 4.47 | 189.9 | 128.5 | 6.00 |
| | 275×200 | 273 | 199 | 9 | 14 | 13 | 51.90 | 40.7 | 3689 | 924.0 | 8.43 | 4.22 | 180.3 | 92.9 | 6.85 |
| | | 275 | 200 | 10 | 16 | 13 | 58.63 | 46.0 | 4182 | 1070 | 8.45 | 4.28 | 202.9 | 107.2 | 6.89 |
| | 300×200 | 298 | 199 | 10 | 15 | 13 | 58.88 | 46.2 | 5148 | 990.6 | 9.35 | 4.10 | 235.3 | 99.6 | 7.92 |
| | | 300 | 200 | 11 | 17 | 13 | 65.86 | 51.7 | 5779 | 1140 | 9.37 | 4.16 | 262.1 | 114.0 | 7.95 |
| | | 303 | 201 | 12 | 20 | 13 | 74.89 | 58.8 | 6554 | 1361 | 9.36 | 4.26 | 292.4 | 135.4 | 7.88 |
| | 325×300 | 323 | 299 | 10 | 15 | 12 | 76.27 | 59.9 | 7230 | 3346 | 9.74 | 6.62 | 289.0 | 223.8 | 7.28 |
| | | 325 | 300 | 11 | 17 | 13 | 85.61 | 67.2 | 8095 | 3832 | 9.72 | 6.69 | 321.1 | 255.4 | 7.29 |
| | | 328 | 301 | 12 | 20 | 13 | 97.89 | 76.8 | 9139 | 4553 | 9.66 | 6.82 | 357.0 | 302.5 | 7.20 |
| | 350×300 | 346 | 300 | 13 | 20 | 13 | 103.11 | 80.9 | 1126 | 4510 | 10.45 | 6.61 | 425.3 | 300.6 | 8.12 |
| | | 350 | 300 | 13 | 24 | 13 | 115.11 | 90.4 | 1201 | 5410 | 10.22 | 6.86 | 439.5 | 360.6 | 7.65 |
| | 400×300 | 396 | 300 | 14 | 22 | 18 | 119.75 | 94.0 | 1766 | 4970 | 12.14 | 6.44 | 592.1 | 331.3 | 9.77 |
| | | 400 | 300 | 14 | 26 | 18 | 131.75 | 103.4 | 1877 | 5870 | 11.94 | 6.67 | 610.8 | 391.3 | 9.27 |
| | 450×300 | 445 | 299 | 15 | 23 | 18 | 133.46 | 104.8 | 2589 | 5147 | 13.93 | 6.21 | 790.0 | 344.3 | 11.72 |
| | | 450 | 300 | 16 | 28 | 18 | 152.91 | 120.0 | 2922 | 6327 | 13.82 | 6.43 | 868.5 | 421.8 | 11.35 |
| | | 456 | 302 | 18 | 34 | 18 | 180.03 | 141.3 | 3434 | 7838 | 13.81 | 6.60 | 1002 | 519.0 | 11.34 |

APPENDIX F  TYPICAL STEEL SHAPES

## Universal Channel — Table F.4

Symbol: Same as ordinary I-shape, but $W_y$ is the section modulus of the corresponding flange tip

$r_1$ —Chamfer radius of flange tip

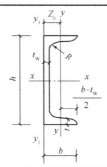

| Type | Section dimension (mm) | | | | | | Sectional area ($cm^2$) | Theoretical weight (kg/m) | Moment of inertia ($cm^4$) | | | Radius of gyration (cm) | | Section modulus ($cm^3$) | | Gravity center (cm) |
|---|---|---|---|---|---|---|---|---|---|---|---|---|---|---|---|---|
| | $h$ | $b$ | $t_w$ | $t$ | $R$ | $r_1$ | | | $I_x$ | $I_y$ | $I_{y1}$ | $i_x$ | $i_y$ | $W_x$ | $W_y$ | $Z_0$ |
| 5 | 50 | 37 | 4.5 | 7.0 | 7.0 | 3.5 | 6.928 | 5.438 | 26.0 | 8.30 | 20.9 | 1.94 | 1.10 | 10.4 | 3.55 | 1.35 |
| 6.3 | 63 | 40 | 4.8 | 7.5 | 7.5 | 3.8 | 8.451 | 6.634 | 50.8 | 11.9 | 28.4 | 2.45 | 1.19 | 16.1 | 4.50 | 1.36 |
| 6.5 | 65 | 40 | 4.3 | 7.5 | 7.5 | 3.8 | 8.547 | 6.709 | 55.2 | 12.0 | 28.3 | 2.54 | 1.19 | 17.0 | 4.59 | 1.38 |
| 8 | 80 | 43 | 5.0 | 8.0 | 8.0 | 4.0 | 10.248 | 8.045 | 101 | 16.6 | 37.4 | 3.15 | 1.27 | 25.3 | 5.79 | 1.43 |
| 10 | 100 | 48 | 5.3 | 8.5 | 8.5 | 4.2 | 12.748 | 10.007 | 198 | 25.6 | 54.9 | 3.95 | 1.41 | 39.7 | 7.80 | 1.52 |
| 12 | 120 | 53 | 5.5 | 9.0 | 9.0 | 4.5 | 15.362 | 12.059 | 346 | 37.4 | 77.7 | 4.75 | 1.56 | 57.7 | 10.2 | 1.62 |
| 12.6 | 126 | 53 | 5.5 | 9.0 | 9.0 | 4.5 | 15.692 | 12.318 | 391 | 38.0 | 77.1 | 4.95 | 1.57 | 62.1 | 10.2 | 1.59 |
| 14a | 140 | 58 | 6.0 | 9.5 | 9.5 | 4.8 | 18.516 | 14.535 | 564 | 53.2 | 107 | 5.52 | 1.70 | 80.5 | 13.0 | 1.71 |
| 14b | 140 | 60 | 8.0 | 9.5 | 9.5 | 4.8 | 21.316 | 16.733 | 609 | 61.1 | 121 | 5.35 | 1.69 | 87.1 | 14.1 | 1.67 |
| 16a | 160 | 63 | 6.5 | 10.0 | 10.0 | 5.0 | 21.962 | 17.241 | 866 | 73.3 | 144 | 6.28 | 1.8 | 108 | 16.3 | 1.80 |
| 16b | 160 | 65 | 8.5 | 10.0 | 10.0 | 5.0 | 25.162 | 9.752 | 935 | 83.4 | 161 | 6.10 | 1.82 | 117 | 17.6 | 1.75 |
| 18a | 180 | 68 | 7.0 | 10.5 | 10.5 | 5.2 | 25.699 | 20.174 | 1270 | 98.6 | 190 | 7.04 | 1.96 | 141 | 20.0 | 1.88 |
| 18b | 180 | 70 | 9.0 | 10.5 | 10.5 | 5.2 | 29.299 | 23.000 | 1370 | 111 | 210 | 6.84 | 1.95 | 152 | 21.5 | 1.84 |
| 20a | 200 | 73 | 7.0 | 11.0 | 11.0 | 5.5 | 28.837 | 22.637 | 1780 | 128 | 244 | 7.86 | 2.11 | 178 | 24.2 | 2.01 |
| 20b | 200 | 75 | 9.0 | 11.0 | 11.0 | 5.5 | 32.837 | 25.777 | 1910 | 144 | 268 | 7.64 | 2.09 | 191 | 25.9 | 1.95 |
| 22a | 220 | 77 | 7.0 | 11.5 | 11.5 | 5.8 | 31.846 | 24.999 | 2390 | 158 | 298 | 8.67 | 2.23 | 218 | 28.2 | 2.10 |
| 22b | 220 | 79 | 9.0 | 11.5 | 11.5 | 5.8 | 36.246 | 28.453 | 2570 | 176 | 326 | 8.42 | 2.21 | 234 | 30.1 | 2.03 |
| 24a | 240 | 78 | 7.0 | 12.0 | 12.0 | 6.0 | 34.217 | 26.860 | 3050 | 174 | 325 | 9.45 | 2.25 | 254 | 30.5 | 2.10 |
| 24b | 240 | 80 | 9.0 | 12.0 | 12.0 | 6.0 | 39.017 | 30.628 | 3280 | 194 | 355 | 9.17 | 2.23 | 274 | 32.5 | 2.03 |
| 24c | 240 | 82 | 11.0 | 12.0 | 12.0 | 6.0 | 43.817 | 34.396 | 3510 | 213 | 388 | 8.96 | 2.21 | 293 | 34.4 | 2.00 |
| 25a | 250 | 82 | 7.0 | 12.0 | 12.0 | 6.0 | 34.917 | 27.410 | 3370 | 176 | 322 | 9.82 | 2.24 | 270 | 30.6 | 2.07 |
| 25b | 250 | 84 | 9.0 | 12.0 | 12.0 | 6.0 | 39.917 | 31.335 | 3530 | 196 | 353 | 9.41 | 2.22 | 282 | 32.7 | 1.98 |
| 25c | 250 | 86 | 11.0 | 12.0 | 12.0 | 6.0 | 44.917 | 35.260 | 3690 | 218 | 384 | 9.07 | 2.21 | 295 | 35.7 | 1.92 |
| 27a | 270 | 7.5 | 7.5 | 12.5 | 12.5 | 6.2 | 39.284 | 30.838 | 4360 | 216 | 393 | 10.5 | 2.34 | 323 | 35.5 | 2.13 |
| 27b | 270 | 9.5 | 9.5 | 12.5 | 12.5 | 6.2 | 44.684 | 35.077 | 4690 | 239 | 428 | 10.3 | 2.31 | 347 | 37.7 | 2.06 |
| 27c | 270 | 11.5 | 11.5 | 12.5 | 12.5 | 6.2 | 50.084 | 39.316 | 5020 | 261 | 467 | 10.1 | 2.28 | 372 | 39.8 | 2.03 |
| 28a | 280 | 7.5 | 7.5 | 12.5 | 12.5 | 6.2 | 40.034 | 31.427 | 4760 | 218 | 388 | 10.9 | 2.33 | 340 | 35.7 | 2.10 |
| 28b | 280 | 9.5 | 9.5 | 12.5 | 12.5 | 6.2 | 45.634 | 35.823 | 5130 | 242 | 428 | 10.6 | 2.30 | 366 | 37.9 | 2.02 |
| 28c | 280 | 11.5 | 11.5 | 12.5 | 12.5 | 6.2 | 51.234 | 40.219 | 5500 | 268 | 463 | 10.4 | 2.29 | 393 | 40.3 | 1.95 |

# APPENDIX F  TYPICAL STEEL SHAPES

continued

| Type | Section dimension(mm) | | | | | | Sectional area ($cm^2$) | Theoretical weight (kg/m) | Moment of inertia ($cm^4$) | | | Radius of gyration (cm) | | Section modulus ($cm^3$) | | Gravity center (cm) |
|---|---|---|---|---|---|---|---|---|---|---|---|---|---|---|---|---|
| | $h$ | $b$ | $t$ | $t_w$ | $R$ | $r_1$ | | | $I_x$ | $I_y$ | $I_{y1}$ | $i_x$ | $i_y$ | $W_x$ | $W_y$ | $Z_0$ |
| 30a | | 7.5 | 7.5 | | | | 43.902 | 34.463 | 6050 | 260 | 467 | 11.7 | 2.43 | 403 | 41.1 | 2.17 |
| 30b | 300 | 9.5 | 9.5 | 13.5 | 13.5 | 6.8 | 49.902 | 39.173 | 6500 | 289 | 515 | 11.4 | 2.41 | 433 | 44.0 | 2.13 |
| 30c | | 11.5 | 11.5 | | | | 55.902 | 43.883 | 6950 | 316 | 560 | 11.2 | 2.38 | 463 | 46.4 | 2.09 |
| 32a | | 8.0 | 8.0 | | | | 48.513 | 38.083 | 7600 | 305 | 552 | 12.5 | 2.50 | 475 | 46.5 | 2.24 |
| 32b | 320 | 10.0 | 10.0 | 14.0 | 14.0 | 7.0 | 54.913 | 43.107 | 8140 | 336 | 593 | 12.2 | 2.47 | 509 | 49.2 | 2.16 |
| 32c | | 12.0 | 12.0 | | | | 61.313 | 48.131 | 8690 | 374 | 643 | 11.9 | 2.47 | 543 | 52.6 | 2.09 |
| 36a | | 9.0 | 9.0 | | | | 60.910 | 47.814 | 11900 | 455 | 818 | 14.0 | 2.73 | 660 | 63.5 | 2.44 |
| 36b | 360 | 11.0 | 11.0 | 16.0 | 16.0 | 8.0 | 68.110 | 53.466 | 12700 | 497 | 880 | 13.6 | 2.70 | 703 | 66.9 | 2.37 |
| 36c | | 13.0 | 13.0 | | | | 75.310 | 59.118 | 13400 | 536 | 948 | 13.4 | 2.67 | 746 | 70.0 | 2.34 |
| 40a | | 10.5 | 10.5 | | | | 75.068 | 58.928 | 17600 | 592 | 1070 | 15.3 | 2.81 | 879 | 78.8 | 2.49 |
| 40b | 400 | 12.5 | 12.5 | 18.0 | 18.0 | 9.0 | 83.068 | 65.208 | 18600 | 640 | 114 | 15.0 | 2.78 | 932 | 82.5 | 2.44 |
| 40c | | 14.5 | 14.5 | | | | 91.068 | 71.488 | 19700 | 688 | 1220 | 14.7 | 2.75 | 986 | 86.2 | 2.42 |

Note: The data of $R$ and $r_1$ are used for pass design, not as delivery conditions.

**Equal leg Angle**  Table F.5

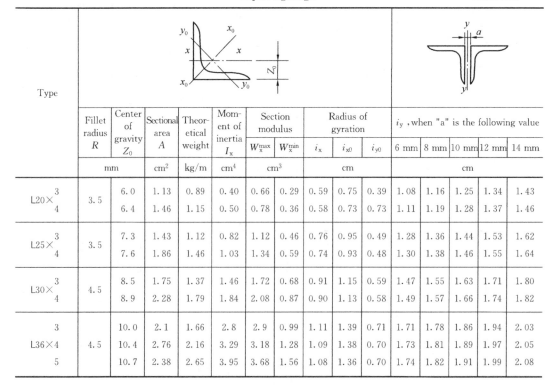

| Type | Fillet radius $R$ | Center of gravity $Z_0$ | Sectional area $A$ | Theoretical weight | Moment of inertia $I_x$ | Section modulus | | Radius of gyration | | | $i_y$, when "a" is the following value | | | | |
|---|---|---|---|---|---|---|---|---|---|---|---|---|---|---|---|
| | | | | | | $W_x^{max}$ | $W_x^{min}$ | $i_x$ | $i_{x0}$ | $i_{y0}$ | 6 mm | 8 mm | 10 mm | 12 mm | 14 mm |
| | mm | | $cm^2$ | kg/m | $cm^4$ | $cm^3$ | | cm | | | cm | | | | |
| L20× 3 | 3.5 | 6.0 | 1.13 | 0.89 | 0.40 | 0.66 | 0.29 | 0.59 | 0.75 | 0.39 | 1.08 | 1.16 | 1.25 | 1.34 | 1.43 |
| L20× 4 | | 6.4 | 1.46 | 1.15 | 0.50 | 0.78 | 0.36 | 0.58 | 0.73 | 0.73 | 1.11 | 1.19 | 1.28 | 1.37 | 1.46 |
| L25× 3 | 3.5 | 7.3 | 1.43 | 1.12 | 0.82 | 1.12 | 0.46 | 0.76 | 0.95 | 0.49 | 1.28 | 1.36 | 1.44 | 1.53 | 1.62 |
| L25× 4 | | 7.6 | 1.86 | 1.46 | 1.03 | 1.34 | 0.59 | 0.74 | 0.93 | 0.48 | 1.30 | 1.38 | 1.46 | 1.55 | 1.64 |
| L30× 3 | 4.5 | 8.5 | 1.75 | 1.37 | 1.46 | 1.72 | 0.68 | 0.91 | 1.15 | 0.59 | 1.47 | 1.55 | 1.63 | 1.71 | 1.80 |
| L30× 4 | | 8.9 | 2.28 | 1.79 | 1.84 | 2.08 | 0.87 | 0.90 | 1.13 | 0.58 | 1.49 | 1.57 | 1.66 | 1.74 | 1.82 |
| L36×3 | 4.5 | 10.0 | 2.1 | 1.66 | 2.8 | 2.9 | 0.99 | 1.11 | 1.39 | 0.71 | 1.71 | 1.78 | 1.86 | 1.94 | 2.03 |
| L36×4 | | 10.4 | 2.76 | 2.16 | 3.29 | 3.18 | 1.28 | 1.09 | 1.38 | 0.70 | 1.73 | 1.81 | 1.89 | 1.97 | 2.05 |
| L36×5 | | 10.7 | 2.38 | 2.65 | 3.95 | 3.68 | 1.56 | 1.08 | 1.36 | 0.70 | 1.74 | 1.82 | 1.91 | 1.99 | 2.08 |

APPENDIX F  TYPICAL STEEL SHAPES

continued

| Type | Fillet radius $R$ | Center of gravity $Z_0$ | Sectional area $A$ | Theoretical weight | Moment of inertia $I_x$ | Section modulus | | Radius of gyration | | | $i_y$, when "$a$" is the following value | | | | |
|---|---|---|---|---|---|---|---|---|---|---|---|---|---|---|---|
| | | | | | | $W_x^{max}$ | $W_x^{min}$ | $i_x$ | $i_{x0}$ | $i_{y0}$ | 6 mm | 8 mm | 10 mm | 12 mm | 14 mm |
| | mm | | cm² | kg/m | cm⁴ | cm³ | | cm | | | cm | | | | |
| L40×3,4,5 | 5 | 10.0 / 11.3 / 11.7 | 2.36 / 3.09 / 3.79 | 1.85 / 2.42 / 2.98 | 3.59 / 4.60 / 5.63 | 3.28 / 4.05 / 4.72 | 1.23 / 1.60 / 1.96 | 1.23 / 1.22 / 1.21 | 1.55 / 1.54 / 1.52 | 0.79 / 0.79 / 0.78 | 1.86 / 1.88 / 1.90 | 1.93 / 1.96 / 1.98 | 2.01 / 2.04 / 2.06 | 2.09 / 2.12 / 2.14 | 2.18 / 2.20 / 2.23 |
| L45×3,4,5,6 | 5 | 12.2 / 12.6 / 13.0 / 13.3 | 2.66 / 3.49 / 4.29 / 5.08 | 2.09 / 2.74 / 3.37 / 3.99 | 5.17 / 6.65 / 8.04 / 9.33 | 4.25 / 5.29 / 6.20 / 6.99 | 1.58 / 2.05 / 2.51 / 2.95 | 1.39 / 1.38 / 1.37 / 1.36 | 1.76 / 1.74 / 1.72 / 1.70 | 0.89 / 0.89 / 0.88 / 0.88 | 2.06 / 2.08 / 2.11 / 2.12 | 2.14 / 2.16 / 2.18 / 2.20 | 2.21 / 2.24 / 2.26 / 2.28 | 2.29 / 2.32 / 2.34 / 2.36 | 2.37 / 2.40 / 2.42 / 2.44 |
| L50×3,4,5,6 | 5.5 | 13.4 / 13.8 / 14.2 / 14.6 | 2.97 / 3.90 / 4.80 / 5.69 | 2.33 / 3.06 / 3.77 / 4.46 | 7.18 / 9.26 / 11.21 / 13.05 | 5.36 / 6.70 / 7.90 / 8.95 | 1.96 / 2.56 / 3.13 / 3.68 | 1.55 / 1.54 / 1.53 / 1.52 | 1.96 / 1.94 / 1.92 / 1.91 | 1.00 / 0.99 / 0.98 / 0.98 | 2.26 / 2.28 / 2.30 / 2.32 | 2.33 / 2.35 / 2.38 / 2.40 | 2.41 / 2.43 / 2.45 / 2.48 | 2.48 / 2.51 / 2.53 / 2.56 | 2.56 / 2.59 / 2.61 / 2.64 |
| L56×3,4,5,6,7,8 | 6 | 14.8 / 15.3 / 15.7 / 16.1 / 16.4 / 16.8 | 3.34 / 4.39 / 5.42 / 6.42 / 7.40 / 8.37 | 2.62 / 3.45 / 4.25 / 5.04 / 5.81 / 6.57 | 10.19 / 13.18 / 16.02 / 18.09 / 21.23 / 23.63 | 6.86 / 8.63 / 10.22 / 14.24 / 12.95 / 14.06 | 2.48 / 3.24 / 3.97 / 4.68 / 5.36 / 6.03 | 1.75 / 1.73 / 1.72 / 1.71 / 1.69 / 1.68 | 2.20 / 2.18 / 2.17 / 2.15 / 2.13 / 2.11 | 1.13 / 1.11 / 1.10 / 1.10 / 1.09 / 1.09 | 2.49 / 2.52 / 2.54 / 2.56 / 2.58 / 2.60 | 2.57 / 2.59 / 2.62 / 2.64 / 2.65 / 2.67 | 2.64 / 2.67 / 2.69 / 2.71 / 2.73 / 2.75 | 2.72 / 2.74 / 2.77 / 2.79 / 2.81 / 2.83 | 2.80 / 2.82 / 2.85 / 2.87 / 2.89 / 2.91 |
| L60×5,6,7,8 | 6.5 | 16.7 / 17.0 / 17.4 / 17.8 | 5.82 / 6.91 / 7.98 / 9.02 | 4.58 / 5.43 / 6.26 / 7.08 | 19.89 / 23.25 / 26.44 / 29.47 | 10.22 / 13.68 / 15.20 / 16.56 | 4.59 / 5.41 / 6.21 / 6.98 | 1.85 / 1.83 / 1.82 / 1.81 | 2.33 / 2.31 / 2.29 / 2.27 | 1.19 / 1.18 / 1.17 / 1.17 | 2.70 / 2.71 / 2.73 / 2.76 | 2.77 / 2.79 / 2.81 / 2.83 | 2.85 / 2.86 / 2.89 / 2.91 | 2.93 / 2.94 / 2.96 / 2.99 | 3.00 / 3.02 / 3.04 / 3.07 |
| L63×4,5,6,7 | 7 | 17.0 / 17.4 / 17.8 / 18.2 | 4.98 / 6.14 / 7.29 / 8.41 | 3.91 / 4.82 / 5.72 / 6.60 | 19.03 / 23.17 / 27.12 / 30.87 | 11.22 / 13.33 / 15.26 / 18.59 | 4.13 / 5.08 / 6.00 / 6.88 | 1.96 / 1.94 / 1.93 / 1.92 | 2.46 / 2.45 / 2.43 / 2.41 | 1.26 / 1.25 / 1.24 / 1.23 | 2.80 / 2.82 / 2.84 / 2.86 | 2.87 / 2.89 / 2.91 / 2.93 | 2.94 / 2.96 / 2.99 / 3.01 | 3.02 / 3.04 / 3.06 / 3.09 | 3.10 / 3.12 / 3.14 / 3.17 |
| L70×4,5,6,7,8 | 8 | 18.6 / 19.1 / 19.5 / 19.9 / 20.3 | 5.57 / 6.88 / 8.16 / 9.42 / 10.67 | 4.37 / 5.40 / 6.41 / 7.40 / 8.37 | 26.39 / 32.21 / 37.77 / 43.09 / 48.17 | 14.16 / 16.89 / 19.39 / 21.68 / 23.79 | 5.14 / 6.32 / 7.48 / 8.59 / 9.68 | 2.18 / 2.16 / 2.15 / 2.14 / 2.12 | 2.74 / 2.73 / 2.71 / 2.69 / 2.68 | 1.40 / 1.39 / 1.38 / 1.38 / 1.37 | 3.07 / 3.09 / 3.11 / 3.13 / 3.15 | 3.14 / 3.17 / 3.19 / 3.21 / 3.23 | 3.21 / 3.24 / 3.26 / 3.28 / 3.30 | 3.28 / 3.31 / 3.34 / 3.36 / 3.38 | 3.36 / 3.39 / 3.41 / 3.44 / 3.46 |
| L75×5,6,7,8,9,10 | 9 | 20.4 / 20.7 / 21.1 / 21.5 / 21.8 / 22.2 | 7.41 / 8.80 / 10.16 / 11.50 / 12.83 / 14.13 | 5.82 / 6.91 / 7.89 / 9.03 / 10.07 / 11.09 | 39.96 / 46.91 / 53.57 / 59.96 / 66.10 / 71.98 | 19.73 / 22.69 / 25.42 / 27.93 / 30.32 / 32.40 | 7.32 / 8.64 / 9.93 / 11.20 / 12.43 / 13.64 | 2.33 / 2.31 / 2.30 / 2.28 / 2.27 / 2.26 | 2.92 / 2.90 / 2.89 / 2.87 / 2.86 / 2.84 | 1.50 / 1.49 / 1.48 / 1.47 / 1.46 / 1.46 | 3.30 / 3.31 / 3.33 / 3.35 / 3.36 / 3.38 | 3.37 / 3.38 / 3.40 / 3.42 / 3.44 / 3.46 | 3.44 / 3.46 / 3.48 / 3.50 / 3.51 / 3.54 | 3.52 / 3.53 / 3.55 / 3.57 / 3.59 / 3.61 | 3.59 / 3.61 / 3.63 / 3.65 / 3.67 / 3.69 |

# APPENDIX F  TYPICAL STEEL SHAPES

continued

| Type | Fillet radius $R$ | Center of gravity $Z_0$ | Sectional area $A$ | Theoretical weight | Moment of inertia $I_x$ | Section modulus | | Radius of gyration | | | $i_y$, when "$a$" is the following value | | | | |
|---|---|---|---|---|---|---|---|---|---|---|---|---|---|---|---|
| | | | | | | $W_x^{max}$ | $W_x^{min}$ | $i_x$ | $i_{x0}$ | $i_{y0}$ | 6 mm | 8 mm | 10 mm | 12 mm | 14 mm |
| | mm | | cm² | kg/m | cm⁴ | cm³ | | cm | | | cm | | | | |
| L80× 5 | 9 | 21.5 | 7.91 | 6.21 | 48.79 | 22.70 | 8.34 | 2.48 | 3.13 | 1.60 | 3.49 | 3.56 | 3.63 | 3.71 | 3.78 |
| 6 | | 21.9 | 9.40 | 7.38 | 57.35 | 26.16 | 9.87 | 2.47 | 3.11 | 1.59 | 3.51 | 3.58 | 3.65 | 3.73 | 3.80 |
| 7 | | 22.3 | 10.86 | 8.53 | 65.58 | 29.38 | 11.37 | 2.46 | 3.10 | 1.58 | 3.53 | 3.60 | 3.67 | 3.75 | 3.82 |
| 8 | | 22.7 | 12.30 | 9.66 | 73.50 | 32.36 | 12.83 | 2.44 | 3.08 | 1.57 | 3.55 | 3.62 | 3.69 | 3.77 | 3.85 |
| 9 | | 23.1 | 13.73 | 10.77 | 81.11 | 35.11 | 14.25 | 2.43 | 3.06 | 1.56 | 3.57 | 3.64 | 3.72 | 3.79 | 3.87 |
| 10 | | 23.5 | 15.13 | 11.87 | 88.43 | 37.68 | 15.64 | 2.42 | 3.04 | 1.56 | 3.57 | 3.66 | 3.74 | 3.81 | 3.89 |
| L90×6 | 10 | 24.4 | 10.64 | 8.35 | 82.77 | 33.99 | 12.61 | 2.79 | 3.51 | 1.80 | 3.91 | 3.98 | 4.05 | 4.12 | 4.20 |
| 7 | | 24.8 | 12.30 | 9.66 | 94.83 | 38.28 | 14.54 | 2.78 | 3.50 | 1.78 | 3.93 | 4.00 | 4.07 | 4.14 | 4.22 |
| 8 | | 25.2 | 13.94 | 10.95 | 106.5 | 42.30 | 16.42 | 2.76 | 3.48 | 1.78 | 3.95 | 4.02 | 4.09 | 4.17 | 4.24 |
| 9 | | 25.9 | 17.17 | 13.48 | 128.6 | 49.57 | 20.07 | 2.74 | 3.45 | 1.76 | 3.98 | 4.05 | 4.13 | 4.21 | 4.28 |
| 10 | | 26.7 | 20.31 | 15.94 | 149.2 | 55.93 | 23.57 | 2.71 | 3.41 | 1.75 | 4.02 | 4.10 | 4.17 | 4.25 | 4.32 |
| L100×6 | 12 | 26.7 | 11.93 | 9.37 | 115.0 | 43.04 | 15.68 | 3.10 | 3.09 | 2.00 | 4.30 | 4.37 | 4.44 | 4.51 | 4.58 |
| 7 | | 27.1 | 13.80 | 10.83 | 131.9 | 48.57 | 18.10 | 3.09 | 3.89 | 1.99 | 4.31 | 4.39 | 4.46 | 4.53 | 4.60 |
| 8 | | 27.6 | 15.64 | 12.28 | 148.2 | 53.78 | 20.47 | 3.08 | 3.88 | 1.98 | 4.34 | 4.41 | 4.48 | 4.56 | 4.63 |
| 10 | | 28.4 | 19.26 | 15.12 | 179.5 | 63.29 | 25.06 | 3.05 | 3.84 | 1.96 | 4.38 | 4.45 | 4.52 | 4.60 | 4.67 |
| 12 | | 29.1 | 22.80 | 17.90 | 208.9 | 71.72 | 29.48 | 3.03 | 3.81 | 1.95 | 4.41 | 4.49 | 4.56 | 4.63 | 4.71 |
| 14 | | 29.9 | 26.26 | 20.61 | 236.5 | 79.19 | 33.73 | 3.00 | 3.77 | 1.94 | 4.45 | 4.53 | 4.60 | 4.68 | 4.76 |
| 16 | | 26.7 | 11.93 | 9.37 | 115.0 | 43.04 | 15.68 | 3.10 | 3.09 | 2.00 | 4.30 | 4.37 | 4.44 | 4.51 | 4.58 |
| L110×7 | 12 | 29.6 | 15.20 | 11.93 | 177.2 | 59.78 | 22.05 | 3.41 | 4.30 | 2.20 | 4.72 | 4.79 | 4.86 | 4.93 | 5.01 |
| 8 | | 30.1 | 17.24 | 13.54 | 199.5 | 66.36 | 24.95 | 3.40 | 4.28 | 2.19 | 4.75 | 4.81 | 4.89 | 4.96 | 5.03 |
| 10 | | 30.9 | 21.26 | 16.69 | 242.2 | 78.48 | 30.60 | 3.38 | 4.25 | 2.17 | 4.78 | 4.86 | 4.93 | 5.00 | 5.07 |
| 12 | | 31.6 | 25.20 | 19.78 | 282.6 | 89.34 | 36.05 | 3.35 | 4.22 | 2.15 | 4.81 | 4.89 | 4.96 | 5.03 | 5.11 |
| 14 | | 32.4 | 29.06 | 22.81 | 320.7 | 99.07 | 41.31 | 3.32 | 4.18 | 2.14 | 4.85 | 4.93 | 5.00 | 5.08 | 5.15 |
| L125× 8 | 14 | 33.7 | 19.75 | 15.50 | 297.0 | 88.20 | 32.52 | 3.88 | 4.88 | 2.50 | 5.34 | 5.41 | 5.48 | 5.55 | 5.62 |
| 10 | | 34.5 | 24.37 | 19.13 | 361.7 | 104.8 | 39.97 | 3.85 | 4.85 | 2.48 | 5.38 | 5.45 | 5.52 | 5.59 | 5.66 |
| 12 | | 35.3 | 28.91 | 22.70 | 423.2 | 119.9 | 47.17 | 3.83 | 4.82 | 2.46 | 5.41 | 5.48 | 5.56 | 5.63 | 5.70 |
| 14 | | 36.1 | 33.37 | 26.19 | 481.7 | 133.6 | 54.16 | 3.80 | 4.78 | 2.45 | 5.45 | 5.52 | 5.60 | 5.67 | 5.74 |
| L140×10 | 14 | 38.2 | 27.37 | 21.49 | 514.7 | 134.6 | 50.58 | 4.34 | 5.46 | 2.78 | 5.98 | 6.05 | 6.12 | 6.19 | 6.26 |
| 12 | | 39.0 | 32.51 | 25.52 | 603.7 | 154.6 | 59.80 | 4.31 | 5.43 | 2.76 | 6.02 | 6.09 | 6.16 | 6.23 | 6.30 |
| 14 | | 39.8 | 37.57 | 29.49 | 688.8 | 173.0 | 68.75 | 4.28 | 5.40 | 2.75 | 6.06 | 6.13 | 6.20 | 6.27 | 6.34 |
| 16 | | 40.6 | 42.54 | 33.39 | 770.2 | 189.9 | 77.46 | 4.26 | 5.36 | 2.74 | 6.09 | 6.16 | 6.24 | 6.31 | 6.38 |

APPENDIX F  TYPICAL STEEL SHAPES

continued

| Type | Fillet radius $R$ | Center of gravity $Z_0$ | Sectional area $A$ | Theoretical weight | Moment of inertia $I_x$ | Section modulus | | Radius of gyration | | | $i_y$, when "$a$" is the following value | | | | |
|---|---|---|---|---|---|---|---|---|---|---|---|---|---|---|---|
| | | | | | | $W_x^{max}$ | $W_x^{min}$ | $i_x$ | $i_{x0}$ | $i_{y0}$ | 6 mm | 8 mm | 10 mm | 12 mm | 14 mm |
| | mm | | cm² | kg/m | cm⁴ | cm³ | | cm | | | cm | | | | |
| L160× 10 | 16 | 43.1 | 31.50 | 24.73 | 779.5 | 180.8 | 66.70 | 4.97 | 6.27 | 3.20 | 6.78 | 6.85 | 6.92 | 6.99 | 7.06 |
| L160× 12 | 16 | 43.9 | 37.44 | 29.39 | 916.6 | 208.6 | 78.98 | 4.95 | 6.24 | 3.18 | 6.82 | 6.89 | 6.96 | 7.03 | 7.10 |
| L160× 14 | 16 | 44.7 | 43.30 | 33.99 | 1048 | 234.4 | 90.95 | 4.92 | 6.20 | 3.16 | 6.85 | 6.93 | 6.99 | 7.07 | 7.14 |
| L160× 16 | 16 | 45.5 | 49.07 | 38.52 | 1175 | 258.3 | 102.6 | 4.89 | 6.17 | 3.14 | 6.89 | 6.96 | 7.03 | 7.10 | 7.18 |
| L180× 12 | 16 | 48.9 | 42.24 | 33.16 | 1321 | 270.0 | 100.82 | 5.59 | 7.05 | 3.58 | 7.63 | 7.73 | 7.77 | 7.84 | 7.91 |
| L180× 14 | 16 | 49.7 | 48.90 | 38.38 | 1514 | 304.6 | 116.25 | 5.57 | 7.02 | 3.56 | 7.66 | 7.73 | 7.80 | 7.87 | 7.94 |
| L180× 16 | 16 | 50.5 | 55.47 | 43.54 | 1701 | 336.9 | 131.13 | 5.54 | 6.98 | 3.55 | 7.70 | 7.77 | 7.84 | 7.91 | 7.98 |
| L180× 18 | 16 | 51.3 | 61.95 | 48.63 | 1881 | 367.1 | 146.64 | 5.51 | 6.94 | 3.53 | 7.76 | 7.83 | 7.90 | 7.97 | 8.04 |
| L200× 14 | 18 | 54.6 | 54.64 | 42.89 | 2104 | 385.1 | 144.70 | 6.20 | 7.82 | 3.98 | 8.47 | 8.53 | 8.60 | 8.67 | 8.74 |
| L200× 16 | 18 | 55.4 | 62.01 | 48.68 | 2366 | 427.0 | 163.65 | 6.18 | 7.79 | 3.96 | 8.50 | 8.57 | 8.64 | 8.71 | 8.78 |
| L200× 18 | 18 | 56.2 | 69.30 | 54.40 | 2621 | 466.5 | 182.22 | 6.15 | 7.75 | 3.94 | 8.54 | 8.61 | 8.68 | 8.75 | 8.82 |
| L200× 20 | 18 | 56.9 | 76.50 | 60.06 | 2867 | 503.6 | 200.42 | 6.12 | 7.72 | 3.93 | 8.56 | 8.64 | 8.71 | 8.78 | 8.85 |
| L200× 24 | 18 | 58.4 | 90.66 | 71.17 | 3338 | 571.5 | 236.17 | 6.07 | 7.64 | 3.90 | 8.65 | 8.73 | 8.80 | 8.87 | 8.94 |
| L220× 16 | 21 | 60.3 | 68.66 | 53.90 | 3187.36 | 528.58 | 199.55 | 6.81 | 8.59 | 4.37 | 9.30 | 9.37 | 9.44 | 9.51 | 9.58 |
| L220× 18 | 21 | 61.1 | 76.75 | 60.25 | 3534.30 | 578.45 | 222.37 | 6.79 | 8.55 | 4.35 | 9.33 | 9.40 | 9.47 | 9.54 | 9.61 |
| L220× 20 | 21 | 61.8 | 84.75 | 66.53 | 3871.49 | 626.45 | 244.77 | 6.76 | 8.52 | 4.34 | 9.36 | 9.43 | 9.50 | 9.57 | 9.64 |
| L220× 22 | 21 | 62.6 | 92.68 | 72.75 | 4199.23 | 670.80 | 266.78 | 6.73 | 8.48 | 4.32 | 9.40 | 9.47 | 9.54 | 9.61 | 9.68 |
| L220× 24 | 21 | 63.3 | 100.51 | 78.90 | 4517.83 | 713.72 | 288.39 | 6.70 | 8.45 | 4.31 | 9.43 | 9.50 | 9.57 | 9.64 | 9.71 |
| L220× 26 | 21 | 64.1 | 108.26 | 84.99 | 4827.58 | 753.13 | 309.62 | 6.68 | 8.41 | 4.30 | 9.47 | 9.54 | 9.61 | 9.68 | 9.75 |
| L250× 18 | 22 | 68.4 | 87.84 | 68.96 | 5268.22 | 770.21 | 290.12 | 7.74 | 9.76 | 4.97 | 10.53 | 10.60 | 10.67 | 10.74 | 10.81 |
| L250× 20 | 22 | 69.2 | 97.05 | 76.18 | 5779.34 | 835.16 | 319.66 | 7.72 | 9.73 | 4.95 | 10.57 | 10.64 | 10.71 | 10.78 | 10.85 |
| L250× 24 | 22 | 70.7 | 115.20 | 90.43 | 6763.93 | 956.71 | 377.34 | 7.66 | 9.66 | 4.92 | 10.63 | 10.70 | 10.77 | 10.84 | 10.91 |
| L250× 26 | 22 | 71.5 | 124.15 | 97.46 | 7238.08 | 1012.32 | 405.50 | 7.63 | 9.62 | 4.90 | 10.67 | 10.74 | 10.81 | 10.88 | 10.95 |
| L250× 28 | 22 | 72.2 | 133.02 | 104.42 | 7700.60 | 1066.57 | 433.22 | 7.61 | 9.58 | 4.89 | 10.70 | 10.77 | 10.84 | 10.91 | 10.98 |
| L250× 30 | 22 | 73.0 | 141.81 | 111.32 | 8151.80 | 1116.69 | 460.51 | 7.58 | 9.55 | 4.88 | 10.74 | 10.81 | 10.88 | 10.95 | 11.02 |
| L250× 32 | 22 | 73.7 | 150.51 | 118.15 | 8592.01 | 1165.81 | 487.39 | 7.56 | 9.51 | 4.87 | 10.77 | 10.84 | 10.91 | 10.98 | 11.05 |
| L250× 35 | 22 | 74.8 | 163.40 | 128.27 | 9232.44 | 1234.28 | 526.97 | 7.52 | 9.46 | 4.86 | 10.82 | 10.89 | 10.96 | 11.04 | 11.11 |

# APPENDIX F  TYPICAL STEEL SHAPES

**Unequal leg Angle**  Table F. 6

| Type $B\times b\times t$ | | Fillet radius $R$ | Center of gravity | | Sectional area $A$ | Theoretical weight | Radius of gyration | | | $i_{y1}$, when a is the following value | | | | $i_{y2}$, when a is the following value | | | |
|---|---|---|---|---|---|---|---|---|---|---|---|---|---|---|---|---|---|
| | | | $Z_x$ | $Z_y$ | | | $i_x$ | $i_y$ | $i_{y0}$ | 6 mm | 8 mm | 10 mm | 12 mm | 6 mm | 8 mm | 10 mm | 12 mm |
| | | mm | mm | | cm² | kg/m | cm | | | cm | | | | cm | | | |
| L25×16× | 3 | 3.5 | 4.2 | 8.6 | 1.16 | 0.91 | 0.44 | 0.78 | 0.34 | 0.84 | 0.93 | 1.02 | 1.11 | 1.40 | 1.48 | 1.57 | 1.65 |
| | 4 | | 4.6 | 10.4 | 1.50 | 1.18 | 0.43 | 0.77 | 0.34 | 0.87 | 0.96 | 1.05 | 1.14 | 1.54 | 1.63 | 1.72 | 1.81 |
| L32×20× | 3 | 3.5 | 4.9 | 9.0 | 1.49 | 1.17 | 0.55 | 1.01 | 0.43 | 0.97 | 1.05 | 1.14 | 1.22 | 1.57 | 1.65 | 1.73 | 1.81 |
| | 4 | | 5.3 | 10.8 | 1.94 | 1.52 | 0.54 | 1.00 | 0.42 | 0.99 | 1.08 | 1.16 | 1.25 | 1.70 | 1.78 | 1.87 | 1.95 |
| L40×25× | 3 | 4 | 5.9 | 11.2 | 1.89 | 1.48 | 0.70 | 1.28 | 0.54 | 1.13 | 1.21 | 1.30 | 1.38 | 1.91 | 1.98 | 2.06 | 2.14 |
| | 4 | | 6.3 | 13.2 | 2.47 | 1.94 | 0.69 | 1.36 | 0.54 | 1.16 | 1.24 | 1.32 | 1.41 | 2.05 | 2.13 | 2.21 | 2.30 |
| L45×28× | 3 | 5 | 6.4 | 13.7 | 2.15 | 1.69 | 0.79 | 1.44 | 0.61 | 1.23 | 1.31 | 1.39 | 1.47 | 2.20 | 2.28 | 2.36 | 2.44 |
| | 4 | | 6.8 | 14.7 | 2.81 | 2.20 | 0.78 | 1.42 | 0.60 | 1.25 | 1.33 | 1.41 | 1.50 | 2.27 | 2.35 | 2.43 | 2.51 |
| L50×32× | 3 | 5.5 | 7.3 | 15.1 | 2.43 | 1.91 | 0.91 | 1.60 | 0.70 | 1.38 | 1.45 | 1.53 | 1.61 | 2.42 | 2.49 | 2.57 | 2.65 |
| | 4 | | 7.7 | 16.0 | 3.18 | 2.49 | 0.90 | 1.59 | 0.69 | 1.40 | 1.48 | 1.56 | 1.64 | 2.48 | 2.55 | 2.63 | 2.71 |
| L56×36× | 3 | 6 | 8.0 | 16.5 | 2.74 | 2.15 | 1.03 | 1.80 | 0.79 | 1.51 | 1.58 | 1.66 | 1.74 | 2.65 | 2.73 | 2.80 | 2.88 |
| | 4 | | 8.5 | 17.8 | 3.59 | 2.82 | 1.02 | 1.79 | 0.79 | 1.54 | 1.62 | 1.69 | 1.77 | 2.74 | 2.82 | 2.90 | 2.98 |
| | 5 | | 8.8 | 18.2 | 4.42 | 3.47 | 1.01 | 1.77 | 0.78 | 1.55 | 1.63 | 1.71 | 1.79 | 2.76 | 2.84 | 2.92 | 3.00 |
| L63×40× | 4 | 7 | 9.2 | 18.7 | 4.06 | 3.19 | 1.14 | 2.02 | 0.88 | 1.67 | 1.74 | 1.82 | 1.90 | 3.96 | 3.04 | 3.11 | 3.19 |
| | 5 | | 9.5 | 20.4 | 4.99 | 3.92 | 1.12 | 2.00 | 0.87 | 1.68 | 1.76 | 1.83 | 1.91 | 3.08 | 3.16 | 3.23 | 3.31 |
| | 6 | | 9.9 | 20.8 | 5.91 | 4.64 | 1.11 | 1.96 | 0.86 | 1.70 | 1.78 | 1.86 | 1.94 | 3.10 | 3.18 | 3.26 | 3.34 |
| | 7 | | 10.3 | 21.2 | 6.80 | 5.34 | 1.10 | 1.98 | 0.86 | 1.73 | 1.80 | 1.88 | 1.97 | 3.12 | 3.20 | 3.28 | 3.36 |
| L70×45× | 4 | 7.5 | 10.2 | 21.5 | 4.55 | 3.57 | 1.29 | 2.26 | 0.98 | 1.84 | 1.92 | 1.99 | 2.07 | 3.33 | 3.41 | 3.48 | 3.56 |
| | 5 | | 10.6 | 22.4 | 5.61 | 4.40 | 1.28 | 2.23 | 0.98 | 1.86 | 1.94 | 2.02 | 2.09 | 3.38 | 3.46 | 3.53 | 3.61 |
| | 6 | | 10.9 | 22.8 | 6.64 | 5.22 | 1.26 | 2.21 | 0.98 | 1.88 | 1.95 | 2.03 | 2.11 | 3.40 | 3.48 | 3.55 | 3.63 |
| | 7 | | 11.3 | 23.2 | 7.66 | 6.01 | 1.25 | 2.20 | 0.97 | 1.90 | 1.98 | 2.06 | 2.14 | 3.42 | 3.50 | 3.58 | 3.66 |
| L75×50× | 5 | 8 | 11.7 | 23.6 | 6.13 | 4.81 | 1.44 | 2.39 | 1.10 | 2.05 | 2.13 | 2.20 | 2.28 | 3.57 | 3.65 | 3.72 | 3.80 |
| | 6 | | 12.1 | 24.0 | 7.26 | 5.70 | 1.42 | 2.38 | 1.08 | 2.07 | 2.15 | 2.22 | 2.30 | 3.60 | 3.67 | 3.75 | 3.83 |
| | 8 | | 12.9 | 24.4 | 9.47 | 7.43 | 1.40 | 2.35 | 1.07 | 2.12 | 2.19 | 2.27 | 2.35 | 3.61 | 3.69 | 3.77 | 3.84 |
| | 10 | | 13.6 | 25.2 | 11.59 | 9.10 | 1.38 | 2.33 | 1.06 | 2.16 | 2.23 | 2.31 | 2.40 | 3.66 | 3.73 | 3.81 | 3.89 |
| L80×50× | 5 | 8 | 11.4 | 26.0 | 6.38 | 5.00 | 1.42 | 2.56 | 1.10 | 2.02 | 2.09 | 2.17 | 2.24 | 3.87 | 3.95 | 4.02 | 4.10 |
| | 6 | | 11.8 | 26.5 | 7.56 | 5.94 | 1.41 | 2.56 | 1.08 | 2.04 | 2.12 | 2.19 | 2.27 | 3.90 | 3.98 | 4.06 | 4.14 |
| | 7 | | 12.1 | 26.9 | 8.72 | 6.85 | 1.39 | 2.54 | 1.08 | 2.77 | 2.82 | 2.88 | 2.94 | 3.92 | 4.00 | 4.08 | 4.15 |
| | 8 | | 12.5 | 27.3 | 9.87 | 7.75 | 1.38 | 2.52 | 1.07 | 2.08 | 2.15 | 2.23 | 2.31 | 3.94 | 4.02 | 4.10 | 4.18 |
| L90×56× | 4 | 9 | 12.5 | 29.1 | 7.21 | 5.66 | 1.59 | 2.90 | 1.23 | 2.22 | 2.29 | 2.37 | 2.44 | 4.32 | 4.40 | 4.47 | 4.55 |
| | 5 | | 12.9 | 29.5 | 8.56 | 6.72 | 1.58 | 2.88 | 1.23 | 2.24 | 2.32 | 2.39 | 2.46 | 4.34 | 4.42 | 4.49 | 4.57 |
| | 6 | | 13.3 | 30.0 | 9.88 | 7.76 | 1.57 | 2.86 | 1.22 | 2.26 | 2.34 | 2.41 | 2.49 | 4.37 | 4.45 | 4.52 | 4.60 |
| | 7 | | 13.6 | 30.4 | 11.18 | 8.78 | 1.56 | 2.85 | 1.21 | 2.28 | 2.35 | 2.43 | 2.50 | 4.39 | 4.47 | 4.55 | 4.62 |

APPENDIX F  TYPICAL STEEL SHAPES

continued

| Type $B\times b\times t$ | | Single angle | | | | | | | | Double angle | | | | | | | |
|---|---|---|---|---|---|---|---|---|---|---|---|---|---|---|---|---|---|
| | | Fillet radius $R$ | Center of gravity | | Sectional area $A$ | Theoretical weight | Radius of gyration | | | $i_{y1}$, when a is the following value | | | | $i_{y2}$, when a is the following value | | | |
| | | | $Z_x$ | $Z_y$ | | | $i_x$ | $i_y$ | $i_{y0}$ | 6 mm | 8 mm | 10 mm | 12 mm | 6 mm | 8 mm | 10 mm | 12 mm |
| | | mm | | | cm² | kg/m | cm | | | cm | | | | cm | | | |
| L100×63× | 6 | 10 | 14.3 | 32.4 | 9.62 | 7.55 | 1.79 | 3.21 | 1.38 | 2.49 | 2.56 | 2.63 | 2.71 | 4.78 | 4.85 | 4.93 | 5.00 |
| | 7 | | 14.7 | 32.8 | 11.10 | 8.72 | 1.78 | 3.20 | 1.38 | 2.51 | 2.58 | 2.66 | 2.73 | 4.80 | 4.87 | 4.95 | 5.03 |
| | 8 | | 15.0 | 33.2 | 12.53 | 9.88 | 1.77 | 3.18 | 1.37 | 2.53 | 2.60 | 2.67 | 2.75 | 4.82 | 4.90 | 4.98 | 5.05 |
| | 10 | | 15.8 | 34.0 | 15.47 | 12.14 | 1.74 | 3.15 | 1.35 | 2.57 | 2.64 | 2.72 | 2.79 | 4.86 | 4.94 | 5.02 | 5.09 |
| L100×80× | 6 | 10 | 19.7 | 29.5 | 10.64 | 8.35 | 2.40 | 3.17 | 1.72 | 3.30 | 3.37 | 3.44 | 3.52 | 4.54 | 4.61 | 4.69 | 4.76 |
| | 7 | | 20.1 | 30.0 | 12.30 | 9.66 | 2.39 | 3.16 | 1.72 | 3.32 | 3.39 | 3.46 | 3.54 | 4.57 | 4.64 | 4.71 | 4.79 |
| | 8 | | 20.5 | 30.4 | 13.94 | 10.95 | 2.37 | 3.14 | 1.71 | 3.34 | 3.41 | 3.48 | 3.56 | 4.59 | 4.66 | 4.74 | 4.81 |
| | 10 | | 21.3 | 31.2 | 17.17 | 13.48 | 2.35 | 3.12 | 1.69 | 3.38 | 3.45 | 3.53 | 3.60 | 4.63 | 4.70 | 4.78 | 4.85 |
| L110×70× | 6 | 10 | 15.7 | 35.3 | 10.64 | 8.35 | 2.01 | 3.54 | 1.54 | 2.74 | 2.81 | 2.88 | 2.96 | 5.22 | 5.29 | 5.36 | 5.44 |
| | 7 | | 16.1 | 35.7 | 12.30 | 9.66 | 2.00 | 3.53 | 1.53 | 2.76 | 2.83 | 2.90 | 2.98 | 5.24 | 5.31 | 5.39 | 5.46 |
| | 8 | | 16.5 | 36.2 | 13.94 | 10.95 | 1.98 | 3.51 | 1.53 | 2.78 | 2.85 | 2.93 | 3.00 | 5.26 | 5.34 | 5.41 | 5.49 |
| | 10 | | 17.2 | 37.0 | 17.17 | 13.48 | 1.96 | 3.48 | 1.51 | 2.81 | 2.89 | 2.96 | 3.04 | 5.30 | 5.38 | 5.46 | 5.53 |
| L125×80× | 7 | 11 | 18.0 | 40.1 | 14.10 | 11.07 | 2.30 | 4.02 | 1.76 | 3.11 | 3.18 | 3.25 | 3.32 | 5.89 | 5.97 | 6.04 | 6.12 |
| | 8 | | 18.4 | 40.6 | 15.99 | 12.55 | 2.28 | 4.01 | 1.75 | 3.13 | 3.20 | 3.27 | 3.34 | 5.92 | 6.00 | 6.07 | 6.15 |
| | 10 | | 19.2 | 41.4 | 19.71 | 15.47 | 2.26 | 3.98 | 1.74 | 3.17 | 3.24 | 3.31 | 3.38 | 5.96 | 6.04 | 6.11 | 6.19 |
| | 12 | | 20.0 | 42.2 | 23.35 | 18.33 | 2.24 | 3.95 | 1.72 | 3.21 | 3.28 | 3.35 | 3.43 | 6.00 | 6.08 | 6.16 | 6.23 |
| L140×90× | 8 | 12 | 20.4 | 45.0 | 18.04 | 14.16 | 2.59 | 4.50 | 1.98 | 3.49 | 3.56 | 3.63 | 3.70 | 6.58 | 6.65 | 6.73 | 6.80 |
| | 10 | | 21.2 | 45.8 | 22.26 | 17.48 | 2.56 | 4.47 | 1.96 | 3.49 | 3.56 | 3.63 | 3.70 | 6.62 | 6.69 | 6.77 | 6.84 |
| | 12 | | 21.9 | 46.6 | 26.40 | 20.72 | 2.54 | 4.44 | 1.95 | 3.55 | 3.62 | 3.70 | 3.77 | 6.66 | 6.74 | 6.81 | 6.89 |
| | 14 | | 22.7 | 47.4 | 30.46 | 23.91 | 2.51 | 4.42 | 1.94 | 3.59 | 3.67 | 3.74 | 3.81 | 6.70 | 6.78 | 6.85 | 6.93 |
| L150×90× | 8 | 12 | 19.7 | 49.2 | 18.84 | 14.79 | 2.55 | 4.84 | 1.98 | 3.42 | 3.48 | 3.55 | 3.62 | 7.12 | 7.19 | 7.27 | 7.34 |
| | 10 | | 20.5 | 50.1 | 23.26 | 18.26 | 2.53 | 4.81 | 1.97 | 3.45 | 3.52 | 3.59 | 3.66 | 7.17 | 7.24 | 7.32 | 7.39 |
| | 12 | | 21.2 | 50.9 | 27.60 | 21.67 | 2.50 | 4.79 | 1.95 | 3.48 | 3.55 | 3.62 | 3.70 | 7.21 | 7.28 | 7.36 | 7.43 |
| | 14 | | 22.0 | 51.7 | 31.86 | 25.01 | 2.48 | 4.76 | 1.94 | 3.52 | 3.59 | 3.66 | 3.74 | 7.25 | 7.32 | 7.40 | 7.48 |
| | 15 | | 22.4 | 52.1 | 33.95 | 26.65 | 2.47 | 4.74 | 1.93 | 3.54 | 3.61 | 3.69 | 3.76 | 7.27 | 7.35 | 7.42 | 7.50 |
| | 16 | | 22.7 | 52.5 | 36.03 | 28.28 | 2.45 | 4.73 | 1.93 | 3.55 | 3.63 | 3.70 | 3.78 | 7.29 | 7.37 | 7.44 | 7.52 |
| L160×100× | 10 | 13 | 22.8 | 52.4 | 25.32 | 19.87 | 2.85 | 5.14 | 2.19 | 3.84 | 3.91 | 3.98 | 4.05 | 7.56 | 7.63 | 7.70 | 7.78 |
| | 12 | | 23.6 | 53.2 | 30.05 | 23.59 | 2.82 | 5.11 | 2.17 | 3.88 | 3.95 | 4.02 | 4.09 | 7.60 | 7.67 | 7.75 | 7.82 |
| | 14 | | 24.3 | 54.0 | 34.71 | 27.25 | 2.80 | 5.08 | 2.16 | 3.91 | 3.98 | 4.05 | 4.12 | 7.64 | 7.71 | 7.79 | 7.86 |
| | 16 | | 25.1 | 54.8 | 39.28 | 30.84 | 2.77 | 5.05 | 2.16 | 3.95 | 4.02 | 4.09 | 4.17 | 7.68 | 7.75 | 7.83 | 7.91 |
| L180×110× | 10 | 14 | 24.4 | 58.9 | 28.37 | 22.27 | 3.13 | 5.80 | 2.42 | 4.16 | 4.23 | 4.29 | 4.36 | 8.49 | 8.56 | 8.63 | 8.71 |
| | 12 | | 25.2 | 59.8 | 33.71 | 26.44 | 3.10 | 5.78 | 2.40 | 4.19 | 4.26 | 4.33 | 4.40 | 8.53 | 8.61 | 8.68 | 8.76 |
| | 14 | | 25.9 | 60.6 | 38.97 | 30.59 | 3.08 | 5.75 | 2.39 | 4.22 | 4.29 | 4.36 | 4.43 | 8.57 | 8.65 | 8.72 | 8.80 |
| | 16 | | 26.7 | 61.4 | 44.14 | 34.65 | 3.06 | 5.72 | 2.38 | 4.26 | 4.33 | 4.40 | 4.47 | 8.61 | 8.69 | 8.76 | 8.84 |
| L200×125× | 12 | 14 | 28.3 | 65.4 | 37.91 | 29.76 | 3.57 | 6.44 | 2.74 | 4.75 | 4.81 | 4.88 | 4.95 | 9.39 | 9.47 | 9.54 | 9.61 |
| | 14 | | 29.1 | 66.2 | 43.69 | 34.44 | 3.54 | 6.41 | 2.73 | 4.79 | 4.85 | 4.92 | 4.99 | 9.44 | 9.51 | 9.59 | 9.66 |
| | 16 | | 29.9 | 67.0 | 49.74 | 39.05 | 3.52 | 6.38 | 2.71 | 4.82 | 4.89 | 4.95 | 5.03 | 9.47 | 9.54 | 9.62 | 9.69 |
| | 18 | | 30.6 | 67.8 | 55.53 | 43.59 | 3.49 | 6.35 | 2.70 | 4.85 | 4.92 | 4.99 | 5.06 | 9.51 | 9.58 | 9.66 | 9.74 |

# APPENDIX F  TYPICAL STEEL SHAPES

**Structural steel pipe**  Table F. 7

Symbol: $I$ — Moment of inertia
  $W$ — Sectional modulus
  $i$ — Radius of gyration

| External diameter $D$ (mm) | Thickness $t$ Seamless | Thickness $t$ Welded | Sectional area (cm²) | Theoretical weight (kg/m) | Surface area (m²/m) | Moment of inertia $I$ (cm⁴) | Sectional modulus $W$ (cm³) | Radius of gyration $i$ (cm) |
|---|---|---|---|---|---|---|---|---|
| 30.0 | — | 2.0 | 1.76 | 1.38 | 0.09 | 1.73 | 1.16 | 0.99 |
|      | — | 2.5 | 2.16 | 1.70 | 0.09 | 2.06 | 1.37 | 0.98 |
| 34.0 | — | 2.0 | 2.01 | 1.58 | 0.11 | 2.58 | 1.52 | 1.13 |
|      | — | 2.5 | 2.47 | 1.94 | 0.11 | 3.09 | 1.82 | 1.12 |
| 38.0 | — | 2.0 | 2.26 | 1.78 | 0.12 | 3.68 | 1.93 | 1.27 |
|      | 2.5 | 2.5 | 2.79 | 2.19 | 0.12 | 4.41 | 2.32 | 1.26 |
|      | 3.0 | — | 3.30 | 2.59 | 0.12 | 5.09 | 2.68 | 1.24 |
|      | 3.5 | — | 3.79 | 2.98 | 0.12 | 5.70 | 3.00 | 1.23 |
| 40.0 | — | 2.0 | 2.39 | 1.87 | 0.13 | 4.32 | 2.16 | 1.35 |
|      | — | 2.5 | 2.95 | 2.31 | 0.13 | 5.20 | 2.60 | 1.33 |
| 42.0 | — | 2.0 | 2.51 | 1.97 | 0.13 | 5.04 | 2.40 | 1.42 |
|      | 2.5 | 2.5 | 3.10 | 2.44 | 0.13 | 6.07 | 2.89 | 1.40 |
|      | 3.0 | — | 3.68 | 2.89 | 0.13 | 7.03 | 3.35 | 1.38 |
|      | 3.5 | — | 4.23 | 3.32 | 0.13 | 7.91 | 3.77 | 1.37 |
|      | 4.0 | — | 4.78 | 3.75 | 0.13 | 8.71 | 4.15 | 1.35 |
| 45.0 | — | 2.0 | 2.70 | 2.12 | 0.14 | 6.26 | 2.78 | 1.52 |
|      | 2.5 | 2.5 | 3.34 | 2.62 | 0.14 | 7.56 | 3.36 | 1.51 |
|      | 3.0 | 3.0 | 3.96 | 3.11 | 0.14 | 8.77 | 3.90 | 1.49 |
|      | 3.5 | — | 4.56 | 3.58 | 0.14 | 9.89 | 4.40 | 1.47 |
|      | 4.0 | — | 5.15 | 4.04 | 0.14 | 10.93 | 4.86 | 1.46 |
| 50.0 | 2.5 | — | 3.73 | 2.93 | 0.16 | 10.55 | 4.22 | 1.68 |
|      | 3.0 | — | 4.43 | 3.48 | 0.16 | 12.28 | 4.91 | 1.67 |
|      | 3.5 | — | 5.11 | 4.01 | 0.16 | 13.90 | 5.56 | 1.65 |
|      | 4.0 | — | 5.78 | 4.54 | 0.16 | 15.41 | 6.16 | 1.63 |
|      | 4.5 | — | 6.43 | 5.05 | 0.16 | 16.81 | 6.72 | 1.62 |
|      | 5.0 | — | 7.07 | 5.55 | 0.16 | 18.11 | 7.25 | 1.60 |
| 51.0 | — | 2.0 | 3.08 | 2.42 | 0.16 | 9.26 | 8.63 | 1.73 |
|      | — | 2.5 | 3.81 | 2.99 | 0.16 | 11.23 | 4.40 | 1.72 |
|      | — | 3.0 | 4.52 | 3.55 | 0.16 | 13.08 | 5.13 | 1.70 |
|      | — | 3.5 | 5.22 | 4.10 | 0.16 | 14.81 | 5.81 | 1.68 |
| 54.0 | — | 2.0 | 3.27 | 2.56 | 0.17 | 11.06 | 4.10 | 1.84 |
|      | — | 2.5 | 4.04 | 3.18 | 0.17 | 13.44 | 4.98 | 1.82 |
|      | 3.0 | 3.0 | 4.81 | 3.77 | 0.17 | 15.68 | 5.81 | 1.81 |
|      | 3.5 | 3.5 | 5.55 | 4.36 | 0.17 | 17.79 | 6.59 | 1.79 |
|      | 4.0 | — | 6.28 | 4.93 | 0.17 | 19.76 | 7.32 | 1.77 |
|      | 4.5 | — | 7.00 | 5.49 | 0.17 | 21.61 | 8.00 | 1.76 |
|      | 5.0 | — | 7.70 | 6.04 | 0.17 | 23.34 | 8.64 | 1.74 |

APPENDIX F  TYPICAL STEEL SHAPES

continued

| External diameter $D$ (mm) | Thickness $t$ Seamless | Thickness $t$ Welded | Sectional area (cm$^2$) | Theoretical weight (kg/m) | Surface area (m$^2$/m) | Section characteristics Moment of inertia $I$ (cm$^4$) | Section characteristics Sectional modulus $W$ (cm$^3$) | Section characteristics Radius of gyration $i$ (cm) |
|---|---|---|---|---|---|---|---|---|
| 57.0 | — | 2.0 | 3.46 | 2.71 | 0.18 | 13.08 | 4.59 | 1.95 |
| | — | 2.5 | 4.28 | 3.36 | 0.18 | 15.93 | 5.59 | 1.93 |
| | 3.0 | 3.0 | 5.09 | 4.00 | 0.18 | 18.61 | 6.53 | 1.91 |
| | 3.5 | 3.5 | 5.88 | 4.62 | 0.18 | 21.14 | 7.42 | 1.90 |
| | 4.0 | — | 6.66 | 5.23 | 0.18 | 23.52 | 8.25 | 1.88 |
| | 4.5 | — | 7.42 | 5.83 | 0.18 | 25.76 | 9.04 | 1.86 |
| | 5.0 | — | 8.17 | 6.41 | 0.18 | 27.86 | 9.78 | 1.85 |
| | 5.5 | — | 8.90 | 6.99 | 0.18 | 29.84 | 10.47 | 1.83 |
| 60.0 | — | 2.0 | 3.64 | 2.86 | 0.19 | 15.34 | 5.11 | 2.05 |
| | — | 2.5 | 4.52 | 3.55 | 0.19 | 18.70 | 6.23 | 2.03 |
| | 3.0 | 3.0 | 5.37 | 4.22 | 0.19 | 21.88 | 7.29 | 2.02 |
| | 3.5 | 3.5 | 6.21 | 4.88 | 0.19 | 24.88 | 8.29 | 2.00 |
| | 4.0 | — | 7.04 | 5.52 | 0.19 | 27.73 | 9.24 | 1.98 |
| | 4.5 | — | 7.85 | 6.16 | 0.19 | 30.41 | 10.14 | 1.97 |
| 63.5 | — | 2.0 | 3.86 | 3.03 | 0.20 | 18.29 | 5.76 | 2.18 |
| | — | 2.5 | 4.79 | 3.76 | 0.20 | 22.32 | 7.03 | 2.16 |
| | 3.0 | 3.0 | 5.70 | 4.48 | 0.20 | 26.15 | 8.24 | 2.14 |
| | 3.5 | 3.5 | 6.60 | 5.18 | 0.20 | 29.79 | 9.38 | 2.12 |
| | 4.0 | — | 7.48 | 5.87 | 0.20 | 33.24 | 10.47 | 2.11 |
| | 4.5 | — | 8.34 | 6.55 | 0.20 | 36.50 | 11.50 | 2.09 |
| | 5.0 | — | 9.19 | 7.21 | 0.20 | 39.60 | 12.47 | 2.08 |
| | 5.5 | — | 10.02 | 7.87 | 0.20 | 42.52 | 13.39 | 2.06 |
| | 6.0 | — | 10.84 | 8.51 | 0.20 | 45.28 | 14.26 | 2.04 |
| 68.0 | 3.0 | — | 6.13 | 4.81 | 0.21 | 32.42 | 9.54 | 2.30 |
| | 3.5 | — | 7.09 | 5.57 | 0.21 | 36.99 | 10.88 | 2.28 |
| | 4.0 | — | 8.04 | 6.31 | 0.21 | 41.34 | 12.16 | 2.27 |
| | 4.5 | — | 8.98 | 7.05 | 0.21 | 45.47 | 13.37 | 2.25 |
| | 5.0 | — | 9.90 | 7.77 | 0.21 | 49.41 | 14.53 | 2.23 |
| | 5.5 | — | 10.80 | 8.48 | 0.21 | 53.14 | 15.63 | 2.22 |
| | 6.0 | — | 11.69 | 9.17 | 0.21 | 56.68 | 16.67 | 2.20 |
| 70.0 | — | 2.0 | 4.27 | 3.35 | 0.22 | 24.72 | 7.06 | 2.41 |
| | — | 2.5 | 5.30 | 4.16 | 0.22 | 30.23 | 8.64 | 2.39 |
| | 3.0 | 3.0 | 6.31 | 4.96 | 0.22 | 35.50 | 10.14 | 2.37 |
| | 3.5 | 3.5 | 7.31 | 5.74 | 0.22 | 40.53 | 11.58 | 2.35 |
| | 4.0 | — | 8.29 | 6.51 | 0.22 | 45.33 | 12.95 | 2.34 |
| | 4.5 | 4.5 | 9.26 | 7.27 | 0.22 | 49.89 | 14.26 | 2.32 |
| | 5.0 | — | 10.21 | 8.01 | 0.22 | 54.24 | 15.50 | 2.30 |
| | 5.5 | — | 11.14 | 8.75 | 0.22 | 58.38 | 16.68 | 2.29 |
| | 6.0 | — | 12.06 | 9.47 | 0.22 | 62.31 | 17.80 | 2.27 |
| | 7.0 | — | 13.85 | 10.88 | 0.22 | 69.58 | 19.88 | 2.24 |

## APPENDIX F  TYPICAL STEEL SHAPES

continued

| External diameter $D$ (mm) | Thickness $t$ Seamless | Thickness $t$ Welded | Sectional area (cm²) | Theoretical weight (kg/m) | Surface area (m²/m) | Section characteristics Moment of inertia $I$ (cm⁴) | Section characteristics Sectional modulus $W$ (cm³) | Section characteristics Radius of gyration $i$ (cm) |
|---|---|---|---|---|---|---|---|---|
| 73.0 | 3.0 | — | 6.60 | 5.18 | 0.23 | 40.48 | 11.09 | 2.48 |
| | 3.5 | — | 7.64 | 6.00 | 0.23 | 46.26 | 12.67 | 2.46 |
| | 4.0 | — | 8.67 | 6.81 | 0.23 | 51.78 | 14.19 | 2.44 |
| | 4.5 | — | 9.68 | 7.60 | 0.23 | 57.04 | 15.63 | 2.43 |
| | 5.0 | — | 10.68 | 8.38 | 0.23 | 62.07 | 17.01 | 2.41 |
| | 5.5 | — | 11.66 | 9.16 | 0.23 | 66.87 | 18.32 | 2.39 |
| | 6.0 | — | 12.63 | 9.91 | 0.23 | 71.43 | 19.57 | 2.38 |
| | 7.0 | — | 14.51 | 11.39 | 0.23 | 79.92 | 21.90 | 2.35 |
| 76.0 | — | 2.0 | 4.65 | 3.65 | 0.24 | 31.58 | 8.38 | 2.62 |
| | — | 2.5 | 5.77 | 4.53 | 0.24 | 39.03 | 10.27 | 2.60 |
| | 3.0 | 3.0 | 6.88 | 5.40 | 0.24 | 45.91 | 12.08 | 2.58 |
| | 3.5 | 3.5 | 7.97 | 6.26 | 0.24 | 52.50 | 13.82 | 2.57 |
| | 4.0 | 4.0 | 9.05 | 7.10 | 0.24 | 58.81 | 15.48 | 2.55 |
| | 4.5 | 4.5 | 10.11 | 7.93 | 0.24 | 64.85 | 17.07 | 2.53 |
| | 5.0 | — | 11.15 | 8.75 | 0.24 | 70.62 | 18.59 | 2.52 |
| | 5.5 | — | 12.18 | 9.56 | 0.24 | 76.14 | 20.04 | 2.50 |
| | 6.0 | — | 13.19 | 10.36 | 0.24 | 81.41 | 21.42 | 2.48 |
| | 7.0 | — | 15.17 | 11.91 | 0.24 | 91.23 | 24.01 | 2.45 |
| 83.0 | — | 2.0 | 5.09 | 4.00 | 0.26 | 41.76 | 10.06 | 2.86 |
| | — | 2.5 | 6.32 | 4.96 | 0.26 | 51.26 | 12.35 | 2.85 |
| | — | 3.0 | 7.54 | 5.92 | 0.26 | 60.40 | 14.56 | 2.83 |
| | 3.5 | 3.5 | 8.74 | 6.86 | 0.26 | 69.19 | 16.67 | 2.81 |
| | 4.0 | 4.0 | 9.93 | 7.79 | 0.26 | 77.64 | 18.71 | 2.80 |
| | 4.5 | 4.5 | 11.10 | 8.71 | 0.26 | 85.76 | 20.67 | 2.78 |
| | 5.0 | — | 12.25 | 9.62 | 0.26 | 93.56 | 22.54 | 2.76 |
| | 5.5 | — | 13.39 | 10.51 | 0.26 | 101.04 | 24.35 | 2.75 |
| | 6.0 | — | 14.51 | 11.39 | 0.26 | 108.22 | 26.08 | 2.73 |
| | 7.0 | — | 16.71 | 13.12 | 0.26 | 121.69 | 29.32 | 2.70 |
| | 8.0 | — | 18.85 | 14.80 | 0.26 | 134.04 | 32.30 | 2.67 |
| 89.0 | — | 2.0 | 5.47 | 4.29 | 0.28 | 51.75 | 11.63 | 3.08 |
| | — | 2.5 | 6.79 | 5.33 | 0.28 | 63.59 | 14.29 | 3.06 |
| | — | 3.0 | 8.11 | 6.36 | 0.28 | 75.02 | 16.86 | 3.04 |
| | 3.5 | 3.5 | 9.40 | 7.38 | 0.28 | 86.05 | 19.34 | 3.03 |
| | 4.0 | 4.0 | 10.68 | 8.38 | 0.28 | 96.68 | 21.73 | 3.01 |
| | 4.5 | 4.5 | 11.95 | 9.38 | 0.28 | 106.92 | 24.03 | 2.99 |
| | 5.0 | — | 13.19 | 10.36 | 0.28 | 116.79 | 26.24 | 2.98 |
| | 5.5 | — | 14.43 | 11.33 | 0.28 | 126.29 | 28.38 | 2.96 |
| | 6.0 | — | 15.65 | 12.28 | 0.28 | 135.43 | 30.43 | 2.94 |
| | 7.0 | — | 18.03 | 14.16 | 0.28 | 152.67 | 34.31 | 2.91 |
| | 8.0 | — | 20.36 | 15.98 | 0.28 | 168.59 | 37.88 | 2.88 |

## APPENDIX F  TYPICAL STEEL SHAPES

continued

| External diameter $D$ (mm) | Thickness $t$ Seamless | Thickness $t$ Welded | Sectional area ($cm^2$) | Theoretical weight (kg/m) | Surface area ($m^2/m$) | Section characteristics Moment of inertia $I$ ($cm^4$) | Section characteristics Sectional modulus $W$ ($cm^3$) | Section characteristics Radius of gyration $i$ (cm) |
|---|---|---|---|---|---|---|---|---|
| 95.0 | — | 2.0 | 5.84 | 4.59 | 0.30 | 63.20 | 13.31 | 3.29 |
| | — | 2.5 | 7.26 | 5.70 | 0.30 | 77.76 | 16.37 | 3.27 |
| | — | 3.0 | 8.67 | 6.81 | 0.30 | 91.83 | 19.33 | 3.25 |
| | 3.5 | 3.5 | 10.06 | 7.90 | 0.30 | 105.45 | 22.20 | 3.24 |
| | 4.0 | — | 11.44 | 8.98 | 0.30 | 118.60 | 24.97 | 3.22 |
| | 4.5 | — | 12.79 | 10.04 | 0.30 | 131.31 | 27.64 | 3.20 |
| | 5.0 | — | 14.14 | 11.10 | 0.30 | 143.58 | 30.23 | 3.19 |
| | 5.5 | — | 15.46 | 12.14 | 0.30 | 155.43 | 32.72 | 3.17 |
| | 6.0 | — | 16.78 | 13.17 | 0.30 | 166.86 | 35.13 | 3.15 |
| | 7.0 | — | 19.35 | 15.19 | 0.30 | 188.51 | 39.69 | 3.12 |
| | 8.0 | — | 21.87 | 17.16 | 0.30 | 208.62 | 43.92 | 3.09 |
| 102.0 | — | 2.0 | 6.28 | 4.93 | 0.32 | 78.57 | 15.41 | 3.54 |
| | — | 2.5 | 7.81 | 6.13 | 0.32 | 96.77 | 18.97 | 3.52 |
| | — | 3.0 | 9.33 | 7.32 | 0.32 | 114.42 | 22.43 | 3.50 |
| | 3.5 | 3.5 | 10.83 | 8.50 | 0.32 | 131.52 | 25.79 | 3.48 |
| | 4.0 | 4.0 | 12.32 | 9.67 | 0.32 | 148.09 | 29.04 | 3.47 |
| | 4.5 | 4.5 | 13.78 | 10.82 | 0.32 | 164.14 | 32.18 | 3.45 |
| | 5.0 | 5.0 | 15.24 | 11.96 | 0.32 | 179.68 | 35.23 | 3.43 |
| | 5.5 | — | 16.67 | 13.09 | 0.32 | 194.72 | 38.18 | 3.42 |
| | 6.0 | — | 18.10 | 14.21 | 0.32 | 209.28 | 41.03 | 3.40 |
| | 7.0 | — | 20.89 | 16.40 | 0.32 | 236.96 | 46.46 | 3.37 |
| | 8.0 | — | 23.62 | 18.55 | 0.32 | 262.83 | 51.53 | 3.34 |
| | 10.0 | — | 28.90 | 22.69 | 0.32 | 309.40 | 60.67 | 3.27 |
| 108.0 | — | 3.0 | 9.90 | 7.77 | 0.34 | 136.49 | 25.28 | 3.71 |
| | — | 3.5 | 11.49 | 9.02 | 0.34 | 157.02 | 29.08 | 3.70 |
| | 4.0 | 4.0 | 13.07 | 10.26 | 0.34 | 176.95 | 32.77 | 3.68 |
| | 4.5 | — | 14.63 | 11.49 | 0.34 | 196.30 | 36.35 | 3.66 |
| | 5.0 | — | 16.18 | 12.70 | 0.34 | 215.06 | 39.83 | 3.65 |
| | 5.5 | — | 17.71 | 13.90 | 0.34 | 233.26 | 43.20 | 3.63 |
| | 6.0 | — | 19.23 | 15.09 | 0.34 | 250.91 | 46.46 | 3.61 |
| | 7.0 | — | 22.21 | 17.44 | 0.34 | 284.58 | 52.70 | 3.58 |
| | 8.0 | — | 25.13 | 19.73 | 0.34 | 316.17 | 58.55 | 3.55 |
| | 10.0 | — | 30.79 | 24.17 | 0.34 | 373.45 | 69.16 | 3.48 |
| 114.0 | — | 3.0 | 10.46 | 8.21 | 0.36 | 161.24 | 28.29 | 3.93 |
| | — | 3.5 | 12.15 | 9.54 | 0.36 | 185.63 | 32.57 | 3.91 |
| | 4.0 | 4.0 | 13.82 | 10.85 | 0.36 | 209.35 | 36.73 | 3.89 |
| | 4.5 | 4.5 | 15.48 | 12.15 | 0.36 | 232.41 | 40.77 | 3.87 |
| | 5.0 | 5.0 | 17.12 | 13.44 | 0.36 | 254.81 | 44.70 | 3.86 |
| | 5.5 | — | 18.75 | 14.72 | 0.36 | 276.58 | 48.52 | 3.84 |
| | 6.0 | — | 20.36 | 15.98 | 0.36 | 297.73 | 52.23 | 3.82 |
| | 7.0 | — | 23.53 | 18.47 | 0.36 | 338.19 | 59.33 | 3.79 |
| | 8.0 | — | 26.64 | 20.91 | 0.36 | 376.30 | 66.02 | 3.76 |
| | 10.0 | — | 32.67 | 25.65 | 0.36 | 445.82 | 78.21 | 3.69 |

## APPENDIX F  TYPICAL STEEL SHAPES

continued

| External diameter $D$ (mm) | Thickness $t$ | | Sectional area ($cm^2$) | Theoretical weight (kg/m) | Surface area ($m^2/m$) | Section characteristics | | |
|---|---|---|---|---|---|---|---|---|
| | Seamless | Welded | | | | Moment of inertia $I$ ($cm^4$) | Sectional modulus $W$ ($cm^3$) | Radius of gyration $i$ (cm) |
| 121.0 | — | 3.0 | 11.12 | 8.73 | 0.38 | 193.69 | 32.01 | 4.17 |
| | — | 3.5 | 12.92 | 10.14 | 0.38 | 223.17 | 36.89 | 4.16 |
| | 4.0 | 4.0 | 14.70 | 11.54 | 0.38 | 251.87 | 41.63 | 4.14 |
| | 4.5 | — | 16.47 | 12.93 | 0.38 | 279.83 | 46.25 | 4.12 |
| | 5.0 | — | 18.22 | 14.30 | 0.38 | 307.05 | 50.75 | 4.11 |
| | 5.5 | — | 19.96 | 15.67 | 0.38 | 333.54 | 55.13 | 4.09 |
| | 6.0 | — | 21.68 | 17.02 | 0.38 | 359.32 | 59.39 | 4.07 |
| | 7.0 | — | 25.07 | 19.68 | 0.38 | 408.80 | 67.57 | 4.04 |
| | 8.0 | — | 28.40 | 22.29 | 0.38 | 455.57 | 75.30 | 4.01 |
| | 10.0 | — | 34.87 | 27.37 | 0.38 | 541.43 | 89.49 | 3.94 |
| 127.0 | — | 3.0 | 11.69 | 9.17 | 0.40 | 224.75 | 35.39 | 4.39 |
| | — | 3.5 | 13.58 | 10.66 | 0.40 | 259.11 | 40.80 | 4.37 |
| | 4.0 | 4.0 | 15.46 | 12.13 | 0.40 | 292.61 | 46.08 | 4.35 |
| | 4.5 | 4.5 | 17.32 | 13.59 | 0.40 | 325.29 | 51.23 | 4.33 |
| | 5.0 | 5.0 | 19.16 | 15.04 | 0.40 | 357.14 | 56.24 | 4.32 |
| | 5.5 | — | 20.99 | 16.48 | 0.40 | 388.19 | 61.13 | 4.30 |
| | 6.0 | — | 22.81 | 17.90 | 0.40 | 418.44 | 65.90 | 4.28 |
| | 7.0 | — | 26.39 | 20.72 | 0.40 | 476.63 | 75.06 | 4.25 |
| | 8.0 | — | 29.91 | 23.48 | 0.40 | 531.80 | 83.75 | 4.22 |
| | 10.0 | — | 36.76 | 28.85 | 0.40 | 663.55 | 99.77 | 4.15 |
| | 12.0 | — | 43.35 | 34.03 | 0.40 | 724.50 | 114.09 | 4.09 |
| 133.0 | 4.0 | 4.0 | 16.21 | 12.73 | 0.42 | 337.53 | 50.76 | 4.56 |
| | 4.5 | 4.5 | 18.17 | 14.26 | 0.42 | 376.42 | 56.45 | 4.55 |
| | 5.0 | 5.0 | 20.11 | 15.78 | 0.42 | 412.40 | 62.02 | 4.53 |
| | 5.5 | — | 22.03 | 17.29 | 0.42 | 448.50 | 67.44 | 4.51 |
| | 6.0 | — | 23.94 | 18.79 | 0.42 | 483.72 | 72.74 | 4.50 |
| | 7.0 | — | 27.71 | 21.75 | 0.42 | 551.58 | 82.94 | 4.46 |
| | 8.0 | — | 31.42 | 24.66 | 0.42 | 616.11 | 92.65 | 4.43 |
| | 10.0 | — | 38.64 | 30.33 | 0.42 | 735.59 | 110.62 | 4.36 |
| | 12.0 | — | 45.62 | 35.81 | 0.42 | 843.04 | 126.77 | 4.30 |
| 140.0 | — | 4.0 | 17.09 | 13.42 | 0.44 | 395.47 | 56.50 | 4.81 |
| | 4.5 | 4.5 | 19.16 | 15.04 | 0.44 | 440.12 | 62.87 | 4.79 |
| | 5.0 | 5.0 | 21.21 | 16.65 | 0.44 | 483.76 | 69.11 | 4.78 |
| | 5.5 | 5.5 | 23.24 | 18.24 | 0.44 | 526.40 | 75.20 | 4.76 |
| | 6.0 | — | 25.26 | 19.83 | 0.44 | 568.06 | 81.15 | 4.74 |
| | 7.0 | — | 29.25 | 22.96 | 0.44 | 648.51 | 92.64 | 4.71 |
| | 8.0 | — | 33.18 | 26.04 | 0.44 | 725.21 | 103.60 | 4.68 |
| | 10.0 | — | 40.84 | 32.06 | 0.44 | 867.86 | 123.98 | 4.61 |
| | 12.0 | — | 48.25 | 37.88 | 0.44 | 996.95 | 142.42 | 4.55 |
| | 14.0 | — | 55.42 | 43.50 | 0.44 | 1113.34 | 159.05 | 4.48 |

APPENDIX F  TYPICAL STEEL SHAPES

continued

| External diameter D | Thickness t | | Sectional area (cm²) | Theoretical weight (kg/m) | Surface area (m²/m) | Section characteristics | | |
|---|---|---|---|---|---|---|---|---|
| | Seamless | Welded | | | | Moment of inertia I (cm⁴) | Sectional modulus W (cm³) | Radius of gyration i (cm) |
| (mm) | | | | | | | | |
| 146.0 | 5.0 | — | 22.15 | 17.39 | 0.46 | 551.10 | 75.49 | 4.99 |
| | 5.5 | — | 24.28 | 19.06 | 0.46 | 599.95 | 82.19 | 4.97 |
| | 6.0 | — | 26.39 | 20.72 | 0.46 | 647.73 | 88.73 | 4.95 |
| | 7.0 | — | 30.57 | 24.00 | 0.46 | 740.12 | 101.39 | 4.92 |
| | 8.0 | — | 34.68 | 27.23 | 0.46 | 828.41 | 113.48 | 4.89 |
| | 10.0 | — | 42.73 | 33.54 | 0.46 | 993.16 | 136.05 | 4.82 |
| | 12.0 | — | 50.52 | 39.66 | 0.46 | 1142.94 | 156.57 | 4.76 |
| | 14.0 | — | 58.06 | 45.57 | 0.46 | 1278.70 | 175.16 | 4.69 |
| 152.0 | 5.0 | 5.0 | 23.09 | 18.13 | 0.48 | 624.43 | 82.16 | 5.20 |
| | 5.5 | 5.5 | 25.31 | 19.87 | 0.48 | 680.06 | 89.48 | 5.18 |
| | 6.0 | — | 27.52 | 21.60 | 0.48 | 734.52 | 96.65 | 5.17 |
| | 7.0 | — | 31.89 | 25.03 | 0.48 | 839.99 | 110.52 | 5.13 |
| | 8.0 | — | 36.19 | 28.41 | 0.48 | 940.97 | 123.81 | 5.10 |
| | 10.0 | — | 44.61 | 35.02 | 0.48 | 1129.99 | 148.68 | 5.03 |
| | 12.0 | — | 52.78 | 41.43 | 0.48 | 1302.58 | 171.39 | 4.97 |
| | 14.0 | — | 60.70 | 47.65 | 0.48 | 1459.73 | 192.07 | 4.90 |
| 159.0 | 5.0 | — | 24.19 | 18.99 | 0.50 | 717.88 | 90.30 | 5.45 |
| | 6.0 | — | 28.84 | 22.64 | 0.50 | 845.19 | 106.31 | 5.41 |
| | 7.0 | — | 33.43 | 26.24 | 0.50 | 967.41 | 121.69 | 5.38 |
| | 8.0 | — | 37.95 | 29.79 | 0.50 | 1084.67 | 136.44 | 5.35 |
| | 10.0 | — | 46.81 | 36.75 | 0.50 | 1304.88 | 164.14 | 5.28 |
| | 12.0 | — | 55.42 | 43.50 | 0.50 | 1506.88 | 189.54 | 5.21 |
| | 14.0 | — | 63.77 | 50.06 | 0.50 | 1691.69 | 212.79 | 5.15 |
| 168.0 | 5.0 | — | 25.60 | 20.10 | 0.53 | 851.14 | 101.33 | 5.77 |
| | 6.0 | — | 30.54 | 23.97 | 0.53 | 1003.12 | 119.42 | 5.73 |
| | 7.0 | — | 35.41 | 27.79 | 0.53 | 1149.36 | 136.83 | 5.70 |
| | 8.0 | — | 40.21 | 31.57 | 0.53 | 1290.01 | 153.57 | 5.66 |
| | 10.0 | — | 49.64 | 38.97 | 0.53 | 1555.13 | 185.13 | 5.60 |
| | 12.0 | — | 58.81 | 46.17 | 0.53 | 1799.60 | 214.24 | 5.53 |
| | 14.0 | — | 67.73 | 53.17 | 0.53 | 2024.53 | 241.02 | 5.47 |
| | 16.0 | — | 76.40 | 59.98 | 0.53 | 2230.98 | 265.59 | 5.40 |
| 180.0 | 5.0 | — | 27.49 | 21.58 | 0.57 | 1053.17 | 117.02 | 6.19 |
| | 6.0 | — | 32.80 | 25.75 | 0.57 | 1242.72 | 138.08 | 6.16 |
| | 7.0 | — | 38.04 | 29.87 | 0.57 | 1425.63 | 158.40 | 6.12 |
| | 8.0 | — | 43.23 | 33.93 | 0.57 | 1602.04 | 178.00 | 6.09 |
| | 10.0 | — | 53.41 | 41.92 | 0.57 | 1936.01 | 215.11 | 6.02 |
| | 12.0 | — | 63.33 | 49.72 | 0.57 | 2245.84 | 249.54 | 5.95 |
| | 14.0 | — | 73.01 | 57.31 | 0.57 | 2532.74 | 281.42 | 5.89 |
| | 16.0 | — | 82.44 | 64.71 | 0.57 | 2797.86 | 310.87 | 5.83 |

## APPENDIX F  TYPICAL STEEL SHAPES

continued

| External diameter $D$ (mm) | Thickness $t$ Seamless (mm) | Thickness $t$ Welded (mm) | Sectional area (cm$^2$) | Theoretical weight (kg/m) | Surface area (m$^2$/m) | Moment of inertia $I$ (cm$^4$) | Sectional modulus $W$ (cm$^3$) | Radius of gyration $i$ (cm) |
|---|---|---|---|---|---|---|---|---|
| 194.0 | 5.0 | — | 29.69 | 23.31 | 0.61 | 1326.54 | 136.76 | 6.68 |
| | 6.0 | — | 35.44 | 27.82 | 0.61 | 1567.21 | 161.57 | 6.65 |
| | 7.0 | — | 41.12 | 32.28 | 0.61 | 1800.08 | 185.57 | 6.62 |
| | 8.0 | — | 46.75 | 36.70 | 0.61 | 2025.31 | 208.79 | 6.58 |
| | 10.0 | — | 57.81 | 45.38 | 0.61 | 2453.55 | 252.94 | 6.51 |
| | 12.0 | — | 68.61 | 53.86 | 0.61 | 2853.25 | 294.15 | 6.45 |
| | 14.0 | — | 79.17 | 62.15 | 0.61 | 3225.71 | 332.55 | 6.38 |
| | 16.0 | — | 89.47 | 70.24 | 0.61 | 3572.19 | 368.27 | 6.32 |
| | 18.0 | — | 99.53 | 78.13 | 0.61 | 3893.94 | 401.44 | 6.25 |
| 203.0 | 6.0 | — | 37.13 | 29.15 | 0.64 | 1803.07 | 177.64 | 6.97 |
| | 8.0 | — | 49.01 | 38.47 | 0.64 | 2333.37 | 229.89 | 6.90 |
| | 10.0 | — | 60.63 | 47.60 | 0.64 | 2830.72 | 278.89 | 6.83 |
| | 12.0 | — | 72.01 | 56.52 | 0.64 | 3296.49 | 324.78 | 6.77 |
| | 14.0 | — | 83.13 | 65.25 | 0.64 | 3732.07 | 367.69 | 6.70 |
| | 16.0 | — | 94.00 | 73.79 | 0.64 | 4133.78 | 407.76 | 6.64 |
| | 18.0 | — | 104.62 | 82.12 | 0.64 | 4517.93 | 445.12 | 6.57 |
| 219.0 | 6.0 | — | 40.15 | 31.52 | 0.69 | 2278.74 | 208.10 | 7.53 |
| | 8.0 | — | 53.03 | 41.63 | 0.69 | 2955.43 | 269.90 | 7.47 |
| | 10.0 | — | 65.66 | 51.54 | 0.69 | 3593.29 | 328.15 | 7.40 |
| | 12.0 | — | 78.04 | 61.26 | 0.69 | 4193.81 | 383.00 | 7.33 |
| | 14.0 | — | 90.16 | 70.78 | 0.69 | 4758.50 | 434.57 | 7.26 |
| | 16.0 | — | 102.04 | 80.10 | 0.69 | 5288.81 | 483.00 | 7.20 |
| | 18.0 | — | 113.66 | 89.23 | 0.69 | 5786.15 | 528.42 | 7.13 |
| | 20.0 | — | 125.04 | 98.15 | 0.69 | 6251.93 | 570.95 | 7.07 |
| 245.0 | 7.0 | — | 52.34 | 41.09 | 0.77 | 3709.06 | 302.78 | 8.42 |
| | 8.0 | — | 59.56 | 46.76 | 0.77 | 4186.87 | 341.79 | 8.38 |
| | 10.0 | — | 73.83 | 57.95 | 0.77 | 5105.63 | 416.79 | 8.32 |
| | 12.0 | — | 87.84 | 68.95 | 0.77 | 5976.67 | 487.89 | 8.25 |
| | 14.0 | — | 101.60 | 79.76 | 0.77 | 6801.68 | 555.24 | 8.18 |
| | 16.0 | — | 115.11 | 90.36 | 0.77 | 7582.30 | 618.96 | 8.12 |
| | 18.0 | — | 128.37 | 100.77 | 0.77 | 8320.17 | 679.20 | 8.05 |
| | 20.0 | — | 141.37 | 110.98 | 0.77 | 9016.86 | 736.07 | 7.99 |
| 273.0 | 8.0 | — | 66.60 | 52.28 | 0.86 | 5351.71 | 428.70 | 9.37 |
| | 10.0 | — | 82.62 | 64.86 | 0.86 | 7154.09 | 524.11 | 9.31 |
| | 12.0 | — | 98.39 | 77.24 | 0.86 | 8396.14 | 615.10 | 9.24 |
| | 14.0 | — | 113.91 | 89.42 | 0.86 | 9579.75 | 701.81 | 9.17 |
| | 16.0 | — | 129.18 | 101.41 | 0.86 | 10706.79 | 784.38 | 9.10 |
| | 18.0 | — | 144.20 | 113.20 | 0.86 | 11779.08 | 862.94 | 9.04 |
| | 20.0 | — | 158.96 | 124.79 | 0.86 | 12798.44 | 937.61 | 8.97 |

APPENDIX F  TYPICAL STEEL SHAPES

continued

| External diameter $D$ | Thickness $t$ | | Sectional area ($cm^2$) | Theoretical weight (kg/m) | Surface area ($m^2/m$) | Section characteristics | | |
|---|---|---|---|---|---|---|---|---|
| | Seamless | Welded | | | | Moment of inertia $I$ ($cm^4$) | Sectional modulus $W$ ($cm^3$) | Radius of gyration $i$ (cm) |
| (mm) | | | | | | | | |
| 299.0 | 8.0 | — | 73.14 | 57.41 | 0.94 | 7747.42 | 518.22 | 10.29 |
| | 10.0 | — | 90.79 | 71.27 | 0.94 | 9490.15 | 634.79 | 10.22 |
| | 12.0 | — | 108.20 | 84.93 | 0.94 | 11159.52 | 746.46 | 10.16 |
| | 14.0 | — | 125.35 | 98.40 | 0.94 | 12757.61 | 853.35 | 10.09 |
| | 16.0 | — | 142.25 | 111.67 | 0.94 | 14286.48 | 955.62 | 10.02 |
| | 18.0 | — | 158.90 | 124.74 | 0.94 | 15748.16 | 1053.39 | 9.96 |
| | 20.0 | — | 175.30 | 137.61 | 0.94 | 17144.64 | 1146.80 | 9.89 |

Steel plates use in Civil Engineering (classified by thickness)　　Table F.8

| Thickness | 0.35-4 | 4.5-20 | 22-60 | >60 |
|---|---|---|---|---|
| Steel plate | Thin | Medium | Thick | Heavy |

Types and specifications of steel plates used in Civil Engineering　　Table F.9

| Type | Executive standard | Specifications |
|---|---|---|
| Plain carbon steel plate | GB/T 3274—2017 | Thickness: 4.5-200 mm |
| Plain carbon steel killed steel plate | GB/T 3274—2017 | Thickness: 4.5-200 mm |
| Low alloy structural steel plate | GB/T 3274—2017 | Thickness: 4.5-200 mm |
| Hot rolled steel plate for building structure | GB/T 19879—2015 | Q345GJ thickness 6-200mm |
| Carbon steel and ordinary low alloy steel plate for bridge | YB 168—1970 | Thickness: 6-50 mm; Width: 1.0-2.4 mm; Length: 2.0-16 mm |

# REFERENCES

[1] Ministry of Housing and Urban-Rural Development of the People's Republic of China. Standard for Design of Steel Structures GB 50017—2017[S]. Beijing: China Architecture & Building Press, 2018.

[2] Ministry of Housing and Urban-Rural Development of the People's Republic of China. Unified Standard for Reliability Design of Building Structures GB 50068—2018[S]. Beijing: China Architecture & Building Press, 2019.

[3] Ministry of Housing and Urban-Rural Development of the People's Republic of China. Load Code for the Design of Building Structures GB 50009—2012[S]. Beijing: China Architecture & Building Press, 2012.

[4] Ministry of Construction of the People's Republic of China. Standard for Acceptance of Construction Quality of Steel Structures GB 50205—2001[S]. Beijing: China Planning Press, 2002.

[5] Ministry of Construction of the People's Republic of China. Technical Code of Cold-formed Thin-wall Steel Structures GB 50018—2002[S]. Beijing: China Standard Press, 2003.

[6] Ministry of Housing and Urban-Rural Development of the People's Republic of China. Standard for General Terms Used in Design of Engineering Structures GB/T 50083—2014[S]. Beijing: China Architecture & Building Press, 2015.

[7] Ministry of Housing and Urban-Rural Development of the People's Republic of China. Technical Code for Steel Structure of Light-weight Building with Gabled Frames GB 51022—2015[S]. Beijing: China Architecture & building Press, 2016.

[8] Ministry of Housing and Urban-Rural Development of the People's Republic of China. Code for Seismic Design of Buildings GB 50011—2010[S]. Beijing: China Architecture & Building Press, 2016.

[9] Ministry of Housing and Urban-Rural Development of the People's Republic of China. Code for Fire Safety of Steel Structures in Buildings GB 51249—2017[S]. Beijing: China Planning Press, 2018.

[10] China Engineering Construction Standardization Association. Code for Anti-collapse Design of Building Structures CECS 392—2014[S]. Beijing: China Planning Press, 2015.

[11] EN 1991-1-7. Eurocode 1-Actions on Structures Part 1.7: General actions-accidental actions due to impact and explosions[S]. Brussels, 2006.

[12] Dan Zeyi. Steel Structure Design Manual (4$^{th}$ Edition) [M]. Beijing: China Architecture & Building Press, 2019.

[13] Chen Shaofan. Principles of Steel Structure Design (4$^{th}$ Edition) [M]. Beijing: Science Press, 2016.

[14] He Ruoquan, et al. Basic Principles of Steel Structures (2$^{nd}$ Edition) [M]. Beijing: China Architecture & Building Press, 2018.

[15] Dai Guoxin. Steel Structure (5$^{th}$ Edition) [M]. Wuhan: Wuhan University of Tecnology Press, 2019.

[16] Leonard Spiegel, George F, Limbrunner. Applied Structural Steel Design (3$^{rd}$ Edition) [M]. Upper Saddle River, New Jersey, Prentice Hall, 2001.

[17] Yao Jian, Xia Zhibin. Steel Structure Behavior and Design (2$^{nd}$ Edition) [M]. Beijing: China Architecture & Building Press, 2011.

[18] Zhang Yaochun, et al. Principles of Steel Structure Design [M]. Beijing: Higher Education Press, 2011.